The Routledge Companion to the Suburbs

The Routledge Companion to the Suburbs provides one of the most comprehensive examinations available to date of the suburbs around the world. International in scope and interdisciplinary in nature, this volume will serve as the definitive reference for scholars and students of the suburbs.

This volume brings together the leading scholars of the suburbs researching in different parts of the world to better understand how and why suburbs and their communities grow, decline, and regenerate. The volume sets out four goals: 1) to provide a synthesis and critical appraisal of the historical and current state of understanding about the development of suburbs in the world; 2) to provide a forum for a comprehensive examination into the conceptual, theoretical, spatial, and empirical discontents of suburbanization; 3) to engage in a scholarly conversation about the transformation of suburbs that is interdisciplinary in nature and bridges the divide between the Global North and the Global South; and 4) to reflect on the implications of the socioeconomic, cultural, and political transformations of the suburbs for policymakers and planners. *The Routledge Companion to the Suburbs* is composed of original, scholarly contributions from the leading scholars of the study of how and why suburbs grow, decline, and transform. Special attention is paid to the global nature of suburbanization and its regional variations, with a focus on comparative analysis of suburbs through regions across the world in the Global North and the Global South.

Articulated in a common voice, the volume is integrated by the very nature of the concept of a suburb as the unit of analysis, offering multidisciplinary perspectives from the fields of economics, geography, planning, political science, sociology, and urban studies.

Bernadette Hanlon is Associate Professor and Chair of the Bachelor of Science Program in City and Regional Planning in the Knowlton School of Architecture at The Ohio State University in Columbus, Ohio, U.S.A. She specializes in political economy of cities and suburbs, environmental sustainability, immigration, and urban environmental policy and planning. She is the author or co-author of three books, including *Once the American Dream: Inner-Ring Suburbs in the Metropolitan United States* (2012).

Thomas J. Vicino is Professor and Chair of the Department of Political Science in the College of Social Sciences and Humanities at Northeastern University in Boston, Massachusetts, U.S.A. He specializes in the political economy of cities and suburbs, focusing on issues of metropolitan development, housing, and demographic analysis. He is the author or co-author of four books, including *Suburban Crossroads: The Fight for Local Control of Immigration Policy* (2013).

The Routledge Companion to the Suburbs

Edited by Bernadette Hanlon and Thomas J. Vicino

Routledge
Taylor & Francis Group

LONDON AND NEW YORK

First published 2019
by Routledge
2 Park Square, Milton Park, Abingdon, Oxon OX14 4RN

and by Routledge
605 Third Avenue, New York, NY 10017

First issued in paperback 2020

Routledge is an imprint of the Taylor & Francis Group, an informa business

British Library Cataloguing-in-Publication Data
A catalogue record for this book is available from the British Library

Library of Congress Cataloging-in-Publication Data
Names: Hanlon, Bernadette, 1969– editor. | Vicino, Thomas J., editor.
Title: The Routledge companion to the suburbs / edited by Bernadette Hanlon and Thomas J. Vicino.
Description: Abingdon, Oxon ; New York, NY : Routledge, 2019. | Includes bibliographical references and index.
Identifiers: LCCN 2018018777 | ISBN 9781138290235 (hbk : alk. paper) | ISBN 9781315266442 (ebk)
Subjects: LCSH: Suburbs. | Suburban life.
Classification: LCC HT351 .R685 2019 | DDC 307.74—dc23
LC record available at https://lccn.loc.gov/2018018777

ISBN 13: 978-0-367-73371-1 (pbk)
ISBN 13: 978-1-138-29023-5 (hbk)

Typeset in Bembo
by Apex CoVantage, LLC

Contents

Contents

PART III
Diversity, exclusion, and poverty in the suburbs   **133**

Figures

Tables

Contributors

Whitney Airgood-Obrycki is Senior Research Analyst at the Joint Center for Housing Studies at Harvard University, U.S.A. Her research interests include affordable and assisted rental housing and suburban neighborhood change and reinvestment.

Katrin B. Anacker is Associate Professor in the Schar School of Policy and Government at George Mason University, U.S.A. Her research interests include housing, urban policy, and race and ethnicity. She is the lead editor of *Introduction to Housing* (2018) and the editor of *The New American Suburb: Poverty, Race, and the Economic Crisis* (2015).

Alison L. Bain is Associate Professor of Geography in the Faculty of Liberal Arts and Professor Studies at York University in Toronto, Canada. Her research examines the complex relationships of cultural workers and LGBTQ+ populations to cities and suburbs. She is the author of the book *Creative Margins: Cultural Production in Canadian Suburbs* (2013) and the co-editor of *Urbanization in a Global Context* (2017).

Lisa Benton-Short is Professor and Chair of the Department of Geography at The George Washington University, U.S.A. Her research focuses on urban sustainability, environmental issues in cities, and public space. She is the author most recently of the book *The National Mall: No Ordinary Public Space* (2016).

Juliet Carpenter is Senior Research Fellow at the School of the Built Environment in the Faculty of Technology, Design, and Environment at Oxford Brookes University, UK. Her research interests include urban regeneration, community engagement, participation in planning, and innovation and competitiveness in urban-regional economies. Her work appears in journals such as *Urban Studies* and *Planning Practice and Research*.

Suzanne Lanyi Charles is Assistant Professor of City and Regional Planning at Cornell University, U.S.A. Her research examines redevelopment and gentrification in post-war suburban neighborhoods, and her work appears in journals such as *Urban Studies* and *Journal of Architectural and Planning Research*.

Chang Gyu Choi is Professor in the Graduate School of Urban Studies at Hanyang University, South Korea. His research interests include city and regional planning and growth management, and his work appears in journals such as *Land Use Policy, Cities, and Urban Studies*.

Magnus Dahlstedt is Professor in the Department of Social and Welfare Studies at Linköping University, Sweden. His research examines the development and changes of citizenship ideals, in the advanced liberal society of today, in the age of international migration. His works appears in journals such as the *Journal of Education and Work.*

Bahar Durmaz-Drinkwater is Assistant Professor at Izmir University of Economics, Turkey, and a Research Associate at Cambridge Architectural Research, UK. Her research focuses on urban architecture, design research, public place, and environmental psychology.

David Ekholm is a Postdoctoral Fellow in the Department for Studies of Social Change and Culture and the Center for Municipality Studies at Linköping University, Sweden. His research examines the role of the welfare state and civil society in confronting social problems, and his work appears in journals such as *Sport in Society* and *Social Inclusion.*

Theresa Enright is Assistant Professor in the Department of Political Science at University of Toronto, Canada. Her research interests include urban and regional studies, critical theory and comparative political economy. She is the author of *The Making of Grand Paris: Metropolitan Urbanism in the Twenty-First Century* (2016).

Anjuli N. Fahlberg is Full-Time Lecturer in the Department of Sociology at Tufts University in Medford, Massachusetts, U.S.A. Her research documents activism in Latin America's gang-controlled neighborhoods. Her recent publications appear in *Politics & Society, Journal of Urban Affairs,* and *Habitat International.*

Fikri Zul Fahmi is Assistant Professor in Urban and Regional Planning in the School of Architecture, Planning and Policy Development at the Institut Teknologi Bandung, Indonesia. His research interests include regional economics and planning, and his work appears in journals such as *Growth & Change.*

Tommy Firman is Professor of Regional Planning, Institut Teknologi Bandung, Indonesia. His research interests include regional planning, urbanization and migration, and decentralization, and his work appears in journals such as *Urban Studies* and *Habitat International.*

Ann Forsyth is Professor of Urban Planning in the Graduate School of Design at Harvard University, U.S.A. Her work examines the social aspects of physical planning and urban development, including how to make more sustainable and healthy cities. Most recently, she is co-editor of the book *Towards 21st Century New Towns: Past, Present, Prospects* (2019).

Robert Freestone is Professor of Planning in the Faculty of Built Environment at The University of New South Wales, Australia. His research interests lie in the development of modern planning theory and practice in Australia, heritage conservation, metropolitan restructuring, and planning education.

Andrew Gorman-Murray is Professor of Geography in the School of Social Sciences at University of Western Sydney, Australia. His research focuses on geographies of gender and sexuality and the politics of belonging, and his work appears in journals such as the *International Journal of Urban and Regional Research* and *Urban Studies.*

Pierre Hamel is Professor in the Department of Sociology at Université de Montréal, Canada. His research specializes in urban policy, social justice, and the capacity of urban institutions to regulate economic and social inequalities. His work appears in journals such as *Urban Geography* and *Progress in Planning*.

Richard Harris is Professor in the School of Geography and Earth Sciences at McMaster University, Canada. His research interests include housing, suburban development, urban social geography, urban historical geography, Canadian and U.S. cities in the twentieth century. Most recently, he is the author of the book *Building a Market: The Rise of the Home Improvement industry, 1914–1960* (2012).

Lawrence A. Herzog is Professor in the Graduate Program in City Planning in School of Public Affairs at San Diego State University, U.S.A. His research specializes in planning, sustainable development and urban design in the United States, Mexico, and Latin America. Most recently, he is the author of the book *Global Suburbs: Urban Sprawl from the Rio Grande to Rio de Janeiro* (2014).

Justin B. Hollander is Associate Professor of Urban and Environmental Policy and Planning at Tufts University, U.S.A. His research examines how cities and regions manage physical change during periods of growth and decline and the cognitive, health, and social dimensions of community well-being. Most recently, he is the author of the book *A Research Agenda for Shrinking Cities* (2018).

Erina Iwasaki is Professor in the Department of French Studies of the Faculty of Foreign Studies at Sophia University, Japan. Her research interests include the socioeconomics of urban communities of North Africa and the Middle East, and her work appears in journals such as *African Development Review*.

Roger Keil is York Research Chair in Global Sub/Urban Studies, Faculty of Environmental Studies, York University in Toronto, Canada. He researches global suburbanization, urban political ecology and regional governance and is the Principal Investigator of the Major Collaborative Research Initiative on Global Suburbanisms. Keil is the author of *Suburban Planet* (2017).

Sugie Lee is Professor of Urban Planning and Engineering in the College of Engineering, Hanyang University, South Korea. His research interests include urban form, travel behavior, mobility, walkability, and urban thermal environment, and his work appears in journals such as *Journal of Planning Education and Research and Urban Studies*.

Willow Lung-Amam is Assistant Professor of Urban Studies and Planning Program in the School of Architecture, Planning, and Preservation at University of Maryland at College Park, U.S.A. Her research examines social inequality and the built environment, and she is the author of the book *Trespassers? Asian Americans and the Battle for Suburbia* (2017).

Ruth McManus is Associate Professor in the School of History and Geography at Dublin City University, Ireland. Her research interests include the nature of the urban and suburban landscape and the physical and social development of everyday spaces. She is the author of the book *Dublin 1910–1940: Shaping the City and Suburbs* (2002).

Catherine J. Nash is Professor of Geography at Brock University, St. Catharines, Ontario, Canada. Her research focuses on gender and sexuality, digital cities, and the production of urban space. Most recently, she is the co-author of the book *Human Geography: People, Place and Culture* (2015).

Asli Ceylan Öner is Associate Professor and Chair of Architecture Department at Izmir University of Economics, Turkey. Her research focuses on global urban networks, planning and governance of global cities, urban theory, metropolitan form and built environment, and comparative urbanization.

Nicholas A. Phelps is Professor and Chair of Urban Planning at the University of Melbourne, Australia. His research interests cover economic geography, economic development, urban sociology, geography and planning, and most recently, he is the author of the books *Post-Suburban Europe* (2006) and *An Anatomy of Sprawl: Planning and Politics in Britain* (2012) and editor of *International Perspectives on Suburbanization* (2011) and *Old Europe, New Suburbanization* (2017).

Simon Pinnegar is Associate Professor of Planning in the Faculty of Built Environment at The University of New South Wales, Australia. His research specializes in with interests in metropolitan planning, urban regeneration, and affordable housing policy.

Colin Polsky is Professor of Geosciences and Director of the Center for Environmental Studies at Florida Atlantic University, U.S.A. His research interests include ecological and coastal resilience, urban planning, and spatial analysis. His work appears in journals such as *The Professional Geographer* and the *Annals of the Association of American Geographers*.

Cody R. Price is Research Analyst at the Ohio Housing Finance Agency in Columbus, Ohio, U.S.A. His research interests include affordable housing, housing policy, stigma, design research, survey research, and environmental psychology.

Bill Randolph is Professor and Director of the City Futures Research Center in the Faculty of Built Environment at The University of New South Wales, Australia. His research specializes in housing policy, housing markets and affordability, urban renewal, sustainability, and metropolitan planning policy issues.

Deden Rukmana is a Professor and Coordinator of the Urban Studies and Planning Program at Savannah State University, Georgia, U.S.A. His research interests include poverty, housing and community development, international planning and development, and urbanization.

Dan Runfola is Assistant Professor of Applied Science at William and Mary College, U.S.A. His research focuses on the intersection between large-scale spatial data, machine learning, and international development. His work appears in journals such as the *Journal of Environmental Economics and Management*.

Alex Schafran teaches in the School of Geography in the Faculty of Environment at University of Leeds, UK. His research examines contemporary restructuring and retrofitting of urban regions, with a particular emphasis on the changing dynamics of race, class and segregation across space and place, and his work appears in journals such as *Housing Policy Debate* and *Journal of Transport Geography*.

Annapurna Shaw retired as Professor of Public Policy and Management at the Indian Institute of Management Calcutta, India. Her research interests include sustainable cities, economic development, informal economy, and she is the author of the book *Indian Cities* (2012).

John Rennie Short is Professor of Public Policy in the School of Public Policy at University of Maryland, Baltimore County, U.S.A. His research focuses on urban issues, environmental concerns, globalization, political geography and the history of cartography. He is the author of 40 books, including most recently, *The Unequal City: Urban Resurgence, Displacement and the Making of Inequality in Global Cities* (2017).

Jon C. Teaford is Professor Emeritus of History at Purdue University, U.S.A. His research specializes in American urban history, and he is the author of ten books, including most recently, *The American Suburb: The Basics* (2007) and *The Metropolitan Revolution: The Rise of Post-Urban America* (2006).

Jaap Vos is the Director of the Urban Design Center and the Program Head for Bioregional Planning and Community Design at the University of Idaho, U.S.A. His research focuses on environmental justice, sustainable development, community involvement, and planning education.

Kyle Walker is Associate Professor of Geography and Director of the Center for Urban Studies at Texas Christian University in Fort Worth, Texas, U.S.A. His research interests include U.S. immigration politics, the demography of cities and suburbs, and open data science, and his work appears in journals such as *Growth & Change* and *Urban Geography*.

Dan Zinder is a recent graduate of the Urban and Environmental Planning Program at Tufts University, U.S.A. His research interests include land use policy, declining cities, GIS, and sustainability.

Preface

The Routledge Companion to the Suburbs is the culmination of a collaboration that spans nearly two decades between the editors, Bernadette Hanlon and Thomas J. Vicino. During these years, we have studied the transformation of the suburban landscape – socially, economically, politically, and spatially, with a focus mostly on the United States. Since the publication of our first journal article in *Urban Studies* on the topic, "The New Metropolitan Reality in the U.S.: Rethinking the Traditional Model," we launched a research agenda based on our mutual interests, confronting this central question: how and why do the suburbs grow, decline, and renew? We explored the many facets of the new metropolitan realities: the impact of deindustrialization and economic restructuring; the impact of socioeconomic diversity; and impact of migration and immigration. Together, these changes have reshaped the U.S. metropolis in important and defining ways. It was evident that uneven patterns of growth and decline benefited some suburban communities while harming still others. At the beginning of the twenty-first century, many suburban jurisdictions were under-prepared or ill-equipped to respond to the political, policy, and planning realities of socioeconomic and demographic changes that were reshaping their communities.

While this body of work addressed some of the fundamental transformations of the suburbs in the United States, questions remained unanswered, especially when taking a more global perspective. Scholars of the suburbs increasingly observed that we lacked a comprehensive understanding about the global nature of the suburbanization processes and needed to move beyond the Anglo-American suburban experience. Similarly, the state of knowledge about the suburbs remained disparate, often aligned with traditional disciplinary boundaries between the humanities and the social sciences. The time has come to integrate our collective understanding of the historical and contemporary transformations of the suburbs – their places, their people, their economy, their landscapes, their politics, and their cultures – around the world.

In this volume, we seek to systematically integrate our shared interdisciplinary perspectives on suburban transformation. Indeed, the study of the suburbs requires an interdisciplinary approach, investigating the transformation of suburbs from varied disciplinary perspectives grounded in cultural studies, economics, history, geography, city and regional planning, political science, and sociology. Our integrating framework uses the geography of a "suburb" as the common unit of analysis. Across our disciplines and different parts of the world, we share many definitions of the suburbs – be it cultural, economic, political, social, or spatial. The concept of the suburb acts as a unifier of disciplinary viewpoints and weaves together the theories, concepts, and approaches to the study of urbanization generally. As the boundaries of disciplines increasingly blur, such perspectives are essential for understanding complex phenomena such as the socioeconomic, cultural, and political transformations of the suburbs.

It is our goal that this volume not only serve as an important reference to the study of the suburbs, but also that it introduces new scholars and students alike to the world of metropolitan studies (the study of cities and suburbs) and that it adds to our shared body of knowledge suburbs as the places that most people live, work, and play in. And so, following in the tradition of our field of urban studies, we hope that *The Routledge Companion to the Suburbs* will serve a definitive and lasting reference to the study of global suburbanization.

The volume's contributors would like to acknowledge the following: Ann Forsyth's chapter is adapted and updated from her 2013 article in the *Journal of Planning Literature* 27(3): 270–281. Anjuli N. Fahlberg acknowledges that the research described in the chapter was funded by the National Science Foundation Law and Social Sciences Division's Doctoral Dissertation Research Improvement Grant (NSF 15–514), the Brudnick Center for the Study of Violence and Conflict, and various grants and fellowships from Northeastern University. Andrew Gorman-Murray and Catherine J. Nash Dean acknowledge Dean Mizzi for providing research assistance. Theresa Enright's chapter is adapted and revised from her 2016 chapter "A Thousand Layers of Governance" in *The Making of Grand Paris: Metropolitan Urbanism in the Twenty-First Century*, Cambridge: MIT Press, 159–190. Erina Iwasaki acknowledges the joint research project between Hitotsubashi University in Japan and the Central Agency for Public Mobilization and Statistics in Egypt. Juliet Carpenter acknowledges that this work has been supported by the European Union's Horizon 2020 research and innovation program under the Marie Skłodowska-Curie grant agreement No 734770, through the Co-Creation project. Bahar Durmaz-Drinkwater, Jaap Vos, and Asli Ceylan Öner would like to thank to their colleagues Lale Başarır, Filiz Keyder Özkan, Gaye Bezircioğlu, and Işın Can.

As the editors, we want to acknowledge two important audiences: 1) the contributors to this volume and 2) the larger contributions of our colleagues in the field. First, we want to express our gratitude to the writers of the individual chapters in this book. In any comprehensive approach to a field of study, bringing together a diverse group of scholars from around the world is no easy task. In fact, we were unable to obtain all the chapters we wanted, and so, as astute readers will recognize, some key places are missing. However, this volume does bring together 42 contributors from a dozen countries. Over the course of nearly two years, we had the privilege of working with these esteemed colleagues, including a network of established and emerging scholars. It is especially fulfilling to see our colleagues working in multigenerational teams, introducing new scholars to a dynamic field of study. It is similarly fulfilling to see multidisciplinary teams come together to study novel questions. Second, we want to acknowledge the work of several generations of scholars that has influenced our thinking and understanding about the city and the suburb. We are grateful for the rich body of work that many of our colleagues and collaborators have produced. You have shaped our thinking, understanding, and approaches to the study of the suburbs.

In addition, our professional societies have provided a venue for sharing our earlier versions of our work. In particular, the Urban Affairs Association, which is the international professional organization for urban scholars, researchers, and public service professionals, has provided an important interdisciplinary space for presenting and receiving feedback on our work. This international network of urban scholars has not only welcomed, but warmly embraced, our work, and our colleagues' work, for many years. Likewise, the American Political Science Association, the American Association of Geographers, and the Association of Collegiate Schools of Planning have all welcomed interdisciplinary urban affairs research as well as parts of our work that spoke to the disciplinary fields of urban politics, urban geography, and urban planning. These venues have enriched our contributions and networks.

Each of the editors would like to individually acknowledge the support for this project.

Bernadette Hanlon would like to thank her colleagues in the Knowlton School at The Ohio State University for their support of this project. The Knowlton School offers me the opportunity to grow intellectually while also being a warm, welcoming, and collaborative environment for students and faculty alike. I am thankful to be there. I am particularly grateful for the support of my faculty friends, Gulsah Akar, Kim Burton, Zhenhua Chen, Maria Conroy, Santina Contreras, Kyle Ezell, Rachel Kleit, Jesus Lara, Jason Reece, Kareem Usher, Tjis Van Maasakkers, and Amber Woodburn. Of course, as always, I am so grateful for my love, Kerry McCarthy.

Thomas J. Vicino would like to thank his colleagues in the Department of Political Science and the School of Public Policy and Urban Affairs at Northeastern University. The College of Social Sciences and Humanities has provided a rich intellectual home for the study and practice of the experiential liberal arts in a global context. I am grateful for the continued support of Joseph Aoun, James Bean, and Uta Poiger. I appreciate the constant support of Janet-Louise Joseph, Britain Scott, Rosy Trovato, and Jermichael Young. Research assistants also provided key support during this project, including Ryan Bloom and Cooper Gould. Last, I wish to thank my entire family for the unwavering support and love, especially Charles Galantini and Bella Vicino Galantini.

Finally, we would like to thank the editorial team at Routledge, including Andrew Mould and Egle Zigaite, for their support, guidance, and patience during this project.

Bernadette Hanlon
Columbus, Ohio
Thomas J. Vicino
Boston, Massachusetts
March 2018

Introduction

Bernadette Hanlon and Thomas J. Vicino

In 2016, more than half of the world's population resided in urban areas (Sennett et al., 2018). Future growth will continue to occur in cities (Burdett and Sudjic, 2007, 2011). According to projections, by 2050, more than two-thirds of the world's population will be urban (Sennett et al., 2018). By then, the urban population is expected to grow to 6.4 billion people (Sennett et al., 2018). Despite methodological challenges to measuring urbanization (Cohen, 2004), and arguments against reducing urbanization to a mere datum point (Brenner and Schmid, 2014), there is unquestionable evidence of increased concentrations of populations in rather complex and dense settlements in ways that bring about extraordinary challenges of social formation, environmental sustainability, land use planning, and urban governance. This fast-paced urbanization is occurring at the fastest-pace in the Global South, where three-quarters of the world's population currently lives (Mitlin and Satterthwaite, 2013). Much urban growth will occur in the periphery of cities, leading to the continued expansion of urban territory. Urban growth is necessarily suburban growth, and urban expansion, suburban expansion.

In their work on planetary urbanization, Brenner and Schmid (2014) note that urbanization is a process that unleashes the constant production of new urban configurations and fabricates patterns of varied urban conditions across the globe. Under capitalist production, urbanization is, in this sense, a boundless process, eroding traditional borders of autonomous urban form, extending outwards into new territories, and resulting in a generalized "explosion of spaces." Echoing the work of Lefebvre (2003, 36), scholars of planetary urbanization suggest "complete urbanization is today being actualized, albeit unevenly, on a worldwide scale" (Brenner, 2013, 19).

As Merrifield (2014, p. x) suggests, this urbanization of the world is a process that manifests itself in both the "undergrowth as well as overgrowth, in abandonment as well as overcrowding, in underdevelopment as well as overdevelopment." Urbanization is a multifaceted process that includes the growth of new territory in conjunction with the destruction of existing cities and towns that are, metaphorically speaking, peripheralized, and must seek out new ways to be remade. This point by Merrifield is relevant to the aim of this book. Sure, we are interested in examining suburbs within the context of the explosion of urban territory. Yet, we also see suburbanization as not merely about typical notions of growth and expansion. It is more complicated, and actually underpinned by dual processes of dispersion and concentration, development and abandonment, homogenization and fragmentation; being made and remade. Suburban

1

space expresses different forms, functions, and social characteristics, and is constantly shifting and changing in dynamic ways.

There are many examples of suburban form that include, for instance, low-density single-family housing development and strip malls of the American and Australian models, the high-rise modernist suburbs of say Paris or Toronto, squatter settlements in Asia, and mega-projects developed in Latin America and India, to name but a few. These different forms are continuously modifying, adjusting, and evolving. For instance, high-rise social housing characteristic of the *banlieues* in Paris undergo renewal (Kipfer, 2016), the post-war housing in U.S. suburbs is torn down and replaced with larger housing (Charles, 2013), so-called New Urbanist developments aimed at increasing density and creating more walkability are recreating new suburban living in Australia (Foster et al., 2015), gated communities are cropping up in borderlands across the globe (Webster et al., 2002), and suburban development creates completely new territories. The global suburban form is both varied and dynamic.

In addition, suburbs are socially diverse, and increasingly so. For instance, the *banlieues* in France are characterized as the sole location of social housing projects with poor and immigrant groups. Yet, the *banlieues* are varied; they can be poor (e.g., Clichy-sous-Bois) and rich (e.g., Neuilly-sur-Seine) and, within the *banlieues* themselves, the population is diverse in ways that include immigrants and native French, old and young people, rich and poor populations (Newman, 2015; Packer, 2015). Suburbs of Australian cities like Melbourne and Sydney are home to diverse groups that includes the Chinese, English, Koreans, Indians, Indonesians, Italians, and Vietnamese, for instance. In the United States, there is much concern about growing suburban poverty, immigrants moving directly to suburbs and bypassing central cities in their search for employment, and suburban decline and shrinkage, especially in former industrialized regions. A comparative study of industrial decline in suburbs of Europe and Latin America finds a similar dynamic (Audirac et al., 2012). The impact of post-Fordist economic restructuring on industrial suburbs is ubiquitous.

Suburban diversity can be fragmented, with groups segregated. In the case of the United States, racial and ethnic groups in the suburbs are often living in separate communities, and Blacks in particular find it difficult to integrate into predominately White suburban communities, regardless of income (Cashin, 2004). The periphery is a contested space, especially as the demand for land for more affluent and luxurious suburbs increases. In the case of India, Roy (2015) describes this well in her description of the forcible eviction of several thousand squatters in the informal settlement of Nonadanga on the edges of the Kolkta metropolitan region. A common occurrence in Indian cities, this eviction cleared land that the metropolitan authorities plan to lease to a developer for luxury retail development serving a growing suburban middle-class. Roy (2015, p. 338) suggests that the Nonadanga story reflects on the "variegated landscapes of slum and suburb that today characterize the metropolitan edges of Southern cities."

In the context of the Global South, the "classic" core periphery model often compares "planned areas to spontaneous developments, urbanized core to non-serviced peripheries" (Buire, 2014, p. 242). The evolution of highly planned "new towns," edge cities, and gated residential complexes across the globe undermine the notion that the periphery is an unplanned and spontaneous space with few services relative to the city core. As scholars have documented, gating is a global phenomenon, long-established as an important (sub)urban form in Latin America and South Africa but since the 1990s growing in prominence in North America, Europe, China, and South Asia. Atkinson and Blandy (2005) suggest that the rise of gated communities reinforces a trend toward a more polarized metropolis, particularly as the middle and upper classes withdraw from the public purse and embrace more privatized forms of governance and a growing privatized security infrastructure.

This book is a study of the heterogeneous nature of global suburbanization that considers an array of suburban landscapes, evolutions, people, policy environments, and political processes across different nations and regions. In this book, we aim to bring the periphery to the center of urban studies. We aim to focus attention on the suburban world not only to discuss the particulars of particular suburbs but, in fact, to examine broader topics of urban definition, morphological transformation, social exclusion, and the global political economy: topics pertinent to our understanding of the urban more generally.

Structure of the book

Let us take the first issue of definitions. We tackle suburban definitions and descriptions in Part 1 of the book. We have presented our work on suburbs at many academic conferences over the years, and at almost every presentation, someone in the audience will inevitably ask, "what do you mean by the suburbs?" In the first chapter of this book, Forsyth grapples with this question. She examines the range of suburban definitions that focus on physical, functional, and social dimensions as well as underlying processes pertinent to suburbanization. She, as importantly, asks us to consider why definitions matter. They matter both for practice and for the development of suburban theory, a process made difficult if terms and concepts are not clearly defined. This is similarly true for the urban and urban theory.

In some respects, the challenge to defining the suburbs is, in part, the result of long-lasting suburban stereotypes. In Chapter 2, Harris describes the positive and negative stereotypes associated with the suburbs, and how they have developed over time. Early critics of the suburbs focused on negative stereotypes including physical homogeneity, social conformity, and lack of culture. Positive stereotypes stress an image of suburbia as a place close to nature, a place where one can own a home and raise a family. Harris notes how positive stereotypes have been actively promoted by developers, builders, lenders, and others who profit from suburban development. In many respects, the positive and negative stereotypes associated with the suburbs scrambles our ability to accurately define them. We think in forms and outlines shaped by varied representations of the suburbs, representations that are often dominated by the Anglo-American suburban experience.

In Chapter 3, Phelps continues the discussion around the difficulty of definitions, noting how the city and suburb are deceptive terms. He offers a suggestion that maybe the suburbs are places that are not yet urban, that are on their way to being urbanized. He tackles too the definition of "post-suburban." This term was first delineated by Lucy and Phillips (1997) who noted the emergence of new spatial forms in the metropolis, including sprawling exurban development and suburban economic concentrations alongside the decline of older inner suburbs. The post-suburban era signals the increased recognition of a multiplicity of (sub)urban configurations.

In their research, Taylor and Lang (2004) refer to a rather exhaustive list of 100 new concepts related to recent urban change, noting the proliferation of terms describing new urban spaces including but not limited to "edge city," "mini-city," "metropolitan suburb," "megacenter," and "boomburb." Many of these forms lie against a backdrop of the traditional city and the traditional suburb. Phelps suggests that consideration of the different types of (post)suburb and longitudinal analysis of certain post(suburbs) is necessary to any empirical analyzes of urbanization processes and should help reveal specific contradictions in these processes. In his visionary 1970 work entitled *The Urban Revolution*, Henri Lefebvre (2003, p. 17), when announcing the coming of an "urban society," noted how it is the "analyst's responsibility to identify and describe the various forms of urbanization and explain what happens to the forms, functions, and urban structures that are transformed by the break up of the ancient city and the process of generalized urbanization." We could add to this not only those urban structures that are transformed but indeed those that are newly formed in peripheral space.

As we aim to demonstrate in this book, the suburbs are complex, varied, and global. In Part 2 of the book, we more explicitly tackle the global nature of suburbanization. Chapter 4 acts as an introduction to this part of the book. This chapter is written by Pierre Hamel and Roger Keil. Both scholars have been part of a Major Collaborative Research Initiative Global Suburbanisms: Governance, Land, and Infrastructure in the Twenty-first Century funded by the Canadian Social Sciences and Humanities Research Council. This initiative includes some 50 researchers working on a variety of suburbanization topics across many different countries around the world. In Chapter 4, Hamel and Keil make an argument for the need for global comparative suburban studies to help produce new empirical insights in the variety of suburban life around the world. These insights, they suggest, will facilitate the production of new ideas that can add to urban theory and our overall understanding of urban society. Their work is explored in other places including in their book, *Suburban Governance: A Global View*.

Part 2 of the *Routledge Companion to the Suburbs* includes case studies of suburbanization in particular cities, regions, or countries, across the globe, with researchers providing an historical perspective, and some focused on more contemporary processes. In Chapter 5, Herzog examines suburbanization in Latin America, identifying both vertical gated enclaves and mega-projects that often result in the displacement of squatter communities. Herzog argues that these types of suburban developments alongside sprawling tiny, box-like houses for the working classes result in extensive ecological damage, and problems of segregation and social isolation.

In Chapter 6, Freestone, Randolph, and Pinnegar examine the history of suburbanization in Australia with an eye to contemporary challenges that include the suburbanization of disadvantage, the regeneration of older suburbs, and effective suburban and metropolitan governance. Australia, like the United States, has difficulty increasing density in its auto-dependent and low-density suburban landscape.

In Chapter 7, McManus describes suburbanization in Dublin, recognizing that it was in the early nineteenth century that modern suburban development occurred around the city on a significant scale. Of course, Dublin's early modern suburbs must be understood in the context of Ireland's colonization by Britain. McManus highlights the somewhat unique mixture of class, religion, and politics that encouraged the development of independently governed Tory suburban townships in the south of Dublin by Protestant and Unionist citizens. Many of these townships had unique characteristics with the less powerful eventually being absorbed into the Dublin city boundary. McManus's chapter offers insights into the impacts of local particularities on Dublin's suburbanization process while, in her examination of more contemporary suburban development, she recognizes that Dublin has become a typical Western globalized metropolis, heavily influenced by global capital and the global economy.

There are some interesting parallels between Dublin's suburbanization and that of India. In Chapter 8, Shaw examines India's suburbanization over different periods of time, with early suburban development, like in the Dublin case, influenced by British colonialism. Yet, India's suburbanization process has many unique characteristics including the sheer size of population expansion, and the evolution of both "formal and informal" settlements on the periphery of cities. Shaw explains that these outer fringes include planned residential layouts as well as unplanned layouts and squatter settlements, leading to a heterogeneous mix of social classes in the periphery of Indian cities.

Turning to another part of Asia, in Chapter 9, Rukmana, Fahmi, and Firman note how Jakarta, as the national capital of Indonesia, has grown beyond its municipal boundary to become a metropolitan region of several different administrative districts and municipalities. The economic policies of the central government have been central to the evolution of large-scale industrial activities in this newly emerging post-suburban agglomeration. The periphery of the Jakarta region has become the site of a variety of satellite cities and economic activity centers.

The central government is also a major player in the suburbanization of Seoul. In Chapter 10, Choi and Lee describe how the central government has used a variety of land use regulations and housing policies to house an emerging middle-class in the periphery of Seoul. Opposite to the U.S. model, the government built high-rise apartment complexes for these middle-class families, way beyond the central business district.

The cases studies outlined in Part 2 of the book highlight the fact that, as Hamel and Keil suggest in Chapter 4, there is much local history to global suburbanization. Suburbanization, in this sense, is diverse and there are differences among suburbs both internally and across different regions, cities, and nations. Yet, there are some commonalities to the suburbanization process. State-led procedures matter as demonstrated by the policies and legislative and planning actions that influenced suburbanization across the different cases. In addition, suburbs are central to the global economic sector. Suburban territorial expansion also has tremendous ecological impacts and represents both a threat and an opportunity for more sustainable living in a time of climate change.

In Part 3 of the *Routledge Companion to the Suburbs*, we turn toward issues of exclusion and inequality in the suburbs. Imagined as defenders of race, class, and sexual uniformity and conformity, suburbs are actually important sites of difference. Chapter 11 by Gorman-Murray and Nash demonstrate this in their examination of queer suburbs, and the experiences of LGBT and queer people living in 'mainstream' suburbia. Their chapter explores the shifting landscape of sexualities and recognizes how the intersection of sexuality, gender, class, race, and age shape LGBT spaces in urban and suburban areas.

The inclusion of different groups in the suburbs is constantly being negotiated, and often ethnic and racial groups and the poor are segregated and isolated spatially and politically. This is illustrated by the examination of poverty and inequality in the suburbs of Cairo. In Chapter 12, Iwasaki demonstrates that in the unplanned suburbs, poverty has become more pronounced and deteriorated in the latter part of the 2000s, just before the Arab Spring of 2011. In Chapter 13, Dahlstedt and Ekholm explore the social exclusion and segregation of young ethnic populations in Sweden, a situation that has resulted in riots and protest. In Rio de Janeiro, the people of the favelas are excluded from urban life. In Chapter 14, Fahlberg demonstrates how, despite their physical location close to downtown, the people of the favelas are symbolically and politically peripheralized from the city.

In Part 3 of the book, there are a number of chapters that consider the growing diversity of U.S. suburbs. For instance, in Chapter 17, Walker describes the contemporary settlement of immigrants in the suburban periphery. The traditional image of suburbia as made up of white middle-class families has been undermined by identification of a diversity of suburban types, and a growing number of different types of people living in U.S. suburbs. In Chapter 15, Teaford reminds us that metropolitan regions of the United States are fragmented and divided because of a multiplicity of local governments each vying for their own interests, and each aiming to exclude certain groups and certain types of development.

The balkanized nature of U.S. suburbia has implications for the fiscal capacity of certain suburbs to deal with the problem of suburban poverty. There are challenges to providing adequate social services for a growing population of the suburban poor (Allard, 2009; Murphy and Wallace, 2010). In Chapter 16, Anacker describes three different place-based policies – the Community Development Block Grant (CDBG) program, the Home Investment Partnership (HOME) program, and the Neighborhood Stabilization (NSP) program – that benefit poor suburbs. She warns us that the loss of these programs could be detrimental to declining suburbs and poor people living in these places.

The social transformation of the suburbs and the social exclusion of particular groups creates political tension and division. In his recent book, *Urban Rage*, Dikeç (2018) examines contemporary riots in cities (and suburbs) in the United States, the United Kingdom, France, Sweden, Greece, and Turkey. He highlights as causes the problems of exclusion, inequality, the stigmatization of minority groups, decreased civil liberties, and large-scale structural changes to the economy and political system that began in the 1970s. After reading both Chapter 18 by Lung and Schafran and Chapter 19 by Airgood-Obrycki and Price, urban rage could just as easily be suburban rage. Both these chapters touch on the ways in which new minority groups living in suburbs (in the context of the United States) are excluded from local political processes, are looked upon with suspicion, and suffer the brunt of police brutality and violence.

Similar to Dikeç (2018), Lung and Schafran find that there is often a spark that ignites suburban rage, and this typically involves the violent death of a young black man at the hands of the authorities. In the case of Sanford, Florida, protests occurred after Trevon Martin, an unarmed black teenager, was shot and killed by a neighborhood watch member in a gated community. In the case of Ferguson, Missouri, unrest happened after Michael Brown, another unarmed black teenager, was shot and killed by a Ferguson police officer. Lung and Schafran suggest that a long history of structural racism, racialized policies and practices, and growing suburban poverty are instrumental factors in the unrest that ensued in these two very different suburbs. In Chapter 19, Airgood-Obrycki and Price ask if the social transformation of U.S. suburbs, and protests for suburban social equity will result in the stigmatization of certain suburbs, particularly in light of the growing awareness of suburban poverty and its consequences.

In Part 4, we consider the suburban policy environment, governance, and suburban morphology, particularly in the context of the impact of planning and behavior on the design and evolution of suburbs. Chapter 20 and Chapter 21 focus on the French context. In Chapter 20, Enright describes the challenges to metropolitan governance in Paris. One of these challenges revolves around the divide between the city of Paris and its suburban periphery. In France, the territorial stigmatization of the *banlieues* (Wacquant, 2008, pp. 163–198) is juxtaposed against a much more positive characterization of the city of Paris. However, as Enright outlines, the emergence of intermunicipal partnerships between the city and its cross-border suburbs has helped establish regional relations and new governance structures. In 2016, a new metropolitan government, the Métropole du Grand Paris, emerged to take over several functions from member municipalities including metropolitan development, housing, urban and environmental policy, crime prevention and economic and social development. The administrative area of the Métropole du Grand Paris includes Paris, the three departments of the first ring of suburbanization, *petit couronne* or "small ring," and the municipalities of the *grand couronne* or "large ring" in the second tier. The metropolitan government is responsible for developing a Metropolitan Master Plan and has the ability to generate tax revenue and permits revenue sharing among local municipal members. In many respects, there is increased recognition of the need to integrate politically what is effectively a global city-region. And yet, there is an argument that the new metropolitan structure does not extend far enough since it still privileges the urban, excluding large periphery territories, and thus encouraging spatial division across what is an expansive economic region.

In Chapter 21, Carpenter details attempts to integrate the *banlieues* into French society. The French *banlieues* have become synonymous with marginalization, and the low-income and ethnic populations living in the large housing estates that exist there feel excluded and isolated from mainstream French society. Carpenter critically explores the various urban renewal programs and planning efforts that have recently been put in place to encourage the redesign and social mixing of the *banlieues*, efforts that in large part have, however, proven unable to alleviate the territorial stigmatization of these places.

In Part 4 of the book, three chapters are focused in suburban morphological transformation and development in the United States context. In Chapter 22, Hollander, Polsky, Zinder, and Runfola consider the problem of suburban shrinkage, examining the extent of housing vacancy. In the United States, planning typically focuses on managing suburban growth and is less concerned with the problem of suburban population loss. Hollander, Polsky, Zinder, and Runfola stress the need for suburban shrinkage policies to help deal with the aftermath of the recent housing crisis that impacted different regions of the United States.

In Chapter 23, Benton-Short explores efforts to design suburban space in specific ways and with specific intentions, focusing on the development of new towns and master-planned communities in the United States during the 1960s and 1970s. Benton-Short describes early prototypes of new town suburbs in the United States, and then provides a case study of the planned suburban community of Mission Viejo in California. Mission Viejo's master plan embraced new town principles with, according to Benton-Short, a number of innovative suburban design elements to help produce a sense of place. Heavily marketed, Mission Viejo sought to attract high- to middle-income families. In more recent years, housing in Mission Viejo has become unaffordable to the typical, young middle-income family attracted to this suburb during the days of its early evolution.

In Chapter 24, Charles explores the tear-down and mansionization process taking place in older suburbs of the United States. A process of creative destruction, Charles explores the motivation behind the tearing down and rebuilding of older homes in inner-ring suburbs, noting the emulation of housing consumption patterns of wealthier households. There is a desire to exhibit wealth and status by having a bigger home than your neighbors. Understanding such motivations provides us with a better understanding of the suburban gentrification process.

Consumption patterns are relevant too in the evolution of suburban Turkey. This is described in Chapter 25. Focusing on the case of Urla located in the metropolitan region of İzmir, Durmaz-Drinkwater, Vos, and Öner note the importance of amenity migration in the emergence of Urla as a residential suburb. Because of its environmental assets, historic and cultural amenities, and proximity to İzmir, this once rural town has evolved into an important destination for upper- and middle-income groups either living or working in İzmir. This has been aided by governmental legislation and policies.

The examination of Urla provides an interesting segue into the examination of cultural production in the suburban context in Chapter 26. In this chapter, Bain snuffs out the stereotype of suburbs as bland and suburbanites as banal. She describes the suburbs as culturally rich and complex places, and often the living environment of creative people. She notes that suburban cultural workers make important contributions to place-making but are largely ignored by civic leaders and policymakers. She suggests that cultural policymakers need to better engage with suburban residents and provide more resources for suburban cultural production.

Part 5 concludes the book. In Chapter 27, Short announces the end of the suburbs and *Suburbia*. This, we think, is a fitting end to the book. *Suburbia* is a myth embedded in a post-war American suburban experience when in reality global suburbanization has tremendous heterogeneity and difference. Short asks the important question: can we legitimately speak of suburbs as something different from the rest of the metropolitan region? He suggests we need to develop more sophisticated models and understandings of the complexity of the global metropolis that do not rely on obsolete terms that can no longer capture the vastness of urban expansion.

As editors, we conclude the book in Chapter 28. We suggest that the ubiquitous nature of the suburbanization process has permanently undone any "classic" division between city and suburb, between center and periphery. As Merrifield (2014) suggests, "inside the urban fabric today we see centers and peripheries all over the place, cities, and suburbs within cities and suburbs, centers that are geographically peripheral and peripheries that suddenly become centers." So, what to do? Do we rid ourselves of the suburbs as obsolete, as Short suggests in Chapter 27? Maybe. We argue, though, that the very study of suburbs and suburbanization has helped and continues to help us understand the complexity of urban society. In his recent book, *Suburban Planet*, Keil (2018, p. 41) suggests that "urban theory has traditionally been an exercise of understanding the city from the inside out." Traditional and now outdated models of the metropolis were built on notions of urban centrality and density characteristic of the nineteenth-century city. From Keil's (2018) perspective we can build a critical urban theory and an understanding of urban society from the outside in. This book, we hope, is part of that effort.

References

Allard, S.W. (2009) *Out of Reach: Place, Poverty, and the New American Welfare State*. New Haven, CT: Yale University Press.

Atkinson, R. and Blandy, S. (2005) "International Perspectives on the New Enclavism and the Rise of Gated Communities." *Housing Studies*, 20(2), pp. 177–186.

Audirac, I., Cunningham-Sabot, E., Fol, S., and Torres Moraes, S. (2012) "Declining Suburbs in Europe and Latin America." *International Journal of Urban and Regional Research*, 36(2), pp. 226–244.

Brenner, N. (ed.) (2013) *Implosions/Explosions: Toward a Study of Planetary Urbanization*. Berlin: Jovis Verlag.

Brenner, N. and Schmid, C. (2014) "The 'Urban Age' in Question." *International Journal of Urban and Regional Research*, 38(3), pp. 731–755.

Buire, C. (2014) "Suburbanisms in Africa? Spatial Growth and Social Transformation in New Urban Peripheries." *African Studies*, 73(2), pp. 241–244.

Burdett, R. and Sudjic, D. (eds.) (2007) *The Endless City*. London: Phaidon.

Burdett, R. and Sudjic, D. (eds.) (2011) *Living in the Endless City*. London: Phaidon.

Cashin, S. (2004) *The Failures of Integration: How Race and Class are Undermining the American Dream*. New York: Public Affairs.

Charles, S.L. (2013) "Understanding the Determinants of Single-Family Residential Redevelopment in the Inner-Ring Suburbs of Chicago." *Urban Studies*, 50(8), pp. 1505–1522.

Cohen, B. (2004) "Urban Growth in Developing Countries: A Review of Current Trends and a Caution Regarding Existing Forecasts." *World Development*, 32(1), pp. 23–51.

Dikeç, M. (2018) *Urban Rage: The Revolt of the Excluded*. New Haven, CT: Yale University Press.

Foster, S., Pereira, G., Christian, H., Knuiman, M., Bull, F., and Giles-Corti, B. (2015) "Neighborhood Correlates of Sitting Time for Australian Adults in New Suburbs: Results from Reside." *Environment and Behavior*, 47(8), pp. 902–922.

Keil, R. (2018) *Suburban Planet Making the World Urban from the Outside In*. Cambridge, MA: Polity Press.

Kipfer, S. (2016) "Neocolonial Urbanism? *La Rénovation Urbaine* in Paris." *Antipode*, 48(3), pp. 603–625.

Lefebvre, H. (2003) *The Urban Revolution*. Minneapolis: University of Minnesota Press.

Lucy, W. H. and Phillips, D. L. (1997) "The Post-Suburban Era Comes to Richmond: City Decline, Suburban Transition and Exurban Growth." *Landscape and Urban Planning*, 36, pp. 259–275.

Merrifield, A. (2014) *The New Urban Question*. London: Pluto Press.

Mitlin, D. and Satterthwaite, D. (2013) *Urban Poverty in the Global South: Scale and Nature*. London: Routledge.

Murphy, A. K. and Wallace, D. (2010) "Opportunities for Making Ends Meet and Upward Mobility: Differences in Organizational Deprivation across Urban and Suburban Poor Neighborhoods." *Social Science Quarterly*, 91(5), pp. 1164–1186.

Newman, A. (2015) *Landscape of Discontent: Urban Sustainability in Immigrant Paris*. Minneapolis, MN: University of Minnesota Press.

Packer, G. (2015) "The Other France: Are the Suburbs of Paris Incubators of Terrorism?" *The New Yorker*, August 31.

Roy, A. (2015) "Governing in Postcolonial Suburbs." In: Hamel, P. and Keil, R. (eds.) *Suburban Governance: A Global View*. Toronto: University of Toronto Press, pp. 337–348.

Sennett, R., Burdett, R., Sassen, S., and Clos, J. (2018) *The Quito Papers and the New Urban Agenda.* London: Routledge.

Taylor, P.J. and Lang, R.E. (2004) "The Shock of the New: 100 Concepts Describing Recent Urban Change." *Environment and Planning A,* 36, pp. 951–958.

Wacquant, L. (2008) *Urban Outcasts: A Comparative Sociology of Advanced Marginality.* Cambridge, MA: Polity Press.

Webster, C., Glasze, G., and Frantz, K. (2002) "The Global Spread of Gated Communities." *Environment and Planning B,* 29, pp. 315–320.

Part I
Suburban definitions and descriptions

Part 1
Standard definitions
and descriptions

1

Defining suburbs

Ann Forsyth

The problem of definition

What is a suburb? In the coming decades, billions of people will move to urban areas and many will live in areas that are already considered suburban. How many people, and what that means, depends heavily on how suburbs are defined. As a vibrant literature on framing in planning work suggests, how urbanists, the press, and the public talk and think about the suburbs shapes how they can see such areas being developed and redeveloped in the future (Caplan and Nelson, 1973; Schon and Rein, 1994; Harvard Law Review Association, 2004; Healey, 2009). In the coming decades, as new suburban areas are built and older suburbs head toward redevelopment, clearer definitions, or better alternatives to the term "suburb," can help focus academic and practical debates on important issues.

In 1958, Kurtz and Eichler published an article in *Social Forces* complaining about confusion over "residence categories" and, in particular, the terms "suburb" and "fringe." As they pointed out, when concepts are not clear it is hard to create an adequate theory. The situation has not improved much in subsequent decades. As Lineberry (1975, p. 2) argued in the mid-1970s, "despite the voluminous literature on suburbia, we are no closer than ever to a definition. It is a mere assumption of convenience that we all know what we are talking about, however variegated the pictures in our heads." While Lineberry attempted to provide some clarity, decades later Harris, reviewing the international literature on suburbs, still complained that the field needed to establish a "minimum definition to which suburbs everywhere conform" (Harris, 2010, p. 26). Among urban scholars, then, there is no consensus as to what exactly constitutes a suburb and the confusion expands when one includes popular and media accounts.

This chapter examines the range of suburban definitions. These include definitions focused on the physical, functional, social, and process dimensions as well as others that take a more analytical or critical view. Obviously, definitions of terms such as suburbs are social constructions or deliberate abstractions, focusing attention on some aspects of suburbs and not others. Several related areas in urban studies – primarily urban planning, urban history, urban sociology, and urban geography – have generated many of the definitions, demonstrating disciplinary differences. For example, historians strive to define suburbs in a way that makes sense over time and urban sociologists are particularly concerned about social relations. Those working in low-income countries often use an overlapping term – peri-urban development – that most commonly refers to the urban or suburban fringe but may also refer to closer areas (Adell, 1999; Iaquinta and Drescher, 1999).

Given this multiplicity, one approach is to abandon the word "suburb" and replace it with terms referring to specific types of suburbs or particular features, such as density. This has some advantages in terms of reducing ambiguity. However, it can also be argued that focusing on key dimensions misses the big picture of metropolitan growth and change.

Finally, it should be noted that a number of authors either reject the term "suburb" as obsolete or propose that it is impossible to define suburbs due to their diversity (Archer, 2005, p. 440). The former group has proposed alternative types of "post-suburban" environments such as technoburbs and urban realms (Phelps, 2015; Teaford, 2008, p. x; Webber, 1964). They have a point, particularly if they continue to define suburbs as primarily residential and middle-class. For the purposes of this review, however, I have included these post-suburban environments as types of suburbs and dealt with them under various dimensions. As I note, alternatives to the residential suburbs have been part of the scholarly debate on suburbs for almost a century, so the idea of the non-residential post-suburban environment is really not so new.

For those wishing to define suburbs as a whole, the most practical approaches are based on the suburbs' outer locations in the metropolis and their relative newness. These two dimensions provide a somewhat distinctive range of opportunities and problems in suburbs including potentially limited access to services, newer social networks, closeness to undeveloped or rural areas, and often lower-cost of land.

In identifying definitions, I encountered a difficulty. Surprisingly few people who write about suburbs define them explicitly as a whole – including many classic, influential, and otherwise important works on suburbs. Some focus on specific types of suburbs, defining them quite clearly but not dealing with suburbs more generally – for example, authors may focus on ethnoburbs, technoburbs, suburban master-planned communities, or streetcar suburbs. Some of my own work has taken this form. Others define suburbs through examples by applying the term suburban to particular places or characteristics from which the reader can deduce a definition. However, it is hard to piece together a comprehensive definition from such accounts. Yet, others focus on areas that are clearly suburban by many definitions – for example, new developments of detached housing on the urban fringe. However, they do not pay much attention to articulating whether other kinds of developments are also suburban.

In order to locate more explicit and comprehensive definitions, I started by reviewing sources likely to define suburbs, such as census agency manuals. I looked at books on suburbs, searching via the combined library catalog, Worldcat. In addition, I located literature using Google Scholar, applying key words such as variants on the terms "suburb" and "definition." I also searched using questions such as, "What is a suburb?" I started with Google Scholar because it picks up a wider variety of sources than, for example, Web of Knowledge (ISI/Thompson). One assessment in social work found it located four times the number of disciplinary journals (Hodge and Lacasse, 2011). It also is very simple to search for works that have cited a particular piece. However, I also checked my search against several other databases including Web of Knowledge, Summon (a database aggregator used by libraries – my library had 800 million references in the database), and the Avery Index to Architectural Periodicals. Combined, these other databases added four more relevant references, all from Summon.

In this way, I traced a range of authors who defined suburbs. More important, I examined the articles and books they cited and located works that cited the sources I had found. Some of these also provided definitions. I did not intend to inventory every work defining suburbs but rather to show the range of definitions across the century in which the urban studies literature has grappled with the phenomenon of suburbanization. A number of definitional issues – such as whether suburbs are essentially residential or can contain employment areas – are long-standing, but that situation is not always obvious from recent debates. While drawing mainly on work from the United States, United Kingdom, and Australasia, where possible, this chapter extends more globally within the English-language literature.

Why definitions matter

Why does better defining suburbs matter for urban studies and planning? People have been discussing and studying suburbs for decades without any consistent definition, so perhaps there is no need for one. There are a number of reasons, however, why it is important to define suburbs clearly.

First is the issue of action. As Caplan and Nelson (1973, p. 200) pointed out some decades ago in the context of social problems, "What is done about a problem depends on how it is defined" (Schon and Rein, 1994; Healey, 2009). For example, if suburbs in the United States are seen as essentially white and middle-class or elite, policymakers may pay less attention to the real achievements and problems of African American suburban residents or low-income suburbs. If they are seen as essentially automobile-dependent, the many examples of transit-oriented suburbs may be ignored. In public debates, people may talk past one another.

Second is the problem of research and theory. Conducting empirical research requires adequate definitions of features and concepts being measured. As Kurtz and Eichler (1958) argued in the 1950s, it is difficult to develop an adequate theory of suburbs if terms are not clearly defined. If one study defines suburbs as metropolitan municipalities outside the central city and another as places that are dominated by detached housing, they will be examining different areas, making comparisons and generalizations more difficult. While researchers may themselves be careful about such issues, those using research findings may well miss these subtle differences and misinterpret the implications.

Finally, even if one does not consider that clear definitions matter for theory and practice, it is still worthwhile to review the variety of definitions to help reduce confusion in the field. Such a review provides scholars, students, and practitioners with a roadmap for identifying the perspectives of contributors to debates about suburbs.

Forms of definitions

Analyzing how suburbs are defined is not easy because the methods of constructing those definitions vary quite a bit, even among those authors who define suburbs explicitly. Table 1.1 explains some of these differences.

First, is whether the definition proposes what a suburb is (also called a positive definition) or focuses on what it is not or what it lacks (also called negative). This is not the same as whether the author likes suburbs or not. A positive definition of a suburb may focus on aspects of suburbia seen as problematic by the analyst (e.g., that it is automobile-oriented); similarly, a negative definition of what a suburb lacks may focus on the absence of problems.

Table 1.1 Approaches to defining suburbs explicitly

	Core essence	*Features and types (family resemblance)*
What a suburb is/positive	Example: LOW-DENSITY primarily residential areas	Example: First, second, and third ring suburbs; suburbs as low-density, with detached houses, middle-class families, substantial open space, and scattered employment
What a suburb lacks/negative	Example: Suburbs are within metropolitan areas (not rural) and outside the central cities (not core)	Example: Not cultured, not diverse, unequal, not dense

Second, and cutting across this first issue, is whether the definition focuses on a core essence of "suburbanity," for example, that all suburbs, or the most typical suburbs, have low densities and are primarily residential, or lists a set of features or types of suburbs that hang together with a family resemblance (Wittgenstein, 2009). Many definitions revolving around features and types are quite complex, which is part of the reason they were not reduced to an essence.

Several key topics often appear in definitions of suburbs, whatever the approach. Table 1.2 demonstrates some of these; these reflect definitions of suburbs as a whole. What is obvious from the short listing, however, is how varied the dimensions are and how potentially complicated definitions become when dimensions are combined.

Table 1.2 Key dimensions for defining suburbs with examples of definitions

Dimension	Brief description	Examples
Physical (where, what)		
Location	Where the suburbs are within a metropolitan area	Suburbs as on the outskirts of a town; definition unofficially derived from U.S. census – suburbs as within metropolitan areas but outside of core cities
Built environment characteristics	Key physical features related to development patterns or building and landscape types	Suburbs as having large areas of low-density detached houses
Functional (operations)		
Transportation	How people access and get around in suburbs	Suburbs as automobile-oriented
Activities	Functions and uses of the place	Suburbs as mainly residential developments with segregated uses
Social (who)		
Political places	Defined by municipal or similar boundaries	U.S. definition of suburbs as municipalities outside the core city
Sociocultural	The population character, level of exclusivity, and cultural heritage and tastes	Suburbs as middle-class or exclusive; a suburban way of life
Process (how, when)		
Styles of building, design, and planning	Who builds; the level of planning in terms of amounts of control and scale of planning unit	Suburbs as incremental and speculative developments
Time	Relates to relative newness, or its period of development	Suburbs as areas from the period since World War II
Analytical		
Critical assessments	Suburbs seen as problematic	Suburbs as sprawling, conformist, isolating, elite, locationally disadvantaged, and/or ugly places
Indices and indicators	Defined using criteria combined into some kind of indicator	Sprawl indices, fiscal capacity classifications

Many definitions of suburbs as a whole incorporate several dimensions at once. Some examples below provide a flavor of these combinations; the dimensions they represent are indicated in Table 1.3. While the focus of this chapter is on single dimensions, many definitions combine at least two of them. For example:

- Gober and Behr (1982) used discriminant analysis to check the importance of nine characteristics thought to distinguish suburbs from core cities in the United States, including age and family status, ethnicity, income, density, auto-orientation, housing age, and employment in manufacturing, retail, and services. They found race and ethnicity to be the most important variable distinguishing central cities and suburbs in the U.S. at the time.

  (Gober and Behr, 1982)

- Harris and Larkham (1999) used historical and geographical approaches to note five characteristics of suburbia focusing on North America and the United Kingdom: 1) peripheral location; 2) residential character; 3) low-density with perhaps high levels of owner occupancy; 4) a distinctive way of life; 5) separate identities for communities often at the municipal level.
- Writing a decade later and attempting to provide a truly global definition, Harris recast suburbs as having three dimensions internationally – 1) "peripheral location;" 2) (usually) having "residential densities intermediate between those of the city and the country;" and 3) relative "newness."

  (Harris, 2010, pp. 27, 29)

- Johnson (2006, p. 261), a geographer working in Australia, saw Harris and Larkham's (1999) definition as being relevant but added her own classic definition: "The idea of a single storied, freestanding dwelling on a relatively large allotment, in a mainly residential area, with strong local identity and limited governance, located midway between the city center and rural lands, where women tend to children and community while their husbands journeyed elsewhere for paid work, encapsulates the Australian suburb." The expansion of Australian cities, she pointed out, had provided an array of different types, so that in major Australian cities, suburbs had fundamentally departed from this definition.

  (Johnson, 2006, p. 261)

- Journalist Flint (2006, p. 2), focusing on the situation in the United States, simply described "suburbia – spread-out, drive-thru, car-dependent, newer-the-better suburbia." This simple definition combines the dimension of density, newness, and dominant transportation mode.

Table 1.3 Example definitions categorized by dimensions dealt with

	Locational	Physical characteristics	Transportation	Activities	Political places	Sociocultural	Styles of building, planning, design	Time	Critical assessments	Indices
Gober and Behr (1982)		x	x	x		x	x			x
Harris and Larkham (1999)	x	x		x		x				
Harris (2010)	x	x						x		
Johnson (2006)	x	x	x	x	x	x	x	x	x	
Flint (2006)		x	x			x				

Table 1.4 Example continua of suburban types

Dimension	Extreme 1		Extreme 2
Sociocultural	Socially homogenous		Socially mixed
Sociocultural	Residence by choice for all groups		Economic and/or regulatory exclusion; resettlement
Sociocultural	Low-cost		High-income elite developments
Style	Large-scale developer/builders	Middle sized builders doing small subdivisions	Build by owner
Style	Planned at a neighborhood or town scale		Incremental, informal
Style	Fitting in well-designed regional plan		Not fitting any regional plan, poor regional plan
Style	Highly designed neighborhoods	Design for major buildings/spaces	Little overall design or planning input
Style	Popular aesthetics		Elite aesthetics
Style	Interspersed with or adjacent to "natural" areas		Heavily urbanized

It is obvious that different authors have different emphases related to their substantive interests and the countries they are studying. They also reflect the changing character of suburbs over time – both in one location such as the Australian suburbs described by Johnson, and between places.

Not all definitions are essentialist. Some that can be better seen as locating a family resemblance list features and types of suburbs. Table 1.4 provides a small sampling of these features for just two dimensions. For example, suburbs may vary from socially homogenous to socially mixed in terms of their sociocultural character, and in terms of style may be built at a large-scale by a developer or house by house by an owner. Combining just these two dimensions creates an amazing variety of suburban types, most of which actually exist, e.g., both large-scale homogenous and socially mixed suburbs, and built-by-owner homogenous or mixed suburbs. One form of definition would list these types.

Dimensions

A less complicated way to analyze this situation is to look at it one dimension at a time.

Location

Suburb is not a new word but rather comes from the Latin *suburbium*, or "under the city," with a plural *suburbia*. The Oxford English Dictionary's (OED) first definition of suburb refers to suburbs as a location with examples dating from the late fourteenth century. "The country lying immediately outside a town or city; more particularly, those residential parts belonging to a town or city that lie immediately outside and adjacent to its walls or boundaries" (Oxford English Dictionary, 2011a; McManus and Ethington, 2007, pp. 319–320). Suburb could mean both a literal place as well as the figurative outskirts of something such as a place, idea, or event (examples include the "suburbs" of Lent, of a narration, of sense).

At their most basic, then, suburbs are outside of a town or city but belonging to it (Frost, 1991). But what does this mean? It in part depends on where the boundary or wall between inside and outside a town or city is located, between what is urban and what is rural.

In the field of urban studies, the city or town is typically the whole metropolitan area, which defines urban as being everything that is not rural or wild. The U.S. census has been changing its definition of urban over the years but has a similar characterization. By 2010, the definition of an urban area was

> a densely settled core of census tracts and/or census blocks that meet minimum population density requirements, along with adjacent territory containing non-residential urban land uses as well as territory with low population density included to link outlying densely settled territory with the densely settled core.
>
> *(U.S. Census Bureau, 2010)*

Urban census tracts have less than 3 square miles in area and population density of at least a thousand per square mile, with a minimum population of 2,500 people. Rural areas are all nonurban locations.

While the U.S. census does not define suburbs, many analysts have made a rough approximation by taking the urban (metropolitan) area and subtracting census-defined central city municipalities. This approach works better in areas with many local governments than in places like Texas, where annexations mean that many central cities take up much of the metropolitan areas. According to calculations published by Statistics Canada, if a similar definition were used in Canada, only 8 percent of Calgary's population and 9 percent of Winnipeg's would be suburban, but 73 percent of Vancouver's would be (Turcotte, 2008).

A major difficulty with a locational definition is that urban areas come in a lot of different shapes and sizes. One estimate is that there are 50,000 urban areas in the world – but some are quite small and a suburb in a town of 5,000 is different in character to one in a city of five million (Satterthwaite, 2007a, 2007b). Suburbs have been developed for a very long-time now; is a place that was on the urban edge 50 or 100 years ago still a suburb, even if it has a classic suburban form or is in a non-central municipality?

In part to get around this problem, Fishman (1987, p. 117) distinguishes between a suburb, defined by location, and a "true suburb . . .[that] must embody in its design a 'marriage of town and country,' a distinct zone set apart from by the solid rows of city streets and from rural fields." That is for "true" suburbs, he rejects a purely locational definition in favor of one based on physical characteristics and styles of building, design, and planning. Lineberry (1975) distinguishes between locational definitions from the U.S. census and cultural definitions like those of Johnson (2006) above – related to family type, segregation of home and work, and so on. He points out that, at least up to the 1970s, the two definitions were often "confused" (Lineberry, 1975, p. 3).

For those who want to keep a locational definition but make it more nuanced and useful for comparative study, one approach is to imagine the city as rings around a core with variously inner (or first), middle (or second), and outer (third) ring suburbs (Johnson, 2006; Green Leigh and Lee, 2005). Gans (1968, p. 49) provides such a definition, comparing the inner-city, the outer city of "stable residential areas that house the working-and middle-class tenant and owner," and the suburbs that are "the latest and most modern ring of the outer city, distinguished from it only by yet lower densities and by the often irrelevant fact of the ring's location outside the city limits." It is often unclear how much these outer suburbs overlap with those areas termed the suburban fringe, the peri-urban fringe, or the exurbs (Adell, 1999). In Canada, the statistical agency has tried to operationalize this kind of definition as rings of a certain distance from the census tract containing the city hall of the most central municipality, with rings out a certain number of kilometers – that is, 0–5, 5–9, etc. (Turcotte, 2008, p. 5). However, this raises the issue of what distances to use.

Built environment characteristics

Some people know suburbs when they see them because they have certain features that can be identified with the naked eye – detached housing, single-story factories and warehouses, campus-style low-rise office complexes, strip mall shopping centers, and large-scale shopping malls. For example, Dunham-Jones and Williamson (2009, p. x) describe suburbs as dominated by lower-density, single-use private buildings designed as objects in a landscape and funded or built by short-term investors such as real estate investment trusts and larger scale home builders; transportation is auto-oriented, with a looped and cul-de-sac network. Others discuss a landscape between the city and the country in form as well as location (Harris, 2010; Fishman, 1987).

Density is one favored characteristic for at least partially defining suburbs physically because it seems to be meaningful and easy to measure. Authors, however, differ as to whether to measure population, employment, or housing unit density; and if density gradients are most important (Gans, 2009; Sridhar, 2004; Harris, 2010). Some early suburbs were denser than the center cities, even in the United States; worldwide there are a number such examples today (Borchert, 1996; Harris, 2010). There are also related concepts such as building intensity (e.g., building bulk), building type, and perceived density that might be relevant in definitions (Forsyth et al., 2007).

Statistics Canada has attempted to construct a definition related to density and housing type: "we will refer to a neighborhood as low-density when at least two-thirds of the occupied housing stock comprises single and semi-detached houses and mobile homes, that is, dwellings that take up the most space or area per occupant" (Turcotte, 2008, p. 6). They chose this to deal with large census tracts with lots of non-residential uses that might be high-density in the residential portions but low-density overall. Density-based definitions have strengths in focusing on issues relevant to policymakers (i.e., issues they can regulate), and that are relatively easy to measure (if researchers can agree on what kind of density to assess).

Transportation

Closely related to definitions based on physical features are functional definitions, prominently definitions related to means of gaining access to suburbs. These are related to locational definitions but foreground linkages to the core city. Clapson, in a review of definitions, cites a common version, defining suburbs as "beyond the heart of the town . . . [but] within its urban orbit" (that is within commuting distance), with a "geography . . . intermediate between the town center and the countryside," and also depending on the town center for "shopping, leisure, and other requirements" (Clapson, 2003, p. 2; Thorns, 1972; Douglass, 1925; see review in Schnore, 1957). Earlier, Douglass (1925, p. 8) had defined as suburban lower-density locations from which "the heart of the city can be reached conveniently, quickly, and at low-cost." This is a viable definition in smaller areas, though its underlying assumptions are less tenable in larger polycentric cities.

Others are less worried about what suburbs have in common in terms of transportation but how they are shaped by a dominant mode. For many authors, suburbs are primarily automobile-based, e.g., "spread-out, drive-thru, car-dependent, newer-the-better suburbia" (Flint, 2006, p. 2). However, collective transportation – from railways and buses to vans and shared taxis – has also provided access to and around suburbs (Warner, 1978). Suburbs could then be defined as the sum of different types, e.g., streetcar suburbs plus automobile-based suburbs. This issue obviously overlaps in important ways with critical assessments as many authors (critically) equate suburbia with automobile-based sprawl.

Activities

A common approach when considering activities has been to see suburbs as primarily residential, providing bedroom areas for the larger metropolis. Others add that it must also be "well off, and marked by single-family homes" (Garreau, 1991, p. 149). This view of suburbs as mostly residential was never the whole picture as even residential areas needed shops, schools, the offices of local professionals and tradespeople, faith communities, and such. Many industries suburbanized early and people maintained some rural activities in early suburban areas (Fogelsong, 2005). But for many this definition of the primarily residential suburb was compelling, particularly through the 1960s.

Of course, suburbs are now more varied and a number of non-residential suburban types have become prominent. Over the past century, authors have argued for new terms to identify non-residential or mixed-use suburban areas. In 1925, Douglass distinguished between suburbs of "production" and "consumption" (Douglass, 1925, pp. 74–92; Harris, 1943; Berger, 1960). In the 1950s, Schnore proposed that "residential suburbs" be distinguished from "employing satellites" (Schnore, 1957, p. 122; Berger, 1960). Many of the areas labeled "post-suburbia" are mixed-use areas or job centers (although some are declining suburbs) (Kling et al., 1991; Lucy and Phillips, 2000, pp. 4–6; Wu and Phelps, 2008; Garreau, 1991). Fishman (1987, p. 184) coined the term

> technoburb . . . [for a] peripheral zone, perhaps as large as a county, that has emerged as a viable socioeconomic unit. . . . Its residents look to their immediate surroundings rather than the [core] city for their jobs and other needs; and its industries find not only the employees they need but also the specialized services.

However, others have continued to use the term suburb; for example, Hartshorn and Muller (1992) call edge city type environments "suburban downtowns."

Some authors focus on the mix of activities in suburbs. For example, Duany et al.'s (2000, p. 3) *Suburban Nation* contrasts "suburban sprawl" with "traditional neighborhoods," both in suburban locations and presumably making up a large percentage of suburban areas in the United States. The traditional neighborhood involves "mixed-use, pedestrian-friendly communities of varied population, either standing free as villages or grouped into towns and cities" (Duany et al., 2000, p. 4). Suburban sprawl in contrast is made up of five "homogenous components" – subdivisions, shopping centers, office parks, civic institutions, and roadways – "which can be arranged in almost any way" (Duany et al., 2000, p. 5). As they claim: "They are polar opposites in appearance, function, and character: they look different, act differently, and they affect us in different ways." Overall such activity-based definitions have been evolving quite quickly.

Political places

Suburbs may also be defined as municipalities or neighborhoods with some political or administrative role. The OED has examples of such usage from the mid-fifteenth century: "2. Any of such residential parts, having a definite designation, boundary, or organization" (Oxford English Dictionary, 2011b). In the United States, a suburban municipality is outside the core city. While this definition is dealt with in the locational section, part of the reason this definition has been used so much is it deals with political culture and activities as well. Some authors see this political independence as being core to the concept of suburbs (Beauregard, 2006). As Teaford describes:

> For Americans the notion of city limits has been vital to the concept of suburbia. . . . Because of the strong tradition of local self-rule in the United States, this political distinction between suburb and central city has been vital to discussions of suburban development, lifestyle, and policy. American suburbs are not simply peripheral areas with larger lawns and more trees than districts nearer the historic hub. They are governmentally independent political units that can employ the powers of the state to distinguish themselves from the city.
>
> *(Teaford, 2008, pp. ix–x)*

Sociocultural

The sociocultural dimension is a prominent one in many definitions. As Healy (1994, p. xiii) states in the introduction to a collection of essays on Australian suburbs, "the terms 'suburb' and 'suburbia' have functioned as imagined spaces on to which a vast array of fears, desires, insecurities, obsessions, and yearnings have been projected and displaced."

Some of these definitions overlap a great deal with styles of building. For example, Fava's (1956, p. 34) "Suburbanism as a way of life" pointed out that suburbs share three sets of traits: "more than their disproportionate share of young married couples and their children," "made up largely of families of middle-class status," and with certain physical qualities such as "private" (detached) houses, low densities, and open spaces. As Nicolaides and Wiese (2006, p. 7) describe in their introduction to the *Suburb Reader*, such features support a way of life: "places shaped by elevated values for home ownership, secluded nuclear families, privacy, a distinctive, gendered division of labor, social exclusivity, semirural landscapes, dislike of cities, political home rule, etc." Some scholars emphasize the economic and regulatory tools for creating and maintaining this conformity (Grant and Mittelstead, 2004), but these are rarely posed as definitions (Fogelsong, 2005).

A subset of such sociocultural definitions examines the intersection between social and physical dimensions, focusing on "Western" style suburbs in low- and middle-income countries. For example, King (2004) describes as suburban middle- and upper middle-class villas in Asia, particularly those with European or North American themes (also Fishman, 2003). As Harris (2010) points out, lower-income squatter settlements, even ones in suburban locations in such countries, are rarely described as suburbs. In contrast, in places such as Europe and Australasia, suburban poverty and "locational disadvantage" in suburban public housing and other low-income areas are key concerns (Maher, 1994). Obviously, this is an area demanding more clarity.

Suburbs are also not all socially homogenously middle class. For decades scholars have emphasized how suburbs attract migrants from rural areas who wanted cheap land where they could incrementally build their own houses, grow some of their own food, or work in suburban manufacturing (Berger, 1960; Nicolaides and Wiese, 2006). Others have pointed to growing communities of international migrants, lower-income groups, gay households, and other diverse populations moving to suburbs. Still others propose a mixture related to settlement age – e.g., people residing in villages and towns swallowed up by a metropolitan area. As Teaford remarks (2008), with perhaps a little boosterism:

> American suburbs include some of the nation's most densely populated communities, as well as areas zoned to accommodate more horses than human beings. Suburbia reflects the ethnic

diversity of America more accurately than the central cities, providing homes for Hispanics, Asians, and blacks as well as non-Hispanic whites. It comprises slums as well as mansions, main streets as well as malls, skyscrapers as well as schools. Some suburbs are particularly gay-friendly; others are planned for senior citizens. Some are known for their fine schools; others are examples of educational failure.

(Teaford, 2008, pp. xiii–xiv)

As Nicolaides and Wiese (2006) explain, this diversity within and between suburbs has challenged earlier definitions, particularly in the field of history in the United States:

The most intensive argument has pivoted around questions of class and race: Was a suburb only a suburb when it was white and middle or upper class? Pioneering scholars in the field . . . implied that the answer was yes. . . . By the 1990s, however, suburban 'revisionists' had begun to challenge this 'orthodox' version of suburbia for what it omitted: in particular, industry, multifamily housing, blue-collar workers, ethnic and racial minorities, and the poor.

(Nicolaides and Wiese, 2006, pp. 7–8)

Globally, suburban public housing estates, industrial worker suburbs, and self-build suburbs (including squatter settlements, shantytowns, mobile home parks, and low-cost subdivisions) also make it hard to defend a view of suburbs as essentially white and affluent (Harris, 2010; Forsyth, 2013). The alternative in some fields has been a definition based on suburbanization, or the process of decentralization of jobs and housing, whether voluntary or involuntary (Nicolaides and Wiese, 2006, p. 8). However, as more people are "suburban-born and bred," and many more move directly to suburbs bypassing the core city, such a definition based on decentralization may need to be enlarged (Fava, 1975, p. 10; Harris, 2010).

Styles of building, design, and planning

Key to how new homeowners interact with suburban developments, and how urban planners and designers think about them, is how the developments are designed and built (Forsyth and Crewe, 2009a). Many authors have pointed out how suburbs meld town and countryside in a unique blend (Fishman, 1987; Archer, 2011). This is a central tenet of the garden tradition of suburban design (Howard, 1902).

How this blend is achieved depends on how suburbs get constructed. Many suburban homes are built one by one to the specifications of the owner, or indeed by the owner over time. Others are built by developers and builders in subdivisions of various scales, and with different levels of fit in various neighborhood, municipal, or regional plans. Many now use prefabrication and other standard building techniques to keep costs down.

Criticisms of "suburbia" frequently focus on one or two aspects such as tract housing, which is how they define suburbia as exemplified by these problematic forms. At larger scale the great debate is over sprawl – thus overlapping with the critical assessment type of definition – whether suburbs are unplanned and incremental or planned to contribute positively to the region (Hayden, 2004, pp. 1–2).

Time

Along with density and location, Harris proposes "newness" as a key criterion for defining suburbs (Harris, 2010, p. 29). As he points out, however, some authors such as Clarke (1966) propose that this period of relative newness lasts only a few years, while others such as Whitzman

(2009) give examples lasting many decades. One issue with such definitions is how to deal with older towns and villages that over time are surrounded by new suburban development, which functionally become suburbs.

Other definitions focus on development after a certain time period, for example post-World War II developments. This is often in combination with some other dimension such as location or building type (Forsyth, 2005). Such definitions are most useful in places like the United States where this has been a key period of suburbanization. An alternative is to define types of suburbs as the total of different periods of suburbs (Forsyth and Crewe, 2009b). Such historical classifications are rarely used to explicitly define suburbs, but suburbs are presumably made up of the sum total of the different types of suburbs (Lang et al., 2006; Hayden, 2003).

Critical assessments

Many definitions of suburbs are really catalogs of their ills. According to the OED, this kind of definition has been around at least since the seventeenth century, when suburban areas were seen as more lawless than the core city (Oxford English Dictionary, 2011b, def. 4). By the nineteenth century, the term "suburban" was used for those "having the inferior manners, the narrowness of view, etc., attributed to residents in suburbs" (Oxford English Dictionary, 2011b). As Barker (2009, p. 13) describes in his defense of British suburban life, for many in the press and design professions to "call anything or anyone 'suburban' is to utter a put-down, an anathema, a curse."

In such cases the commentator is frequently focusing on certain kinds of suburbs that have specific problems – social, aesthetic, environmental, or cultural. Other areas that under a locational definition would be classed as suburbs are defined as something else – small towns, employment areas, new towns, and so on. A cluster of urban design critiques point to the lack of urbanity in suburbs, with suburbs defined as lacking positive urban features (Montgomery, 1998).

There is a long and rich history of such critiques, with a number of authors providing substantial reviews of them (Eichler and Kaplan, 1967; Popenoe, 1977; Gans, 1963; Forsyth, 2005; Bruegmann, 2006). Of these more critical assessments, the ones that are most easy to defend are environmental critiques of suburbia as energy inefficient, land grabbing, and water quality eroding. But even these are hard to sustain across all suburbs. For example, any suburban areas are transit-oriented or else blend natural systems with development (Crewe and Forsyth, 2011). Overall, these are important "negative" definitions that specify through critique.

Indices

Finally, there are indices or indicators for defining and distinguishing suburbs – here again, suburbs are the sum total of different types. For example, Orfield (2002) used cluster analysis applied to almost 5,000 suburban places in the United States, distinguishing them according to fiscal capacity and location. He proposed six suburban clusters or types, including three "at risk" types (segregated, older, low-density), two types of suburban jobs centers (best off fiscally), and bedroom suburbs. Similarly, Mikelbank (2004) used cluster analysis to create a typology of U.S. census-defined suburbs based on demographic characteristics, economic variables, physical features, and some aspects of government. Data were drawn from the population, economic, and government censuses. From this, he created ten types of suburbs, ranging from traditional wealthy bedroom suburbs though working-class diverse suburbs (p. 950). Hanlon et al. (2009, p. 261) similarly developed an "index of suburban transformation" related to population, income, and poverty (see also Vicino, 2008).

Sprawl indices that can distinguish suburbs from historic areas by features such as density and street pattern are another of this style (Ewing et al., 2002). These are useful and data rich ways of defining types of suburbs and could be further developed to define suburbs as a whole.

Conclusion

Given all the confusion around the term, one option is to give up on the term suburb. There are two ways to do this. The first is to replace it with more specific environmental types such as post-war subdivision, edge city, and office park. This has some potential among suburban experts who are typically aware that suburbs are quite diverse. In comparative work it may be easier to examine specific suburban types. Such an approach may, however, be a challenge for those working on more general processes of urban growth or speaking to the public. Terms such as neighborhood and community are similarly murky but also hard to give up completely.

A variation on the strategy is to focus on specific features such as location, density, or historical period of development and not use the term suburb. This is in fact quite often done in urban studies. Many research projects on, say, environments and health do not state that specific environments are suburban but rather that their street patterns are dominated by large blocks (and some such areas will of course be located in core cities, but more are likely to be in suburbs). This approach allows research variables to be clearly conceptualized. For example, Knapp and Zhao's (2009) overview of "Smart Growth and Urbanization in China" does not mention the term suburbs, though it is clearly dealing with development in suburban locations and it has a number of very compelling illustrations of outer urban development, freeways, and a low-density subdivision (Song and Ding, 2009). Rather, their paper deals with smart growth dimensions such as land use mix and farmland preservation. I have certainly used this strategy in my own work. However, non-experts can find it hard to interpret such variables, so researchers often have to provide examples using terms such as "suburban apartment area" or "low-density suburb."

Alternatively, it may be possible to keep the term. One way to do this is to better distinguish between types of suburbs – so that all references to suburbs are qualified by an adjective. This makes sense because different types of suburbs will have different problems and different planning needs. Many authors today deal with this situation by focusing on fairly clearly defined types of suburbs such as Fishman's (2003, p. 1) "American-style 'suburbs of prosperity'" or the "ethnoburb" (Li, 1998). If such suburbs are defined with some precision, then they could be the bases of conceptual models, larger theories, and thoughtful practice.

To define suburbs as a whole, rather than types of suburbs, is more complex. Following an extensive international review of suburban scholarship, Harris (2010) proposed a definition based on location, density, and newness. Of these, location and newness can be the most consistently applied. Even this definition raises questions about how far out and how new? A suburban index could, perhaps, help deal with some of these problems.

As urbanization continues, the term suburb represents a long-standing and viable term for describing development beyond the core city (Clapson and Hutchison, 2010). Alternatives such as "peri-urban" are no more clearly defined. Suburbs do have important features in common. Because of their location and relative newness, such locations are likely to have a particular range of functions, transportation modes, social characters, and physical features. More clearly distinguishing suburbs from other kinds of development, and different types of suburban environments, can help both those who want to understand suburbs and those involved in planning and (re)developing them.

Guide to further reading

Harris, R. and Vorms, C. (eds.) (2017) *What's in a Name? Talking about "Suburbs."* Toronto: University of Toronto Press.

Jackson, K. (1987) *Crabgrass Frontier: The Suburbanization of the United States.* New York: Oxford University Press.

Keil, R. (ed.) (2013) *Suburban Constellations: Governance, Land and Infrastructure in the 21st Century.* Berlin: Jovis Verlag.

References

Adell, G. (1999) *Theories and Models of the Peri-Urban Interface: A Changing Conceptual Landscape.* London: Development Planning Unit. Available at: http://discovery.ucl.ac.uk/43/1/DPU_PUI_Adell_THEORIES_MODELS.pdf.

Archer, J. (2005) *Architecture and Suburbia.* Minneapolis, MN: University of Minnesota Press.

Archer, J. (2011) "Suburbs." In: Banerjee, T. and Loukaitou-Sideris, A. (eds.) *Companion to Urban Design.* London: Routledge.

Barker, P. (2009) *The Freedoms of Suburbia.* London: Francis Lincoln.

Beauregard, R. (2006) *When America became Suburban.* Minneapolis, MN: University of Minnesota Press.

Berger, B. (1960) *Working-Class Suburb: A Study of Auto Workers in Suburbia.* Berkeley, CA: University of California Press.

Borchert, J. (1996) "Residential City Suburbs: The Emergence of a New Suburban Type 1880–1930." *Journal of Urban History*, 22, pp. 283–307.

Bruegmann, R. (2006) *Sprawl: A Compact History.* Chicago, IL: University of Chicago Press.

Caplan, N. and Nelson, S. (1973) "On Being Useful: The Nature and Consequences of Psychological Research on Social Problems." *American Psychologist*, 28(3), pp. 199–211.

Clapson, M. (2003) *Suburban Century.* Oxford: Berg.

Clapson, M. and Hutchison, R. (2010) "Introduction: Suburbanization in Global Society." In: Clapson, M. and Hutchison, R. (eds.) *Suburbanization in Global Society.* Bingley, UK: Emerald Publishing.

Clarke, S.D. (1966) *The Suburban Society.* Toronto: University of Toronto Press.

Crewe, K. and Forsyth, A. (2011) "Compactness and Connection in Environmental Design: Insights from Ecoburbs and Ecocities for Design with Nature." *Environment and Planning B*, 38(2), pp. 267–288.

Douglass, H.P. (1925) *The Suburban Trend.* New York: The Century Company.

Duany, A., Plater-Zyberk, E., and Speck, J. (2000) *Suburban Nation: The Rise of Sprawl and the Decline of the American Dream.* New York: North Point Press.

Dunham-Jones, E. and Williamson, J. (2009) *Retrofitting Suburbia.* New York: Wiley.

Eichler, E. and Kaplan, M. (1967) *The Community Builders.* Berkeley, CA: University of California Press.

Ewing, R., Pendall, R., and Chen, D. (2002) *Measuring Sprawl and its Impact.* Washington, DC: Smart Growth America. Available at: www.smartgrowthamerica.org/sprawlindex/MeasuringSprawl.PDF.

Fava, S. (1956) "Suburbanism as a Way of Life." *American Sociological Review*, 21(1), pp. 34–37.

Fava, S. (1975) "Beyond Suburbia." *The Annals of the American Academy of Political and Social Science*, 422, pp. 10–24.

Fishman, R. (1987) *Bourgeois Utopias: The Rise and Fall of Suburbia.* New York: Basic Books.

Fishman, R. (2003) "Global Suburbs." University of Michigan Working Paper, URRC-0301. Available at: http://sitemaker.umich.edu/urrcworkingpapers/all_urrc_working_papers&mode=single&recordID=308464&nextMode=list.

Flint, A. (2006) *This Land: The Battle over Sprawl and the Future of America.* Baltimore, MA: Johns Hopkins University Press.

Fogelsong, R.M. (2005) *Bourgeois Nightmares: Suburbia 1870–1930.* New Haven, CT: Yale University Press.

Forsyth, A. (2005) *Reforming Suburbia.* Berkeley, CA: University of California Press.

Forsyth, A. (2013) "Global Suburbia and the Transition Century: Physical Suburbs in the Long-Term." *Urban Design International*, 9(4), pp. 259–273.

Forsyth, A. and Crewe, K. (2009a) "New Visions for Suburbia: Reassessing Aesthetics and Place-Making in Modernism, Imageability, and New Urbanism." *Journal of Urban Design*, 14(4), pp. 415–438.

Forsyth, A. and Crewe, K. (2009b) "A Typology of Comprehensive Designed Communities since the Second World War." *Landscape Journal*, 28(1), pp. 56–78.

Forsyth, A., Oakes, J.M., Schmitz, K.H., and Hearst, M. (2007) "Does Residential Density Increase Walking and Other Physical Activity?" *Urban Studies*, 44(4), pp. 679–697.

Frost, L. (1991) *The New Urban Frontier*. Kensington: University of New South Wales University Press.

Gans, H. (1963) "Effects of the Move from City to Suburb." In: Duhl, L. (ed.) *The Urban Condition*. New York: Basic Books.

Gans, H. (1968) *People and Plans*. New York: Basic Books.

Gans, H. (2009) "Imagining the Suburban Future." Robert A. Catlin Memorial Lecture, Bloustein School of Planning and Public Policy, Rutgers, 5 February.

Garreau, J. (1991) *Edge City*. New York: Doubleday.

Gober, P. and Behr, M. (1982) "Central Cities and Suburbs as Distinct Place Types: Myth or Fact?" *Economic Geography*, 58(4), pp. 371–385.

Grant, J. and Mittelsteadt, L. (2004) "Types of Gated Communities." *Environment and Planning B*, 31, p. 913.

Green Leigh, N. and Lee, S. (2005) "Philadelphia's Space in between: Inner-Tine Suburb Evolution." *Opolis*, 1(1), pp. 13–32.

Hanlon, B., Short, J.R., and Vicino, T.J. (2009) *Cities and Suburbs: New Metropolitan Realities in the US*. New York: Routledge.

Harris, C.D. (1943) "Suburbs." *American Journal of Sociology*, 49(1), pp. 1–13.

Harris, R. (2010) "Meaningful Types in a World of Suburbs." In: Clapson, M. and Hutchinson, R. (eds.) *Suburbanization in Global Society*. Bingley, UK: Emerald Publishing.

Harris, R. and Larkham, P.J. (1999) "Suburban Foundation, Form and Function." In: Harris, R. and Larkham, P. (eds.) *Changing Suburbs: Foundation, Form and Function*. New York: E. & F.N. Spon.

Hartshorn, T. and Muller, P. (1992) "The Suburban Downtown and Urban Economic Development Today." In: McLean, M. (ed.) *Sources of Metropolitan Growth*. New Brunswick, NJ: Rutgers University Press.

Harvard Law Review Association (2004) "Locating the Suburb." *Harvard Law Review*, 117, pp. 2003–2022.

Hayden, D. (2003) *Building Suburbia: Green Fields and Urban Growth, 1820–2000*. New York: Pantheon Books.

Hayden, D. (2004) *A Field Guide to Sprawl*. New York: Norton.

Healey, P. (2009) "The Pragmatic Tradition in Planning Thought." *Journal of Planning Education and Research*, 28, pp. 277–292.

Healy, C. (1994) "Introduction." In: Ferber, S., Healey, C. and McAuliffe, C. (eds.) *Beasts of Suburbia: Representing Cultures in Australian Suburbs*. Melbourne: Melbourne University Press.

Hodge, D. and Lacasse, J. (2011) "Ranking Disciplinary Journals with the Google Scholar H-Index: A New Tool for Constructing Cases for Tenure, Promotion, and Other Professional Decisions." *Journal of Social Work Education*, 47(3), pp. 579–596.

Howard, E. (1902, orig. 1898) *Garden Cities of Tomorrow*. London: Sonnenschein.

Iaquinta, D.L. and Drescher, A.W. (1999) *Defining the Peri-Urban: Rural-Urban Linkages and Institutional Connections*. Rome: UN Food and Agriculture Organization. Available at: www.fao.org/DOCREP/003/X8050T/x8050t02.htm#P13_2357.

Johnson, L.C. (2006) "Style Wars: Revolution in the Suburbs?" *Australian Geographer*, 37(2), pp. 259–276.

King, A.D. (2004) *Spaces of Global Cultures: Architecture Urbanism Identity*. London: Routledge.

Kling, R., Olin, S., and Poster, M. (eds.) (1991) *Postsuburban California*. Berkeley, CA: University of California Press.

Knapp, G. and Zhao, X. (2009) "Smart Growth and Urbanization in China." In: Song, Y. and Ding, C. (eds.) *Smart Urban Growth for China*. Cambridge, MA: Lincoln Institute for Land Policy.

Kurtz, R. and Eichler, J.B. (1958) "Fringe and Suburb: A Confusion of Concepts." *Social Forces*, 37(1), pp. 32–37.

Lang, R., Le Furgy, J., and Nelson, A.C. (2006) "The Six Suburban Eras of the United States." *Opolis*, 1(2), pp. 65–72.

Li, W. (1998) "Anatomy of a New Ethnic Settlement: The Chinese Ethnoburb in Los Angeles." *Urban Studies*, 35(3), pp. 479–501.

Lineberry, R. (1975) "Suburbia and the Metropolitan Turf." *Annals of the Academy of Political and Social Science*, 422, pp. 1–9.

Lucy, W. and Phillips, D.L. (2000) *Confronting Suburban Decline*. Washington, DC: Island Press.

Maher, C. (1994) "Residential Mobility, Locational Disadvantage, and Spatial Inequality in Australian Cities." *Urban Policy and Research*, 13(3), pp. 185–191.

McManus, R. and Ethington, P.J. (2007) "Suburbs in Transition: New Approaches to Suburban History." *Urban History*, 34(2), pp. 317–337.

Mikelbank, B. (2004) "A Typology of U.S. Suburban Places." *Housing Policy Debate*, 15(4), pp. 935–964.

Montgomery, J. (1998) "Making a City: Urbanity, Vitality, and Urban Design." *Journal of Urban Design*, 3(1), pp. 93–116.

Nicolaides, B. and Wiese, A. (2006) "Introduction." In: Nicolaides, B. and Wiese, A. (eds.) *The Suburb Reader*. New York: Routledge.

Orfield, M. (2002) *American Metropolitics: The New Suburban Reality*. Washington, DC: Brookings Institution Press.

Oxford English Dictionary (2011a) "Suburb." Available at: www.oed.com/viewdictionaryentry/Entry/193229.

Oxford English Dictionary (2011b) "Suburban." Available at: www.oed.com/view/Entry/193230.

Phelps, N. (2015) *Sequel to Suburbia*. Cambridge, MA: MIT Press.

Popenoe, D. (1977) *The Suburban Environment*. Chicago, IL: University of Chicago Press.

Satterthwaite, D. (2007a) "The Scale of Urban Change Worldwide, 1950–2000 and Its Underpinnings." IIED Human Settlements Discussion Paper. Available at: www.iied.org/urban/Urban_Change.html.

Satterthwaite, D. (2007b) "The Transition to a Predominantly Urban World and Its Underpinnings." IIED Human Settlements Discussion Paper. Available at: www.iied.org/pubs/display.php?o=10550IIED.

Schnore, L. (1957) "Satellites and Suburbs." *Social Forces*, 36(2), pp. 121–127.

Schon, D. and Rein, M. (1994) *Frame Reflection: Towards the Resolution, of Intractable Policy Controversies*. New York: Basic Books.

Song, Y. and Ding, C. (eds.) (2009) *Smart Urban Growth for China*. Cambridge, MA: Lincoln Institute for Land Policy.

Sridhar, K. (2004) "Cities with Suburbs: Evidence from India." National Institute of Public Finance and Policy Working Paper. Available at: www.nipfp.org.in/working_paper/wp04_nipfp_023.pdf.

Teaford, J. (2008) *The American Suburb: The Basics*. New York: Routledge.

Thorns, D. (1972) *Suburbia*. London: McGibbon and Kee.

Turcotte, M. (2008) *The City/Suburb Contrast: How Can We Measure It?* Available at: www.statcan.gc.ca/pub/11-008-x/2008001/article/10459-eng.htm#2.

U.S. Census Bureau (2010) *2010 Census Urban and Rural Classification and Urban Area Criteria*. Available at: www.census.gov/geo/www/ua/2010urbanruralclass.html.

Vicino, T.J. (2008) "The Spatial Transformation of First-Tier Suburbs, 1970 to 2000: The Case of Metropolitan Baltimore." *Housing Policy Debate*, 19(3), pp. 479–517.

Warner, S. (1978) *Crabgrass Frontier: The Process of Growth in Boston (1870–1900)*. Cambridge, MA: Harvard University Press.

Webber, M. (1964) "The Urban Place and the Nonplace Urban Realm." In: Webber, M. (ed.) *Explorations into Urban Structure*. Philadelphia, PA: University of Pennsylvania Press.

Whitzman, C. (2009, orig. 1962) *Suburb, Slum, Urban Village: Transformations in Toronto's Parkdale Neighborhood, 1975–2002*. Vancouver: UBC Press.

Wittgenstein, L. (2009, orig. 1953) *Philosophical Investigations*. 4th ed. Edited and Translated by Hacker, P.M.S. and Schulte, J. Oxford: Wiley-Blackwell.

Wu, F. and Phelps, N. (2008) "From Suburbia to Post-Suburbia in China? Aspects of the Transformation of the Beijing and Shanghai Global City Regions." *Built Environment*, 34, pp. 464–481.

2

Suburban stereotypes

Richard Harris

Introduction

Stereotypes have a bad reputation. This is not surprising because, whether of people or places, most have undesirable connotations: think of "slum," "ghetto," or, for Parisians, *les banlieues*. But some are positive. "Neighborhood," for example, evokes warm feelings of community. And then there are names that carry mixed messages. "Suburb" is one: praised by some, damned by others. Indeed, the very same aspects of suburbs may be viewed positively or negatively. As if this were not complicated enough, in some parts of the world, the places that North Americans would call suburbs have no identity and are associated with no stereotype. And so the first, and major, section of this survey explores the diversity, and occasional absence, of suburban stereotypes.

Even positive stereotypes are criticized when they oversimplify and mislead. For example, many neighborhoods in fact lack community. But we should be careful not to criticize a stereotype just for being inaccurate. After all, definitions are no better. There are few places that conform in all respects to the usual American definition of a suburb – residential, low-density, peripheral location, and with a separate political identity – and yet we commonly refer to all sorts of other places as suburbs. It is convenient to do so. The same with stereotypes: they can be damaging, but they can also be useful. Take "the suburban way of life." In North America this phrase implies a lifestyle that is car-dependent and focused on home-owning families. It connotes a focus on everyday domesticity that is welcome to some, anathema to others, but either way stereotyped. Of course, many suburbs and their residents do not fit this template, but some slippage is surely acceptable. After all, it is a convenient way of speaking about a type of experience. And so we might think of "suburban way of life" as a *useful* stereotype. Objections begin when the connotations deviate greatly from reality, when *many* residents are renters who are not living in nuclear families, raising children, or driving cars. At what point do we decide that a line has been crossed between a stereotype that is serviceable and one that is misleading? There is no easy answer, an issue that is discussed in the second major section.

If stereotypes are often inaccurate, why are they so common? The answer differs depending on the stereotype in question and who believes it. The general answer is that they help us make sense of our world, expressing the aspirations of people at particular times, including their desire to set themselves apart. They have been articulated and reproduced in various ways: by the media, and by private agents, including land developers, whose purposes they serve. Even when a stereotype serves no purpose, or is indeed counterproductive, it may persist through sheer inertia. The third and final major section of this survey considers these elements in the life of stereotypes, while a concluding comment offers advice about how we should use them.

The diversity of suburban stereotypes

You probably have some idea in your mind of what suburbs are. You may assume that this is *the* suburban stereotype, the image that everyone else possesses, and many probably do. But others − depending on where they live, how they make a living, and what media they consume − will see things differently. At any time, there may be a dominant stereotype but these vary historically and geographically, while some places have none at all.

Positive stereotypes

The word "suburb" was born in late eighteenth century England where it soon captured people's imagination. That is not surprising: by 1850, Britain had become the first urban nation, in that a majority of people then lived in cities. This trend involved large-scale suburbaniza-tion, a process that continued through the late nineteenth century. Working-class districts in industrial cities were crowded and unhealthy. Those who could escape to the suburbs did so, and reformers extolled the anti-urban virtues of suburban living: a healthier environment that was better for children; more space in and around the home; access to nature, in gardens and parks; a moral, home-centered life, distant from the corrupting influences of the city (Clapson, 2003, pp. 51–78). The purest version was called "the garden suburb" (Whitehand and Carr, 1999). It was an appealing vision, a suburban ideal.

Many elements of this stereotype persisted and were transmuted as they became influential elsewhere in Anglo-America: Australia, New Zealand, Canada, and the United States. These countries inherited anti-urbanism, with its valuation of space, nature, and domesticity (Bunce, 1994). To these, locals added white-settler virtues of home ownership, which after 1945 were seen as part of the American (and Canadian, and Australian) dream (Archer, 2005, pp. 250–289; Jackson, 1986, pp. 68–72). Owning one's home was above all the hope of immigrants and workers, and many were able to realize that aspiration, if necessary by building their own. It is therefore ironic that by the 1950s homeownership had come to be seen as a typically middle-class achievement. In the United States, with its history of slavery, the suburban connotations of exclusivity were as much racial as economic, these being ensured by having a separate political identity (Jackson, 1986). The American suburban stereotype, then, became more specific: white, middle-class, owner-occupied single-family homes, and self-governed.

As suburbanization continued, a more particular, related stereotype emerged and acquired its own name: exurbia. Persistent anti-urbanism and widespread car ownership enabled some people to move to largely rural settings and still commute (Bunce, 1994, pp. 89–101). An early dramati-zation of this ideal was the book and then movie *Mr. Blandings Builds his Dream House* (1948). The image was one of living close to nature, perhaps growing produce, keeping horses, and enjoying rural leisure pursuits such as golf. Over time, it acquired connotations of a commitment to the conservation of the countryside. Exurbia, then, became a landscape and lifestyle ideal that was distinct from that of the more urbanized suburbanite, and in some ways opposed to it (Cadieux and Taylor, 2013; Sandberg et al., 2013).

These positive images of suburbs and exurbs reflected the desires and experience of those who moved out of the city. But there is a different experience and vision of these places, which embodies the aspirations of those who have traveled in the opposite direction. It is a stereotype that hardly exists now in North America, Europe, or Japan, where rural-urban migration has run its course. But in the rest of the world it is a widespread, and sometimes dominant force.

In Europe and North America, until the mid-twentieth century, millions moved from rural areas to cities. Some were dislocated by famine, evictions, or farm mechanization. Others simply sought jobs and a better life. Most ended up in inner cities, crowded into deteriorated housing but close to nearby work. These people had a simple stereotype of urban life: opportunity!

Since World War II, the scale of urbanization elsewhere has been greater than anything experienced earlier. In the region first affected, Latin America, the initial wave of rural migrants also moved to inner cities where the jobs were (Turner, 1968, pp. 356, 359). Those with more secure work could afford to commute from the urban fringe (Turner, 1968, pp. 357–359). But the numbers of people involved soon made this strategy impossible. Increasingly, rural migrants got no further than the suburbs, where they occupied various types of shelters, ranging from squatter settlements to densely redeveloped villages embraced by urban expansion. This pattern has happened on a massive scale in China (Wu, 2013, pp. 169–170) and India (Dupont, 2004, pp. 176–178), as well as in Southeast Asia and sub-Saharan Africa. In Noida, for example, a planned new town on Delhi's periphery, a survey revealed that most residents were migrants, and that in Noida's slums the proportion reached 99 percent (Dupont, 2004, p. 180).

For some migrants, suburban settlement is temporary. Recently in China, the system of *hukuo* residency permits has prevented rural migrants from accessing urban services, including education. For many, then, the urban experience is limited and temporary. For example, on the fringes of Dongguan, a Chinese city of eight million, temporary migrants comprise three-quarters of all residents (Liu, 2007). In parts of southern Africa "circular migration" has developed, whereby workers regularly move between farms and urban homes, mostly located at the urban fringe (Ferguson, 1999; Potts, 2010). More generally, in large parts of the world today, the typical resident of the urban periphery has sought the city, not fled it.

The aspiration of these migrants is clear enough. As Dupont (2004, pp. 181, 185) observed, "Delhi's squatter settlements shelter mostly migrant households attracted by the employment opportunities provided by the city" while their presence in suburban Noida testifies to "the power of attraction of the new industrial center." Migrants often hope to provide their children with better economic and social prospects. There is also a complex gendered dimension. Almost everywhere, employment and educational opportunities for women are better in urban than in rural areas, while attitudes are more progressive there with respects to women's rights. But sometimes only the men move, and commonly the decision to migrate is made by the husband alone and not always with his partner's interests in mind (Haider, 2000).

Beyond plausible speculations about "opportunity," we know little about the nature of the stereotype that rural migrants have of cities or of their fringes: the research has not been done. We do know that, once a pattern starts, migrants are influenced by the experience of friends and kin who have already moved: based on reports and connections, they follow those who went before. This was true of the Irish who moved to Manchester, England, in the nineteenth century, or to American cities in the early twentieth; it was as true of Italians who moved to Toronto after 1945 as it is of Indian and Chinese villagers who now look for work in Mumbai or Shenzhen. Because most migrant destinations are now located at the urban fringe, the stereotype that millions of migrants must have is in that sense "suburban." Whether it is seen as an end in itself, or as a stepping-stone to a more purely urbanized ideal, is an open question.

As many nations in the Global South have prospered, more of their peoples have migrated to the urban fringe in the modern Western way: they have moved from the city. Everywhere, but above all in China, a class of people has emerged who aspire to own property at the urban fringe. Often the dwellings are in medium- or high-rises rather than ground-oriented single-family structures, but their aspirations are recognizably similar to those of their North American counterparts. In India, the aim is "to distance oneself from the city, to be at ease, away from the chaos of the streets" (Brosius, 2010, p. 94). To accomplish that, for example in a gated development, is a mark of prestige (Waldrop, 2008). Status is also a key element in the attraction of similar developments in China (Fleischer, 2010). Here, in the past, many used to "prefer a bed in the central city to a house in the suburbs," but recent years have seen the emergence of "suburban living ideals" (Zhou and Logan, 2008, p. 157; Shen and Wu, 2013, p. 1823). Significantly, in both countries, the names given to many developments echo Western ideals: "Victoria Gardens," "Gulmohar Greens," and "Rose Garden Villas" (Brosius, 2010; Wu, 2013, p. 169). Elsewhere – in Southeast Asia, Latin America, and South Africa, as in the United States – security is also a prominent motive in the growth of gated communities (Caldeira, 1996; Dick and Rimmer, 1998, p. 2317; Herzog, 2015). These places offer variants of a common vision: some include security, some feature lower-density and greenery, but all imply a removal from urban problems, control over living space, and an address that speaks of achievement, if not prestige. They are locally appropriate modifications of the Anglo-American suburban ideal.

Negative stereotypes

These positive stereotypes have had their negative counterpoints. That is obviously true of migrant shantytowns, but it has been no less true of middle-class suburbia, wherever it has developed. And the criticisms have been present almost from the beginning.

Among nineteenth-century cities, London's suburbs were by far the most extensive. They were lampooned for their supposed social and physical mediocrity. As one historian puts it, the "end product was generally viewed with distaste, ridicule or contempt" (Thompson, 1982, p. 3). Part of the reason was that "the suburbs appeared monotonous, featureless, without character . . . infinitely boring to behold." Another part was social: the "wastelands of housing [were seen as] settings for dreary, petty lives without social, cultural or intellectual interests, settings which fostered a pretentious preoccupation with outward appearances." In a phrase, suburbs were seen as bastions of mediocrity.

Just as the positive stereotype evolved when it was exported, so did the negative. Across the English Channel, by the 1920s the lower middle-class *pavillonaires* who moved into the Parisian suburbs were being damned for their small-minded parochiality (Harris and Vorms, 2017, p. 8). Versions of this stereotype persist for current settlement in the *zone périurbain* in the outer fringe. Across the Atlantic, in the 1920s and then more prominently during the post-war suburban boom, critics built on the theme of conformity (Archer, 2005, pp. 283–285, 331–340; Nicolaides, 2006, pp. 81, 91). Grids were replaced by a landscape of loops and lollipops that was no more visually stimulating. With rising prosperity, home, and car ownership, the new conformity was one of mass consumerism: a car in every driveway, a fridge in every kitchen, and a TV in every living room on which everyone watched *I Love Lucy* at the same time every week. This could be seen as an achievement, but also as an enforceable norm. Suburbanites – a new term – were stereotyped as status-conscious joiners, eager to fit in, to match or even outdo their neighbors, but only in minor, acceptable ways. A new, gendered critique emerged (Clapson, 2003, pp. 125–141). Women were seen to bear the brunt of everyday social pressures, notably those of food preparation, childrearing, and home decoration, while being confined to home and neighborhood. Part of the social stereotype, then, was the neurosis that conformity created. Accordingly, suburban life was typed as comfortable but bland, and unhappy because it was devoid of larger meaning.

In recent decades, the package of criticism has changed and seen some significant additions. The charge of social conformity and homogeneity, and the image of suburbs as "a cultural void or desert" have persisted, and can still be found in the entertainment media, but are eroding (Huq, 2013, p. 6). The feminist critique of women's isolation has widened as commentators have pointed out that children and the elderly are often the most disadvantaged by car dependency (Clapson, 2003, pp. 125–141). And new lines of argument have been developed (Bruegmann, 2005; Rome, 2001). Suburbs have been represented as irresponsible. Because they are unfriendly to pedestrians and cyclists, they are seen to have played a part in the rise of obesity, with negative effects on public health. In economic terms, because of their lower densities, they are costly to service. As generous consumers of carbon energy, steel, cement, wood, brick earth, and tarmac, they are damned for destroying farmland and wetlands as well as water and air quality, while contributing more than their share to global warming. The modern negative stereotype of suburbia, then, extends beyond its effects on residents to include the public at large, nationally and globally. Lumped under the label "sprawl," a purely negative term, suburbs are now seen by some as the epitome of how humanity is sending itself to hell in a hand basket.

To some extent the positive and negative stereotypes of Anglo-American suburbs emphasize different things. The environmental critique, for example, has no positive side. But on other points the contrasting pictures involve a different take on the same features (Davison, 2013, p. 10). What, to one observer, appears as social conformity, land hungry sprawl, consumerism, and a lack of culture, may appear to another as responsible citizenship, spaciousness, a decent standard of living, and a healthy emphasis on home and family. Above all, then, the difference is one of point of view.

The same is true for suburban migrant settlements and shack towns. Their promise is one of striving and opportunity, but living conditions are often appalling. As a result, squatter settlements have been condemned and stereotyped as unhealthy firetraps, crime-ridden havens of immorality, fit only for demolition. That is why policies of slum clearance, begun in the Global North in the late nineteenth century, came to be used widely, everywhere, in the twentieth. Even those who are disinclined to blame their residents for the appalling conditions have used slums and squatter settlements as a symbol of failure – of governments, or of the whole global economic system (e.g., Davis, 2006).

Point of view also accounts for the divergence of opinion about the middle-class suburban developments that have become common in the Global South since the 1980s. Given that these developments are situated in countries that have struggled to reduce poverty and to acquire the trappings of Western affluence, criticism of urban fringe developments has been somewhat muted. Even so, they have been subject to negative stereotyping. In China, for example, the critique of sprawl has given birth to a new term, *tan da bing* (making a big pancake) (Fleischer, 2010). The standard charge of suburban blandness has been extended by claiming that suburban developments merely ape North American originals – in their naming, design, and often their auto-dependence (Herzog, 2015). They therefore appear marginal, makers of U.S.-led urbanization. It might seem that no type of suburb has escaped a negative stereotype.

But what about those places whose urban fringes lack a generic name, whether that be "suburb," "exurb," "shack town," *banlieue*, or some Chinese, Hindi, Persian, or Spanish equivalent? This is not an academic question, for there are many such places. People in Teheran and Cairo, for example, have no single name for what North Americans call "suburbs" (Harris and Vorms, 2017). The same is true in India, even though it was long subject to colonial rule. The British introduced the word "suburb" in the nineteenth century, and used it for over a century; after winning independence in 1947, India made English one of its official languages. But only in Mumbai is "suburb" now used in a generic sense, and it has no equivalent in any indigenous language (Harris, 2017). In China, there are generic words, but they differ according to the speaker: urban experts talk of *jiaoqu* and local residents of *nong cun* (" rural," "village") or *jinjiao nong cun* ("close-in villages"); in Germany, experts use terms such as *suburbanisierung* and *umland-gemeinde* while residents prefer *stadtrand* (Harris and Vorms, 2017). In yet other countries, the words come from different languages. In Zambia, Anglophone planners use "peri-urban" to refer to the outer urban fringe but local Swahili-speakers talk of *kiungu* ("the thing attached"). What does exist in some places are names for different *types* of suburbs, whether affluent and gated or poor and ramshackle. In these varied situations, there is no agreed-upon word for what most English speakers call suburbs. Without such a word, we are not likely to find a single stereotype, and perhaps not even several. The concept of a suburban stereotype simply does not apply.

Cold reality

Stereotypes can be useful, but there are many situations where they are employed but should not be. They can arise when people are speaking about the present, but they are even more common when referring to the past.

Hardly any suburb has ever been as desirable as the ideal that prevailed in its day, or as dire as critics claimed. The post-war suburb is a case in point. Because they were new, no suburb was as verdant as it was supposed to be, its neighbors as friendly, its location as convenient. Many were not white and middle-class, or had apartments as well as single-family homes, or failed to conform to any elements of the stereotype except for being new and peripheral. Collectively, Toronto's suburbs are a good example (Harris, 2015). Conversely, even those places that consisted of white, middle-class homeowners were never as bland and homogeneous as their critics supposed. There were men in blue-collar as well as white-collar occupations; some wives worked outside the home for pay; family interests and childrearing practices varied. The residents were, after all, people.

Often, where two opposing stereotypes exist, the truth lies in between. There are always grains of truth: most of the residents of 1950s suburbs were indeed able to own their own homes, raised families, and were rather narrow and conformist in their tastes; the shack towns of Delhi or Lagos are indeed poor places to live but most of their residents are better off than where they came from and many succeed in making meaningful lives; those nearby in gated communities may aspire to Western-style consumerism but will also draw on local values and traditions. And so for those who wish to know where the truth lies, a useful first step is to consider both, or all, of the applicable stereotypes. Collectively, these are likely to point to important features of the place in question.

Exceptions occur when a stereotype is so compelling that it endures beyond its expiration date. This applies to the Anglo-American ideal. The author of a survey of popular culture concluded that representations of suburban, white, affluent, male-headed nuclear families have persisted long after the world has changed, and that these have contributed to the misleading persistence of a mythic image; she also notes that the times are now changing (Huq, 2013, p. 194). More fundamentally, an argument can be made that the very idea of a suburb – simplified very simply as a residential space at the urban fringe – is no longer valid. Fringe areas have always contained a mix of land uses, including workplaces, shops, and roads, but for decades it was reasonable to assume that the people who lived in the suburbs worked in the city. That is no longer true. In most metropolitan areas, there are more jobs in the suburbs than downtown, these being clustered in nodes of varying size. Residents of suburbs may not define themselves or their places of residence in relation to the city center, or think of themselves as suburban (Shanks et al., 2017). In that sense, "suburban" models of any sort have become irrelevant.

At least, when applied to the present, a stereotype can easily be contradicted by looking at census statistics or going for a ride. When used to make sense of the past, however, inaccuracies can easily persist or grow, not least because, in some measure, they always express our own concerns. Part of today's image of the 1950s suburb – good and bad – reflect our current concerns and views: feminists see domestic patriarchy beyond anything that contemporaries perceived; environmentalists highlight effects of sprawl that were overlooked at the time; nostalgic conservatives praise the nuclear families that once were taken for granted. These tweaks to existing stereotypes are not necessarily any more accurate than the originals, and may be less so. Qualifying them, however, requires the sorts of research that few have the desire, or the resources, to undertake.

How stereotypes arise and persist

So if stereotypes are often inaccurate, and become more so over time, why do they persist? To answer this question, we first have to consider how they arise in the first place. Here there is a fundamental difference in the raison d'être of the positive and the negative versions.

Positive stereotypes embody people's aspirations. These may be modest indeed, as with those rural migrants who make their way to the city. They want to be able to eat, to put a roof over their head, and just possibly get ahead. They may not care whether they end up in inner or outer parts of the urban area as long as they can fulfill Maslow's basic needs. Those moving out of the city are better off and aim to satisfy higher needs: comfort, security, and privacy. Since those are more clearly associated with the suburbs, theirs is a more specifically suburban vision. Either way, the stereotype articulates an ideal, a vision, a goal.

This ideal is most forcefully articulated by those who gain from promoting it. These include the agents of land development: developers, builders, lenders, and the professions that serve them, including lawyers and real estate agents. Developers are central, and their advertising – originally in newspapers and on billboards, now more commonly online – is the clearest articulation of the stereotypical ideal, and of its local variations. Brosius (2010) provides examples for Delhi. It is their business to be in tune with local needs while, for marketing purposes, it is in their interests to oversimplify, for example by "extend[ing] the tone of the garden suburb to that of suburbia at large" (Bunce, 1994, p. 168). In other words, their task is to market a stereotype.

The same line of argument does not help explain negative stereotypes. Those who dislike suburbs need not live there, and hardly anyone stands to profit from damning them. But there are rewards other than monetary. Historically, condemnations of the suburbs have been class-based, coming from those who looked down on those whose aspirations were limited to ownership of a comfortable home. Critics were either self-styled, urbane intellectuals whose milieu was the city, or the owners of landed estates beyond city limits (Bruegmann, 2005, pp. 118–119; Thompson, 1982, pp. 2–4). Their modern equivalents are those "university-educated professionals," including journalists and academics, who are actual or potential gentrifiers, as well as prosperous exurbanites who bought extensive acreages in more rural settings (Davison, 2013, p. 15). Both the urbanites and, paradoxically, the exurbanites often have a "self-righteous and condescending" attitude toward the intervening suburbs, dismissing them as featureless sprawl (Davison, 2013, p. 15; Sandberg et al., 2013, p. 19). Negative stereotyping, then, can be a form of one-upmanship.

That is still true, but there are now objective grounds for critique. As the scale of suburbanization has increased, the loss of farmland has become a more serious issue. So has the impact on wildlife and local watersheds, while there is wide agreement that sprawl contributes to global warming. Everyone now has a stake in limiting suburban growth, and raising the density of that which has already occurred. To be effective, however, those who invoke modern, negative stereotypes would be advised to lose the self-righteousness and condescension. That is not the way to win friends and influence people.

Both the negative and, to a lesser extent the positive, stereotypes have persisted because of the influence of the media. Popular forms of entertainment have often invoked and reinforced negative images of the suburbs (Huq, 2013). The same appears to be true of local newspapers (Harris and Hendershott, 2018). This matters because many of those who criticize the suburbs, and those who live there, do not actually know what they are talking about. Nicolaides (2006, p. 90) has noted of Lewis Mumford and Jane Jacobs – two of the most influential urbanists of the past century, and both critics of suburbia – that "neither actually observed the suburbs firsthand." Ignorance, above all, is the basis on which stereotypes thrive.

Conclusion

We must handle stereotypes with care, but we cannot do without them. A recent study found that post-war Toronto has not fit several of the clichéd descriptions discussed previously: a significant minority of its residents never bought into the suburban dream; its suburbs have always been socially diverse; and, perhaps most surprisingly, a number of local, educated professionals have always defended them (Harris, 2015). But, at the same time, those stereotypes do provide a point of departure for understanding the city. They are a scaffolding, eventually to be discarded, without which it would be difficult to construct a coherent account. And what is true for Toronto is surely true for anywhere.

But if we cannot do without them, we should also handle with care. For every positive version, there is a negative and, because the truth often lies in between, we should always consider both. We should pay attention to who is articulating each one, and consider what their stake in the matter might be, whether financial, social, or perhaps even psychological. We should be wary of stereotypes that have been around for a long-time because their relevance may have eroded, and we should be even more skeptical of those of the past, for they may in fact say much more about the present. And we should be alert to the possibility that, in any particular setting, no stereotype might exist, not even the idea of the suburb itself.

Guide to further reading

Brosius, C. (2010) *India's Middle Class: New Forms of Urban Leisure, Consumption and Prosperity.* New Delhi: Routledge.

Bruegmann, R. (2005) *Sprawl: A Compact History.* Chicago, IL: University of Chicago Press.

Davison, G. (2013) "The Suburban Idea and Its Enemies." *Journal of Urban History*, 39, pp. 829–847.

Harris, R. (2015) "Using Toronto to Explore Three Suburban Stereotypes." *Environment and Planning A*, 47, pp. 30–49.

Harris, R. and Vorms, C. (2017) "Introduction." In: Harris, R. and Vorms, C. (eds.) *What's in a Name? Talking about Urban Peripheries.* Toronto: University of Toronto Press, pp. 3–42.

Herzog, L.A. (2015) *Global Suburbs. Urban Sprawl From the Rio Grande to Rio de Janeiro.* New York: Routledge.

Huq, R. (2013) *Making Sense of Suburbia through Popular Culture.* London: Bloomsbury.

Jackson, K. (1986) *Crabgrass Frontier: The Suburbanization of the United States.* New York: Oxford University Press.

Nicolaides, B. (2006) "How Hell Moved from the City to the Suburbs." In: Kruse, K.M. and Sugrue, T.J. (eds.) *The New Suburban History.* Chicago, IL: University of Chicago Press, pp. 80–98.

Shen, J. and Wu, F. (2013) "Moving to the Suburbs: Demand-Side Driving Forces of Suburban Growth in China." *Environment and Planning A*, 45, pp. 1823–1844.

References

Archer, J. (2005) *Architecture and Suburbia: From English Villa to American Dream House, 1690–2000.* Minneapolis, MN: University of Minnesota Press.

Brosius, C. (2010) *India's Middle Class: New Forms of Urban Leisure, Consumption and Prosperity.* New Delhi: Routledge.

Bruegmann, R. (2005) *Sprawl: A Compact History.* Chicago, IL: University of Chicago Press.

Bunce, M. (1994) *The Countryside Ideal.* Toronto: Routledge.

Cadieux, K.V. and Taylor, L. (eds.) (2013) *Landscape and the Ideology of Nature in Exurbia: Green Sprawl.* New York: Routledge.

Caldeira, T. (1996) "Fortified Enclaves: The New Urban Segregation." *Public Culture*, 8, pp. 303–328.

Clapson, M. (2003) *Suburban Century: Social Change and Urban Growth in England and the USA.* Oxford: Berg.

Davis, M. (2006) *Planet of Slums.* London: Verso.

Davison, G. (2013) "The Suburban Idea and Its Enemies." *Journal of Urban History*, 39, pp. 829–847.

Dick, H.W. and Rimmer, P.J. (1998) "Beyond the Third World City: The New Urban Geography of Southeast Asia." *Urban Studies*, 35, pp. 2303–2321.

Dupont, V. (2004) "Urban Development and Population Redistribution in Delhi." In: Champion, A.G. and Hugo, G. (eds.) *New Forms of Urbanization: Beyond the Rural-Urban Dichotomy.* Aldershot: Ashgate, pp. 171–190.

Ferguson, J. (1999) *Expectations of Modernity: Myths and Meanings of Urban Life in the Zambian Copperbelt.* Berkeley, CA: University of California Press.

Fleischer, F. (2010) *Suburban Beijing: Housing and Consumption in Contemporary China.* Minneapolis, MN: University of Minnesota Press.

Haider, F. (2000) "Migrant Women and Urban Experience in a Squatter Settlement." In: Dupont, V., Tarlo, E. and Vidal, D. (eds.) *Delhi: Urban Space and Human Destinies.* New Delhi: Manohar, pp. 29–49.

Harris, R. (2015) "Using Toronto to Explore Three Suburban Stereotypes." *Environment and Planning A*, 47, pp. 30–49.

Harris, R. (2017) "Transnational Urban Meanings: The Passage of 'Suburb' to India, and Its Rough Reception." In: Sandoval-Strausz, A. and Kwak, N. (eds.) *Making Cities Global: The Transnational Turn in Urban History.* Philadelphia, PA: University of Pennsylvania Press (in press).

Harris, R. and Vorms, C. (eds.) (2017) *What's in a Name? Talking about Urban Peripheries.* Toronto: University of Toronto Press.

Harris, R. and Hendershott, K. (2018) "How Newspapers Portray Suburbs: A Paradox." *Journal of Urban Affairs.* DOI: 10.1080/07352166.2018.1431050.

Herzog, L.A. (2015) *Global Suburbs: Urban Sprawl from the Rio Grande to Rio de Janeiro.* New York: Routledge.

Huq, R. (2013) *Making Sense of Suburbia through Popular Culture.* London: Bloomsbury.

Jackson, K. (1986) *Crabgrass Frontier: The Suburbanization of the United States.* New York: Oxford University Press.

Liu, G.C.S. (2007) "Peri-Urbanism in Globalising China: A Study of New Urbanism in Dongguan." *Eurasian Geography and Economics*, 47, pp. 28–53.

Nicolaides, B. (2006) "How Hell Moved from the City to the Suburbs." In: Kruse, K.M. and Sugrue, T.J. (eds.) *The New Suburban History.* Chicago, IL: University of Chicago Press, pp. 80–98.

Potts, D. (2010) *Circular Migration in Zimbabwe and Contemporary Sub-Saharan Africa.* Oxford: James Currey.

Rome, A. (2001) *The Bulldozer in the Countryside: Suburban Sprawl and the Rise of American Environmentalism.* Cambridge, MA: Cambridge University Press.

Sandberg, L.A., Wekerle, G., and Gilbert, L. (2013) *The Oak Ridges Moraine Battles: Development, Sprawl, and Nature Conservation in the Toronto Region.* Toronto: University of Toronto Press.

Shanks, A., Coates, V., and Harris, R. (2017) "Doubts about 'Suburbs' in Canada." In: Harris, R. and Vorms, C. (eds.) *What's in a Name? Talking about Urban Peripheries.* Toronto: University of Toronto Press, pp. 89–111.

Shen, J. and Wu, F. (2013) "Moving to the Suburbs: Demand-Side Driving Forces of Suburban Growth in China." *Environment and Planning A*, 45, pp. 1823–1844.

Thompson, F.M.L. (1982) "Introduction: The Rise of Suburbia." In: Thompson, F.M.L. (ed.) *The Rise of Suburbia.* Leicester: Leicester University Press, pp. 2–25.

Turner, J.F.C. (1968) "Housing Priorities, Settlement Patterns and Urban Development in Modernizing Countries." *Journal of the American Institute of Planners*, 34, pp. 354–363.

Waldrop, A. (2008) "Gating and Class Relations: The Case of a New Delhi Colony." *City and Society*, 16, pp. 93–116.

Whitehand, J.W.R. and Carr, C.M.H. (1999) "England's Garden Suburbs: Development and Change." In: Harris, R. and Larkham, P.J. (eds.) *Changing Suburbs: Foundation, Form and Function*. London: E. & F.N. Spon, pp. 76–90.

Wu, W. (2013) *The Chinese City*. London: Routledge.

Zhou, Y. and Logan, J. (2008) "Growth on the Edge." In: Logan, J.R. (ed.) *Urban China in Transition*. Oxford: Blackwell Publishing, pp. 140–160.

3

In what sense a post-suburban era?

Nicholas A. Phelps

Introduction

In what sense do we live in a post-suburban world? On the one hand, a sense of a clear and distinct break between suburban and post-suburban eras is problematic, not least because of the highly relative nature of the very definition of the word "suburb." At any number of geographical scales across what Neil Smith (1981) depicted as the see-saw of uneven development or what Lefebvre (2003) depicted as the implosion-explosion dialectic of planetary urbanization, there is of course a fundamental unity in the urbanization process. The likes of inner-city gentrification and suburbanization are linked as part of a single process involving investment, disinvestment, and reinvestment. Yet, on the other hand, there is a sense that the post-suburban is necessarily implied in the suburb itself. It is implied in the suburb as a settlement (or more correctly, perhaps, a people and their institutions) in the process of *becoming* urban – in form, in function, and in ways of life. This sense of a settlement characterized by a *process* of becoming doubtless has existed from ancient times but is, I suggest, perhaps most palpable today as we investigate the many housing forms and settlement types that make up the suburbs in an era of planetary urbanization.

In this chapter, I set out three different senses in which we might think of there being a post-suburban era followed with additional reflections on an earlier review (Phelps et al., 2010). First, I discuss ideas of a post-suburban era. Second, I discuss the sense of post-suburbia as settlement form. While there are limits to this view, questions of form nevertheless remain more important than contemporary discourse on planetary urbanization might suggest. Third, I develop the idea that post-suburbs and the processes producing them are implied in the contradictions of suburbs and processes of suburbanization. The third of these is the one I favor. This is because of the difficulties of distinguishing suburbs from post-suburbs in terms of a new era when such considerations are based on static empirical measures and the problems associated with the lack of attention to geographical or morphological variation apparent in the most abstract theoretical formulations of "planetary urbanization" (Lefebvre, 2003). My favored third sense of a post-suburban era is one implied in the modernity of state and private sector-developed residential suburbs of the Global North (and especially Anglophone countries), but here I also briefly discuss whether it has wider applications to the individual and collective aspirations to modernity implied in the self-built shanties, favelas, and kampungs through which suburbanization has occurred across much of the Global South. That is, the emergence of a class of post-suburbs may be an ideological and political mutation of the traditional residential suburb in notably Anglo-American settings – a local, more modest and concrete manifestation of the broader politics of second modernity of which Ulrich Beck and colleagues have spoken. This perspective is not without its own limitations, but I believe it offers a tractable basis for considering and evaluating empirical evidence and can provide valuable insights into the very contradictions that lie at the heart of capitalist urbanization processes.

In the following sections, I first examine definitions of the suburb and processes of suburbanization before going on to consider the adequacy of three different senses in which we can speak of a post-suburban era. In closing, I draw some broader conclusions regarding the value of thinking about what a class of post-suburbs (alongside other settlement types) and processes of post-suburbanization can reveal about the fundamental unity of the capitalist urbanization process.

The suburb and suburbanization defined

Both city and suburb are deceptive terms. They are both difficult to define precisely and are rather more a case of "we know one when we see one." One composite definition developed recently by Harris and Larkham (1999, p. 8) provides a starting point for thinking about suburbs but also post-suburbs and processes of suburbanization and post-suburbanization. Harris and Larkham (1999, p. 8) provide a composite definition of a suburb as a settlement: 1) in a peripheral location relative to a dominant urban center; 2) partly or wholly residential in character; 3) of low-density of development; 4) with a distinctive culture or way of life; and 5) a separate community identity often embodied in a local government. The definition is a starting point as it arguably is exposed, just as definitions of the post-suburban can be, to the very process of urbanization it seeks to shed light on. The first element also forms the basis of Ekers et al. (2012, p. 407) definition of suburbanization as "the combination of non-central population and economic growth with urban spatial expansion." However, the outward expansion of cities renders the first definition highly relative – what was once peripheral comes to be regarded as central by virtue of the ongoing urbanization process. The second ingredient of the composite definition becomes problematic when we move beyond the Anglophone world, let alone beyond the Global North.

The suburb has been an escape from the city in some societal contexts, and an approach to the city in others. Moreover, there are no easy area-based generalizations that can be made with respect to these two aspects of the suburb. Examples of the suburb as approach to and retreat from the city are both to be found within a single complex continent like Europe (Phelps and Vento, 2015). There are important commonalities between the suburbs of southern European cities and those in Latin American cities – and indeed across much of the Global South – in terms of the informality of their mode of development and their being an approach to the opportunities and services of cities.

Perhaps a lowest common denominator that might integrate discourse on the suburb is that it is, in one or more senses, less than urban. If we accept this, we might also – although this is more controversial in some regards – accept the suggestion that, by the actions of their people and institutions, these settlements are in a process of becoming more urban. This is a question that has rarely been touched on in the literature (Bourne, 1996) and remains a largely vacant research agenda (see also McManus and Ethington, 2007). That is, rather than the suburb as a category, it is the process of (post)suburbanization that needs to be considered (Ekers, Hamel and Keil, 2012).

Yet, given the sheer diversity of suburbs and the meanings attached to them, it is also likely that classification of different (post)suburban settlement types is not at all antithetical to understanding the operation of processes – indeed in the face of such diversity, it may be a vital first cut at theoretical understanding and empirical verification.

The post-suburb as new era in urbanization

The first, and perhaps most straightforward, sense of our having entered a post-suburban era in some societal contexts was outlined by Lucy and Phillips (1997) in the case of the state of Virginia in the United States and also appears present in Essex and Brown's (1997) work on Australia. Lucy and Phillips (1997, p. 260) defined the post-suburban era as a "time period which is succeeding the suburban era and which includes several spatial forms, including a sprawling exurban rural pattern, which is of much lower-density than most suburbs." More specifically still, they define this post-suburban era in terms of "inner suburban population loss and relative income decline, suburban employment increase, suburban out commuting reduction, exurban population and income increase and farmland conversion."

One set of problems with this definition stems from the fact that it is oriented primarily to situations found in Anglophone countries where the residential suburb has been a bourgeois retreat from the city (Fishman, 1987). Across much of the Global South and some parts of Europe, for example, self-built residential "suburbs" are an approach to the city by low-income populations, meaning that successively peripheral suburban developments are unlikely to have incomes that are growing faster than elsewhere across metropolitan areas. The limits of this definition are also reached when one recognizes the "temporal disparity" that exists not only within the Global North but also between the Global North and Global South: the rising average income levels and urban transitions that are seen to drive suburbanization as a global residential preference (Bruegmann, 2006) and process have taken place at different times and have yet to happen in many other instances.

A second problem is that this definition of a post-suburban era is also entirely relative within any given national context let alone when considering the divergent trajectories of nations noted earlier. Whether and when a particular settlement or jurisdiction is post-suburban depends entirely on where it is located relative to a historic city. Or, as a process, it is dependent on some minimal extension of cities. To put this slightly differently, and returning to our definition of a suburb outlined above, suburbs can be rendered invisible by the outward development of cities. What was once peripheral becomes ostensibly central, what was once a suburb becomes an inner suburb or hardly distinguishable from what we today take for the historic core of a metro area, not least because suburbs have often been annexed as part of cities.

The missing post-suburbs of planetary urbanization

The majority of the world's population now live in officially defined urban areas. This fact alone might signal that we have entered an era of what Lefebvre (2003) termed "planetary urbanization." More profoundly, what Lefebvre pointed to was the fact that the majority of the world now experienced an urban way of life regardless of whether residing in officially defined urban or rural areas. Lefebvre's work is exciting for the research agenda regarding an "urban revolution" that it opens to view. It is an urban revolution that is in large measure a suburban revolution. Yet, one has to look hard to find any reference to (post)suburbs or processes of (post)suburbanization in the burgeoning body of literature on planetary urbanization.

At a fundamental level, there is a unity to the urbanization process that belies academic deliberation over cities and suburbs. Yet, in other respects, attention to (post)suburbs and processes of (post)suburbanization are set already to be underappreciated aspects of planetary urbanization – perhaps replicating the long-standing lack of interest in and attention to suburbs within urban studies (Harris, 2010). Notwithstanding the difficulties of identifying specifically urban dimensions of socioeconomic processes (Saunders, 1983), the subtly different urban, suburban, and indeed rural questions may disappear from view altogether in this rapidly emerging agenda if all is now urban. Yet, as Ekers et al. (2012) have recently elaborated, it is precisely the explosion – the suburban credentials – of this revolution that are worthy of further investigation, as it is here that the generalized contractions of capitalist planetary urbanization are played out and most visible. For example, to the extent that the urban question is about collective consumption, the less-than-urban properties of suburbs noted previously make them good places to consider the urban question (Phelps et al., 2015). It is the post-suburban transformation of suburbs that provides perhaps the best vantage point to understand the urban question, as suburbs have from their outset posed questions about the making good of deficits of infrastructure, amenities, services, and political representation. In dialectical terms, from the outset, suburbs, and the process of suburbanization contained the seeds of their own negation.

In adhering to the dialectical method, and at an abstract ontological and theoretical level, this lack of reference to (post)suburbs, and the detailed processes of change to which they are subject, *may* of course be entirely correct. Lefebvre's writings provide a highly abstract basis for considering the contemporary urbanization process. With reference to Lefebvre's writings, Nussli (2017) argues against the classification of different settlement types in analysis. However, it is the sheer enormity of this research agenda that really begs for the adoption of approaches that offer tractability, both for proper theoretical elaboration of the potential of this agenda and meaningful empirical validation or exploration of it. Lefebvre's work (2003) may be rather limiting in terms of understanding of urbanization – a process, as I have noted earlier, that in almost all instances involves differentiated settlements and peoples striving for urbanity. Galster et al. (2001), for example, identify eight separate characteristics of urban sprawl as a complex and multifaceted process. The variable geography and morphology of planetary urbanization – its form – must remain an important focus of study within this agenda.

There is a rich tradition of neologisms used in urban studies to depict new forms of urbanization typically in terms of its ever more diffuse and even difficult to delimit nature. Little wonder that Lang and Knox (2009) have argued recently that city and suburb are themselves "zombie" categories. Indeed, earlier work defined the process of post-suburbanization primarily in terms of the new urban-regional scale at which the urbanization process was occurring in places such as North America (Kling et al., 1995). As a result of the mixing of different morphologies and land uses, peripheral settlements became cities in function but not in form (Fishman, 1987). Unless we think that this urban-regional scale of urbanization is exclusive to North America, it is worth noting how the urban-regional scale is one appealed to elsewhere when depicting processes that we might, for convenience's sake, term planetary urbanization. In a European context, particularly in Germany, the term *Zwischenstadt* – or in-between city – resembles such a phenomenon (Sieverts, 2003). Similarly, the term *desakota* (McGee, 1991), which combines the Indonesian words for village and city, has emerged as an influential summary of unique conditions and processes in East Asia. More generally across the Global South, the picture of peripherally located informal settlements found across the Global South has become overlain with gated residential communities, master-planned towns, and industrial parks. At least when viewed at this urban-regional scale, the peripheries of many city-regions of the Global South can look post-suburban, even if there are differences in the context and the balance of forces producing a scale and mix of urbanization similar to parts of the Global North. Thus, in Indonesia, Jakarta's extended urbanization is considered post-suburban (Firman, 2011). In Chile, the periphery of Santiago de Chile is also considered distinctly post-suburban in the mixture of elements (Borsdorf et al., 2007).

However, we should be careful of interpreting the same processes at play in settlement patterns with superficially similar morphology. The geography and morphology of settlement can equally alert us to qualitatively different processes by which suburban communities are becoming urban. Just as questions of urban policy mobility and inter-referencing retain a regional complexion in their relationality (Roy, 2009), so the geography and morphology of (post)suburbanization is regionally distinct in a way that is revealing of potentially different underlying processes of development and further evolution of settlement. Thomas Sieverts' *Zwischenstadt* forms one influential reference point within discourse on planetary urbanization. At one level, the sorts of mixing of urban and rural and uses might prompt comparison to the *desakota* landscape of East Asia. Yet, it would be a mistake to think the processes underlying this "rurban" mixture are anywhere near the same. Despite its ambiguous form, *Zwischenstadt* is fully urban in function in that the vast majority of development and employment is in the formal sector. This contrasts markedly with the fundamental in-betweenity of *desakota* landscapes. *Desakota* captures the betweenity of some instances of planetary urbanization: of communities caught between urban and rural, formal and informal, and modernity and tradition (Phelps, 2017).

The second modernity of post-suburbanization

A third way one might think of post-suburbia's emergence in different societal settings may suffer less from the limitations of the first two senses of post-suburbia noted previously – as a new era and as new settlement space. This is a definition of post-suburb and post-suburbanization implied in the inherent limits or contradictions of suburbs – as settlements in a process of becoming. In the present this is a sense of a post-suburban era borne of the unintended consequences of modern urbanization and in particular to the manner in which modern state interventions have licensed suburban forms of development with their own contradictions. These inherent contradictions of suburbanization are something to which classic suburbanization studies within the Marxist tradition (e.g., Walker, 1981; Harvey, 1985) alert us, though it should be remembered that they do so with the United States in mind. This, then, is a definition of a post-suburban era measured not in terms of a particular period of time or scale of urbanization, but one measured in terms of a transformation in suburban ideology and politics as a response to the inherent limits or contradictions of suburbs.

The process of suburbanization has itself been a "spatial fix" (Harvey, 1985) for capitalist societies – a vast opportunity zone for all those industries reliant on the development of land for their profits (architects, land speculators, developers, construction and house-building companies, real estate brokers) and a vast market outlet for other industries that supplied all the items that filled up the houses and offices (vacuum cleaners, refrigerators, cars) (Walker, 1981). The building of these suburban expanses did not happen of its own accord. It was licensed by modern state interventions, notably in the form of new credit and new roads that systematically distorted geographical patterns of accessibility at the metropolitan scale and beyond. That is, they greatly increased the accessibility of peripheral land. The thriving U.S. economy during the decades after World War II is inseparable from the physical transformation of U.S. cities that took place in that time (Walker, 1981).

However, if in general state interventions have their own contradictions or unintended consequences (Scott and Roweis, 1977), the suburbs facilitated by some of the largest state interventions also have their specific contradictions. Some examples are the extreme separation of land uses, the additional trips and ultimately congestion these generate, and, in some instances, an exclusionary politics that appears likely to prevent future change (Harvey, 1985). The modern suburb is a confluence of government interventions and rationalized corporate production and marketing techniques and similar interventions played a big part in the emergence of sub-urban forms in not just the U.S. but also increasingly elsewhere (in China, Indonesia, and Chile, for example). This might be regarded as a specific instance of the more generalized unintended consequences of modern state interventions as a whole (Beck et al., 2003). It is the unanticipated consequences of a host of modern interventions by states, major corporations, or the technocracy that John Kenneth Galbraith saw making up the merger of the two that have ushered in what Beck refers to as an era of second modernity. Beck concentrates on the unintended environ-mental consequences of modernity that have prompted more individualized and environmentally conscious politics. Although he does not mention the suburbs, one can hardly speak of the poli-tics of energy consumption and associated global warming without reference to the ongoing pro-cesses of suburbanization. That is, modern suburbs contain within them contradictions – the seeds of the post-suburbs of second modernity. The question that arises is whether we can observe anything of Beck's more generalized politics of second modernity in the suburbs?

In all but its most affluent incarnations, the residential suburb was a development model that was flawed from the start (Gallagher, 2014). Even in the United States, its limitations became apparent almost immediately, setting in motion distinctly post-suburban politics as early as the 1950s (Teaford, 1997). By the 1970s, the urbanization of the suburbs (Masotti and Hadden, 1973) and the emergence of outer cities (Muller, 1975) had become apparent. These developments, among others, can be classified into settlements and their trajectories of change as an initial means to consider the differential politics (Phelps and Wood, 2011; Phelps, 2016). The scheme we offered is not exhaustive. Doubtless some categories of settlement we depicted could be subdivided further, perhaps along the lines described by Harris and Larkham (1999). Aspects of the post-suburban transformation that we focused on are also hardly exclusive to middle-income outer suburbs; they are also found in struggling, older, inner, working-class suburbs (Sweeney and Hanlon, 2017). However, our scheme does start to focus attention on 1) some of the continuities apparent as suburban communities transform into post-suburbs, and 2) the qualitatively differ-ent (demographic, economic, and institutional) character, ideology, and politics of settlements, including the variety of suburbs to be found in many extensive city-regions.

We distinguished a class of what might be termed post-suburbs from a class of stable, afflu-ent, residential suburbs – the stereotypical idea of the suburb found in much of the Anglophone and even Global North literature – and from declining inner (often industrial or employment originated) suburbs that may in some instances amount to sub-suburbs. Ostensibly, as Masotti and Hadden (1973) implied and as Teaford (1997) made explicit, the class of post-suburbs in the United States is represented by the communities centered on the numerous "edge cities" that now characterize metropolitan areas (Garreau, 1991). The case of Kendall Dadeland in Miami-Dade County provided one early and only partially successful New Urbanism-inspired fashion-ing of a downtown for sprawling outer suburban residential areas (Phelps, 2016). The past growth and current plans to reshape Tysons Corner in Fairfax County, Virginia, outside Washington, DC, stand as a prime example of political deliberations implied in suburbs transforming into post-suburbs and into something akin to cities in function, even if not in form. Tysons Corner grew in the first instance precisely in recognition of the fiscal limitations on county government expenditure imposed by the predominantly residential nature of development in Fairfax up until the 1960s. The aggressive recruitment and corralling of business into Tysons Corner was the political bargain struck between business and development elites and residents in order to make good the shortfalls in urban amenities and services that had accumulated to that point. By today, Tysons Corner has succeeded in that role but now contradicts itself to the point that recent planning deliberations have sought to further develop this suburban business center into a proper downtown destination for the surrounding residential communities of Fairfax (Phelps, 2012).

If Tysons Corner is perhaps the best example of the emergence of distinctly post-suburban politics, the same sorts of deliberations are increasingly apparent elsewhere in the edge cities and outer suburbs of the United States – in Houston's Energy Corridor and in Philadelphia's King of Prussia to name just two (Dunham-Jones and Williamson, 2009).

It could be argued that the scheme and our writings on post-suburbanization – as with much urban theory – are oriented overly to the Global North and to the North American context in particular. We would not want to deny this, since it was developed from an appreciation of these countries in mind. However, recast in the broader terms of communities seeking to become more urban – more modern even – the process of post-suburbanization is visible, albeit with important differences, for example, in Europe (Phelps et al., 2006; Phelps et al., 2015). It is also not antithetical to the peripherally located settlements of the Global South in which complex and perhaps more multifaceted desires – of states, of individual households, and of corporations alike – for urbanity and modernity coexist.

Across the Global South, states are engaging in the sorts of infrastructure developments seen in the United States with some of the attendant elements of development such as industry and office parks, gated communities and shopping malls. In this regard alone, the challenge of reworking suburbs into something more urban will indeed be salient elsewhere (Forsyth, 2013). At the same time, the sense of modernity inherent in processes of suburbanization across the Global South may run deeper than the appearance of state and corporate infrastructure provision and planning rationalities. It may extend more fundamentally to the individual aspiration to modernity visible within informality as a mode of development. Much of the *desakota* pattern mentioned previously involves the peripheral and ribbon development of self-built kampungs (rural and urban village scale communities). Kampungs themselves are interesting in that they are self-, or rather collectively, built in the first instance for purposes of shelter. However, not only do dwellings evolve to become larger and more permanent over time, the kampung typically also evolves rapidly to become as much a place of work as residence. This happens independently of formal state intervention, but state intervention in the form of housing upgrading programs are an important part of the mix in these distinctive suburban forms.

Conclusion

If the suburbs and processes of suburbanization have been shunned by all except a small band of historians, they stand to be further ignored in contemporary debates about planetary urbanization. This is curious for some of the reasons outlined earlier. While on the most abstract theoretical level, there is little doubting the unity of urbanization processes that render distinctions between settlement types rather meaningless. It is clear that consideration of different types of (post)suburbs and longitudinal studies of particular (post)suburbs is vital to: 1) meaningful empirical analysis of the urbanization processes; 2) revealing specific contradictions of the urbanization process; and 3) unified analytical discussion.

While according to contemporary sensibilities in urban studies the inductive identification and classification of different types of settlements might be ridiculed for the reductionism imputed to such taxonomies, without such imperfect schemes, there is a very real danger that any and all empirical findings can be poured into generalized theoretical meta-containers without 1) real consideration of significant areal differentiation in urban function and form and 2) real investigation of the very contradictions that are the drivers of the unity of urbanization processes. It is in this sense that entertaining a category of post-suburbs, associated processes of post-suburbanization, and even a post-suburban era can be a useful ingredient in dialectical method when applied to understanding contemporary urbanization.

Nicholas A. Phelps

Guide to further reading

Phelps, N.A. (2017) "Suburbs in the Metropolitan Economy." In: Berger, A., Kotkin, J. and Balderas Guzman, T. (eds.) *Infinite Suburbia*. Hudson, NY: Princeton Architectural Press, pp. 314–327.

Walker, A.R. (1981) *Theory of Suburbanization: Capitalism and the Construction of the Urban Space in the United States*. London: Methuen.

References

Beck, U., Bonss, W., and Lau, C. (2003) "The Theory of Reflexive Modernization: Problematic, Hypotheses and Research Programme." *Theory, Culture & Society*, 20, pp. 1–33.

Borsdorf, A., Hidalgo, R., and Sánchez, R. (2007) "A New Model of Urban Development in Latin America: The Gated Communities and Fenced Cities in the Metropolitan Areas of Santiago de Chile and Valparaíso." *Cities*, 24, pp. 365–378.

Bourne, L.S. (1996) "Reinventing the Suburbs: Old Myths and New Realities." *Progress in Planning*, 46, pp. 163–184.

Bruegmann, R. (2006) *Sprawl: A Compact History*. Chicago, IL: University of Chicago Press.

Dunham-Jones, E. and Williamson, J. (2009) *Retrofitting Suburbia: Urban Design Solutions for Redesigning Suburbs*. Chichester: Wiley and Sons.

Ekers, M., Hamel, P., and Keil, R. (2012) "Governing Suburbia: Modalities and Mechanisms of Suburban Governance." *Regional Studies*, 46, pp. 405–422.

Essex, S.J. and Brown, G.P. (1997) "The Emergence of Post-Suburban Landscapes on the North Coast of New South Wales: A Case Study of Contested Space." *International Journal of Urban and Regional Research*, 21, pp. 259–287.

Firman, T. (2011) "Post Suburban Elements in an Asian Extended Metropolitan Region: The Case of Jabodetabek." In: Phelps, N.A. and Wu, F. (eds.) *International Perspectives on Suburbanization: A Post-Suburban World?* Basingstoke: Palgrave Macmillan, pp. 195–209.

Fishman, R. (1987) *Bourgeois Utopias: The Rise and Fall of Suburbia*. New York: Basic Books.

Forsyth, A. (2013) "Suburbs in Global Context: The Challenges of Continued Growth and Retrofitting." *Planning Theory and Practice*, 14, pp. 403–406.

Gallagher, L. (2014) *The End of the Suburbs: Where the American Dream is Moving*. London: Portfolio.

Galster, G., Hanson, R., Ratcliffe, M.R., Wolman, H., Coleman, S., and Freihage, J. (2001) "Wrestling Sprawl to the Ground: Defining and Measuring an Elusive Concept." *Housing Policy Debate*, 12, pp. 681–717.

Garreau, J. (1991) *Edge City: Life on the New Frontier*. New York: Doubleday.

Harris, R. (2010) "Meaningful Types in a World of Suburbs." In: Clapson, M. and Hutchinson, R. (eds.) *Suburbanization in Global Society*. Bingham: Emerald Publishing, pp. 15–47.

Harris, R. and Larkham, P. (1999) "Suburban Foundation, Form and Function." In: Harris, R. and Larkham, P. (eds.) *Changing Suburbs: Foundation, Form and Function*. London: Routledge, pp. 1–31.

Harvey, D. (1985) *The Urbanisation of Capital*. Oxford: Blackwell Publishing.

Kling, R., Olin, S.C., and Poster, M. (1995) *Postsuburban California: The Transformation of Orange County since World War II*. Berkeley, CA: University of California Press.

Lang, R. and Knox, P. (2009) "The New Metropolis: Rethinking Megalopolis." *Regional Studies*, 43, pp. 789–802.

Lefebvre, H. (2003 [1970]) *The Urban Revolution*. Minneapolis, MN: University of Minnesota Press.

Lucy, W.H. and Phillips, D.L. (1997) "The Postsuburban Era Comes to Richmond: City Decline, Suburban Transition and Exurban Growth." *Landscape and Urban Planning*, 36, pp. 259–275.

Masotti, L. and Hadden, J. (1973) *The Urbanization of the Suburbs*. London: Sage.

McGee, T.G. (1991) "The Emergence of Desakota Regions in Asia: Expanding a Hypothesis." In: Ginsburg, N., Koppel, B. and McGee, T. (eds.) *The Extended Metropolis: Settlement Transition in Asia*. Honolulu, HI: University of Hawaii Press, pp. 3–25.

McManus, R. and Ethington, P.J. (2007) "Suburbs in Transition: New Approaches to Suburban History." *Urban History*, 34, pp. 317–337.

Muller, P.O. (1975) "The Outer City: Geographical Consequences of the Urbanization of the Suburbs." Washington, DC: Association of American Geographers Resource Paper No. 75–2.

Nussli, R. (2017) "Between Farming Villages and Hedge-Fund Centres: The Politics of Urbanization in the Border Zone of the Metropolitan Region of Zurich." In: Phelps, N.A. (ed.) *Old Europe, New Suburbanization: Land Infrastructure and Governance in Europe*. Toronto: University of Toronto Press, pp. 207–236.

Phelps, N. (2012) "The Growth Machine Stops? Urban Politics and the Making and Remaking of an Edge City." *Urban Affairs Review*, 48, pp. 670–700.

Phelps, N.A. (2016) *Sequel to Suburbia: Glimpses of America's Post-Suburban Future*. Cambridge, MA: MIT Press.

Phelps, N.A. (2017) *Interplaces: An Economic Geography of the Inter-Urban and International Economies*. Oxford: Oxford University Press.

Phelps, N.A., Parsons, N., Ballas, D., and Dowling, A. (2006) *Post-Suburban Europe: Planning and Politics at the Margins of Europe's Capital Cities*. Basingstoke: Palgrave Macmillan.

Phelps, N. A. and Tarazona Vento, A. (2015) "Suburban Governance in Western Europe." In: Hamel, P. and Keil, R. (eds.) *Suburban Governance: A Global View*. Toronto: University of Toronto Press, pp. 155–176.

Phelps, N.A., Tarazona Vento, A., and Roitman, S. (2015) "The Suburban Question: Grass Roots Politics and Place Making in Spanish Suburbs: The Cases of Badalona and Getafe." *Environment & Planning C*, 33, pp. 512–532.

Phelps, N.A. and Wood, A.M. (2011) "The New Post-Suburban Politics?" *Urban Studies*, 48, pp. 2591–2610.

Phelps, N.A., Wood, A.M., and Valler, D.C. (2010) "A Post-Suburban World? An Outline of a Research Agenda." *Environment & Planning A*, 42, pp. 366–383.

Roy, A. (2009) "The 21st-Century Metropolis: New Geographies of Theory." *Regional Studies*, 43, pp. 819–830.

Saunders, P. (1983) *Social Theory and the Urban Question*. London: Hutchinson.

Scott, A.J. and Roweis, S.T. (1977) "Urban Planning in Theory and Practice: A Reappraisal." *Environment and Planning A*, 9, pp. 1097–1119.

Sieverts, T. (2003) *Cities without Cities: An Interpretation of the Zwischenstadt*. London: Routledge.

Smith, N. (1981) *Uneven Development*. Oxford: Blackwell Publishing.

Sweeney, G. and Hanlon, B. (2017) "From Old Suburb to Post-Suburb: The Politics of Retrofit in the Inner Suburb of Upper Arlington, Ohio." *Journal of Urban Affairs*, 39, pp. 241–259.

Teaford, J.C. (1997) *Post-Suburbia: Government and Politics in the Edge Cities*. Baltimore, MA: Johns Hopkins University Press.

Walker, R.A. (1981) "A Theory of Suburbanization: Capitalism and the Construction of the Urban Space in the United States." In: Dear, M. and Scott, A.J. (eds.) *Urbanization and Urban Planning in Capitalist Societies*. London: Methuen, pp. 383–429.

Part II

Global perspectives on the suburbs

4

Toward a comparative global suburbanism

Pierre Hamel and Roger Keil

The dialectics of extended urbanization at a global scale

Speaking, as we will in this chapter, of comparative global suburbanism, we build predominantly on two relevant theoretical interventions. One goes back to Henri Lefebvre's observation, first articulated in the late 1960s, that humanity had entered an urban revolution or a stage of complete urbanization (2003). In this context, Lefebvre isolated a dialectic of imploding and exploding developments, one in which he contrasts urban concentration with deconcentration. He expressed this most succinctly in this oft-quoted passage:

> [T]he tremendous concentration (of people, activities, wealth, goods, objects, instruments, means and thought) of urban reality and the immense explosion, the projection of numerous, disjunctive fragments (peripheries, suburbs, vacation homes, satellite towns) into space.
> *(Lefebvre, 2003, p. 14)*

There are no easy analogies here of inner and outer cities, of dense downtowns and sprawling suburbs, of skyscrapers here and single-family homes there. As we will see below, sub/urban forms and functions come in wildly different shapes and intensities, but among the exploding antithesis of the imploding center, Lefebvre counts "suburbs." However, while the suburban can be understood as a part of extended urbanization overall (Monte-Mor, 2014a, 2014b), it is also a specific process that can be studied empirically and theorized conceptually. In the first instance, then, when we say "suburbs" or "suburbia," we refer to this aspect of current urbanization dynamics. Specifically, in our chapter here, we take as a starting point that suburbanization "is a combination of non-central population and economic growth with urban spatial expansion" (Ekers et al., 2012, p. 407). Suburbanism(s) refers to suburban ways of life. The definitional simplicity must not overshadow, however, the vast diversity of processes and forms we find in suburbanization and suburbanisms worldwide.

The second theoretical intervention on which we base our argument in this chapter takes its cues from the understanding that urban and suburban research needs to develop new "geographies of theory" (Roy, 2009), where cities and suburbs around the world can neither be studied from any privileged (i.e., normally Western) position nor be reduced to containers of socioeconomic

1

activities, but should be considered instead as sites of experimentation where the urban condition is both limiting and enabling (Robinson, 2006). This intervention can be roughly divided into four related aspects:

1) The first aspect is the very notion of new geographies of theory itself. This mode of viewing the sub/urban world suspends norms that have driven the field for more than a hundred years. From their new vantage point, sub/urban studies defy allegedly predetermined trajectories of urbanization (following in particular the modernizing thrust of Western urban theory and developmental practice), and throw this inquiry into open conceptual and empirical territory.

2) The second aspect here is "urbanism," which needs liberation from a purely normative design strategy. If suburbanism is global suburbanism, it will be important to remember the conventional implications of this term that, following Sheppard et al. (2013, p. 894) "has come to refer to a distinct kind of site (the city), separable from other rural places, and taken to be a hallmark of modernism, progress, development, and the metropole – the opposite of provincialism. At the same time, urbanism is associated with a set of social ills, the dark side of development contrasted with an idyllic rural past. This dissonance implies the need for intervention – urban planning to achieve development while minimizing a social dysfunctionality." Conventionally, this aspect of urbanism (or suburbanism) implies a naturalization of liberal democracy and capitalism, and suggests it presents "ubiquitous norms and [is] capable of overcoming the poverty, inequality, and injustice seen as so pervasive across the global South" (Sheppard et al., 2013, p. 894).

   Ash Amin speaks in this context of "telescopic urbanism" that hides "the myriad hidden connections and relational doings that hold together the contemporary city as an assemblage of many types of spatial formations, from economically interdependent neighborhoods to infrastructures, flows, and organizational arrangements that course through and beyond the city" (2013, p. 484). We ultimately follow Roy here, who has proposed to understand urbanism in four dimensions. Initially, "urbanism refers to the territorial circuits of late capitalism" (2011a, p. 8). Suburban land is created in a capitalist process of the production of space. Then, Roy notes that while capital structures urban space in its image, it cannot structure it freely. Instead, "urbanism indicates a set of social struggles over urban space," a set of claims to the right to the city and the suburb. Also, urbanism appears as a "formally constituted object, one produced through the public apparatus that we may designate as planning." This is closest to the general usage of the term in today's public debates. And finally, Roy argues that "urbanism is inevitably global" (2011a, p. 9), which gets us to the third aspect of this intervention.

3) When we talk about *global* suburbanism in this chapter, we refer to a range of interlocking registers of geographical globality, ubiquity, simultaneity – although both of those latter processes are uneven – and planetarity, harking back to Lefebvre's idea of planetary urbanization (as recently discussed and popularized most prominently by Neil Brenner and Christian Schmid [2015]). As Stuart Elden (2014) has noted, Lefebvre viewed "the planetarization [la planétarisation] of the urban" (Lefebvre, 2014, p. 205) as a threat and the coverage of the actual earth [*terre*] with urban settlement as uneven and full of contradictions.

4) We have now arrived at the fourth and final aspect of this second intervention, the comparative lens. For several reasons, comparisons are required for better understanding (sub)urban processes through their diversified empirical components. From an historical perspective, the central cities and suburbs have always been interrelated. As mentioned previously, however, these links do not borrow unidirectional models. The contradictions brought on by class

relations, social inequalities and unequal access to territorial amenities involved in shaping the (sub)urban worldwide are redefined in different ways according to economical, political, and cultural settings. In that respect, defining global suburbanism means that comparison is called upon.

Global suburbanisms

What have we learned from studying global suburbanization and suburbanisms? (See Keil, 2018, p. 207, for an elaboration on this argument). Generally, this comparative research, engaging multiple sites and perspectives from across the globe, places suburbanization and suburbanisms in the mainstream of urban studies. However, it also opens up the urban periphery up to critical theory. In the first instance, this work points to the simple fact that urbanization today mostly comes in the form of peripheral extension or suburbanization. If ten billion humans eventually populate the earth, we can expect that hundreds of millions who are still on a trajectory of rural-to-urban migration and those who leave established cities for the margins – running a spectrum from forcefully displaced communities to aspirational seekers of isolation and privilege – will live in some form of suburbia. We can now make a few summary observations about this global phenomenon, whereby "global" is understood here in the double sense of generalized and worldwide. If the temptation of subsuming the variation of urban development under the dominant force of economic globalization remains strong, insistence on variations must be emphasized through the multiplicity of components, processes, cultural settings, class relationships and models of suburbanism occurring (Ong, 2011). That said, what are some general empirical trends?

- Suburbanization is a process that tends to produce sameness on site (reproducing houses and housing units in massive developments) and across sites (following international blueprints in many locations) (Easterling, 2014). Suburbanization is itself a major process of "inter-referencing" (Ong, 2011) a domain of sameness across the world (this is true for the forms, the social and class formation processes, transportation patterns, etc.).
- In this sense, the *banlieues à l'américaine* (Le Goix, 2017, p. 255) have provided a model for suburbanization worldwide. Still, the American or Anglo-Saxon model (single-family homes often in master-planned subdivisions) have always been only one among other models of suburbanization (Keil, 2017, chapters 4–6). As the lifestyle of the American model has retained its attraction for aspirational middle-classes across the world, the Global South has developed its own system of suburban inter-referencing and models of peri-urban developments and lifestyles (Bloch, 2015; Gururani and Kose, 2015; Mabin, 2013; McGee, 2013, 2015; Roy, 2015; Wu and Shen, 2015).
- Suburbanization is diverse. This may not be a big surprise given multiple trajectories of building the peripheral across the globe that have existed previously. But today, variety rules. These global processes come in all manner of shapes, forms and institutions, organized and disorganized, formal and informal, gated and squatted, and range from single-family homes to high-rise towers, including massive infrastructures, employment, and commercial zones, conservation areas and greenbelts (Keil, 2018, 2017).
- And speaking about diversity: suburbs, once imagined as bulwarks of class and racial uniformity and experienced that way by their inhabitants, are now a prime location where urban difference is present and being negotiated. Immigrant suburbs, suburbs as a place of poverty, the suburban as a location of racialized populations are now common experiences.

- There is much local history and geography of suburbanization. Yes, it is plausible to identify larger contexts such as, for example, American Fordism, neoliberalization or globalization, but the imprint of these processes will be specific to local histories and geographies, and in turn those local specificities help define the contours of these larger trends. National, regional, and local histories and geographies of suburbanization are important (see, for example, Richard Harris (1996, 2004) on Canada; also Keil et al., 2015, Keil and Addie, 2016; or Paul Knox on London's "metroburbia") (2017, pp. 12–14).

- Suburbs are, further, products of multi-scale, multi-topological processes of production and governance that involve worldwide interactions, money flows, idea transmission and aspirations in a global world. Andre Ortega (2016), for example, demonstrated that Manila's suburbanization was the product of a series of interlocking, global flows of capital, labor, and culture in a transnational chain of interdependencies that have a powerful effect on all participants beyond the actual physical building of suburban space.

- Next, suburbs are the ultimate capitalist commodity: a place of intense use value consumption (albeit realized through immense fetishization of labor practices) and at the same time a prime site for the realization of exchange values, especially in an era of financialization. Suburbanization has become the prime economic sector in many regions and countries, and has made the massive construction of housing, shopping malls and workspaces the source of both national wealth and sometimes crony capitalism (Guney et al., forthcoming).

- Relatedly, then, suburbanization today is a prime process through which "peripheral" urbanization (Caldeira, 2016) is converted to corporate, state-built and more "formal" types of settlement. The state-led processes of building huge suburban tower estates or identical gated communities with golf courses in nations such as China, Korea, Turkey, Vietnam, Brazil, and India stand out here. We have seen similar complexes emerge from the deserts of southern California and the fruit lands of Silicon Valley, where technology companies also serve as suburban builders, planners, and infrastructure providers, having turned classic residential suburbia into an elongated workbench of the world's most aggressive disruptors.

- Suburbanization is characterized by the complexities of post-suburbanization, as common perceptions of center and periphery blur and densities and morphologies are inverted. Knox (2017, p. 6) calls Outer London "at once metropolitan and suburban in character. It is, to borrow from the lexicon of urban studies, metroburban: a multimodal mixture of residential and employment settings, with a fusion of suburban and central city characteristics." The trend Knox observes here may be specific to London, but it certainly has been documented across the world from China to Europe and North America (Charmes and Keil, 2015; Le Goix, 2017; Phelps, 2015; Phelps and Wu, 2011; Sieverts, 2003).

- Physical forms of suburbanization and social processes of suburban expansion are accompanied by distinct suburbanisms that define the spread of the urban tissue (Moos and Walter-Joseph, 2017; Quinby, 2011; Walks, 2013). The dependence on the automobile persists in most suburban locations, but increasing densities and socioeconomic diversities have led to more challenges to the status quo of how one lives in the suburbs. Mobility, nutrition, work, and recreation are reassembled in and through suburban locations.

- A particularly important role is assigned to suburban infrastructures in, of, and for the suburbs (Addie, 2016; Filion and Pulver, 2018), and infrastructure blueprints have begun to stand in for (sub)urbanization more generally (Easterling, 2014, p. 12). The infrastructures contain the DNA of the suburban form; they are the medium that contains the message of a particular form and process of urban extension (Filion, 2013).

- The production of suburban land and the diversity of suburban form are common and ubiquitous features of suburban expansion (Harris and Lehrer, 2018).
- Density deserves its own mention on this list as a category that was once considered a clear indicator of suburbanism: if it is low-density and if it is on the periphery, it must be suburban. Today, such certainties have disappeared as "debates over density and sprawl become not only sterile or semantic, but also increasingly irrelevant, for the majority urban experience" (Tonkiss, 2013, p. 43). Still, densification and intensification – often paired with the notion of compactness – are desired processes of urban planning and development. "Sprawl repair" has become one of the battle cries of climate change discourse (Dunham-Jones and Williamson, 2011; Jessen and Roost, 2015). The enthusiasm for density and compactness would not be controversial except for the complex problems facing the reality of retrofitting in an age of post-suburbanization with its myriad contradictions (Charmes and Keil, 2015; Keil, 2017).
- Suburban political ecologies now need to be understood on their own terms. A good example is the relevance of boundaries to understanding suburbanization, as they bring into focus the (real and imagined) natural environment beyond. It has long been known that the suburban was sold to its users and inhabitants as a place close to nature, while the suburbs represent a threat to sustainability and are a major source of climate change. What is new is the fact that the suburban edge is the location of innovative and potentially path-breaking suburban political ecologies such as greenbelts, conservation areas, and new forms of suburban living emerging at the city's edge.
- Last but not least, the very terminology of suburban research has come under scrutiny, as the notion of the suburban has often been identified with a particular North American phenomenon. As Richard Harris and Charlotte Vorms (2018) have shown, there are near-endless different terms for almost equally diverse suburbanizations worldwide.

Space and spatiality in sub/urbanization processes

Suburban studies have played a crucial role in the definition of a new understanding of the urban. Suburbanization processes are contributing to the general thrust towards urbanization in diverse physical forms and social modes of organization.

However, from a sociology of science standpoint, it may seem pretentious or short-sighted to claim that the field of sub/urban studies is undergoing a paradigmatic shift, as the diverse strands involved in the field of urban research have been quite unstable and its concepts contested ever since its emergence at the end of the nineteenth century (Brenner and Schmid, 2015, p. 163). So, channeling Thomas Kuhn (1962), how can we talk of a paradigmatic shift today?

Critical sub/urban studies in both the political economy and poststructuralist traditions have brought to the fore a new grammar of spatiality with several processes: the multiplicity – and diversity – of urban forms including "new spatial arrangements" (Amin, 2007, p. 102); the exploding/imploding tensions inherent through sub/urban development (Keil, 2017); the emerging new geographies of the urban as "emplaced by heterogeneity" (Roy, 2011b, p. 309); the extension of urbanizing landscapes in connection with "intangible (virtual) and tangible aspects"

(Shields, 2015, p. 59); and the reproduction of class relations combined with recognition issues (Donald, 1999). These components have certainly contributed to highlighting the ways sub/urbanization has entered an era of the "postmetropolis" (Soja, 2000) or "post-suburbia" (Wu and Phelps, 2008). Through emphasizing the multiplicity and competing analytical discourses in the production of spatial configurations, sub/urban studies have shown us that suburbs – from their origins to their diverse typologies in connection with politics – have followed unlikely hetero-geneous trajectories where class, race, and gender collide.

Sub/urban studies are thus open to one of the major social scientific challenges of our times, the search for regularities beyond the overwhelming presence of subjectivities and predomi-nance of social constructivism. In that respect, we think that the suggestion of Charles Tilly to "build systematic knowledge of social construction into superior analysis of social processes" (2008, p. 5) must be taken seriously. This position does not seek to reconcile the competing discourses or analytical frames but instead to confront the competing discourses and ideas. Thus, the "effects" of scientific activity can gain an additional understanding of the variety of processes, practices, perspectives involved in "the constitution of the social environment" (Tilly, 2008).

Understood as a research field, global suburbanism, like the field of urban studies more gener-ally, can be divided in a variety of ways into sub-specialized fields (history, sociology, economics, and political science). Even if these sub-fields are open to an interdisciplinary approach, they remain influenced by the methodology and normative orientations of their discipline (Bowen et al., 2010, p. 199).

Without tracing back all the stages that punctuated its history, one can recall that urban stud-ies were initially defined as a specialized field of sociological inquiry (Perry and Harding, 2002). More importantly, sociology as a general social science converged above all with the study of the city or with what would become urban sociology. For the first European sociologists (Weber, 1966 [1921]; Simmel, 1950 [1903]), and soon after for the American sociologists of the Chi-cago School (Park et al., 1925), cities and especially modern metropolises were research topics of exception. This was because those metropolises were the main places where modernity was experienced. In addition, for them, before being an object of study as such, metropolises were entry points to understanding social relations, social domination, and the new importance gained by subjectivity in the modern age. The deployment of actors' subjectivity was creating new meanings in order to cope with the sweeping changes underway with urbanization and indus-trialization. Today, we make the argument that suburbanization serves as a prominent process through which we can study the crisis and evolution of the modern society that was the original subject of urban studies.

The notion of space, including its substance – spatiality – has been a particularly important dimension through which these processes can be observed and analyzed. Following Doreen Massey (2005), space cannot be separated from the practices involved in its construction. Space can be defined before all as social space pointing towards the relations individuals and communi-ties build together: "instead of being this flat surface it's like a pincushion of a million stories" (Massey, 2013).

Consequently, space implies the recognition of multiplicity with the presence of the other – the social question – or how actors are going to live and make choices, introducing as a consequence the "geography of power." Power relations – defining and contesting time-space relations – are thus inevitable. In the face of the historical path taken by capitalism and the Western World, espe-cially in the neoliberal globalizing era, a new form of resistance emerged through local struggles (Massey, 2005, p. 183).

This vision of space by Massey is consistent with Lefebvre's perspective regarding the "production of space," where he insists on the Gramscian idea of hegemony for building collective action in connection with class interests (2005). From this standpoint, class hegemony – with its race, gender, colonial, and other relevant dimensions – is clearly achieved through space. Suburbanization is part of this historical production of concentrated, extended, and differential space. In post-war America, for example, the emergence of white middle-class suburbia was interpreted as a specific spatial "solution" to the over-accumulation of Fordism's problems and the simultaneous production of a landscape of white privilege excluding racialized populations from a historical process of wealth formation underlying American power differentials to this day. In more recent years, to give another example, the production of suburban space has been seen as coinciding with processes of differentiation that make some suburban areas the domain of – often gated – privilege, while others are portrayed as spaces of exclusion and control. The increasing diversity of suburban productions of space itself necessitates comparative research.

Suburbs and beyond from a critical theory approach

At its inception critical urban theory entails acknowledging the new "geographies of theory" referred to previously. Global generalizations have to face the test of heterogeneity of flux, processes, and practices manifested in particular cities. The implications for urban theory entail awareness of contradictory impulses involved in global suburbanism, where the peripheral suburban landscapes are engaged in diversified relationships of power. Suburban globality is anything but a process in the making, with multiple forms of interconnections between local actors, at times opposing promoters, on other occasions finding accommodating paths.

Beyond the fact researchers disagree about the meaning of sub/urban changes underway, a consensus is emerging concerning the main components of those changes and the fact that through diverse suburban processes, metropolitan areas everywhere are experiencing structural transformations (Hamel and Keil, 2015). Suburbanization is currently a constitutive part of regionalization and as such also subject to debates on regional governance (Keil et al., 2017; Hanlon et al., 2010). The suburbs, once thought of as the problematic Other of metropolitan unity, with self-selected, dispersed, and contrarian constituencies (Hoch, 1985), are the connective tissue of burgeoning regions that have become the operative terrain of globalized industrial, commercial, and infrastructural activities. In that, the suburbs often remain the literal dumping grounds of the social and environmental "bads" of the region, but they also come into their own with identities forged from their new demographic and economic composition (Belina and Keil, 2017).

The indeterminacy of power – consistent with the implementation of democracy, including local democracy – is deeply ingrained in the conduct of sub/urban governance. At a metropolitan or regional scale, the capacity of subordinated or excluded actors to influence resource distribution and promote increasing equality cannot be taken for granted anywhere. If greater equity defined as a renewed challenge in a post-suburban era does not necessarily undermine the fundamental democratic project as sketched by Lefebvre (2003/1970) through the "urban revolution," nonetheless the promises of regionalism have not met expectations and "the (post-)suburban re/insurgence in regional politics is not settled" (Addie and Keil, 2015, p. 17). This open-ended assessment serves as a reminder that the emergence of governance on the political agenda at a regional scale did not solve social conflicts. Coinciding with renewed questions about the nature of the state and the recognition of its historical and cultural character (Bevir and Rhodes, 2010; Pinson, 2015), does governance distance itself from the principle of the sovereign state defined as a monolithic reality and promote openness towards non-state actors in decision-making processes?

The challenge of state sovereignty in pluralized contexts, taking into account modalities and mechanisms of suburban governance as implemented in the world outside the Global North, does not correspond with the dominant representation or the dominant implementation of state governing. Thus, the Western model of sovereignty was challenged and/or reinforced differently. In Latin America, for example, we should recall that the self-led peri-urban expansion during the first half of the twentieth century was largely implemented outside state control and policies. Afterwards, suburban development expanded under capital and/or the initiatives of land and real estate promoters – including international corporations – with state support. At the same time, "the national governments . . . tend to have a much bigger say in the governance of at least the major urban centers than those in North America or Europe" (Heinrichs and Nuissl, 2015, p. 231).

In Asia, and especially in India, economic liberalization has influenced governance, and it is towards land politics that the failure to meet social expectations is most visible. As Shubhra Gururani and Burak Kose underline, land politics and policies are "a messy business" (2015, p. 293). Class and caste divisions remain strong, but private land developers – able to invest in local infrastructures – have the support of central and state governments (Gururani and Kose, 2015). Confronted with this alliance, slum residents are in a difficult situation.

In addition, the specificity of localities remains strong. Resistance and insurgent practices are supported and restrained simultaneously through local culture, experiencing diverse modalities of inhabiting space, reminding urban researchers that cultural appropriation cannot adequately account for slum residents' urban creativity (Roy, 2011b). If suburbanization and suburbanism(s) can take place in "extended metropolitan regions" – where centrality is redefined along "decentred spatial flows" (Firman, 2011, p. 195) – they can also prevail beyond "metrocentric tendencies" (Bunnel and Maringanti, 2010).

What is at stake is a deeper understanding of state restructuring under sub/urban governance defined by a global suburbanism perspective. Generally, urban researchers agree that with governance, despite diversity of interests, actors share a cognitive framework. It is what explains the possibility of regulating conflicts. Bringing diversity and domination issues to the fore in considering governance goes hand in hand with a revised understanding of its repercussions on state restructuring. That issue, above all, is at stake here (Pinson, 2015). And thus far, this has been overlooked in the field of sub/urban studies. Returning to Lefebvre's apprehension of the "planetarization of the urban" and its contradictions, it is necessary to include "post-colonial reading practices" (Robinson, 2016, p. 22). Such a perspective can help overcome and distance ourselves, with the agreed-upon vision of governance as a tool for regulating conflicts, and for dodging the cultural aspects and axiological content of those conflicts.

Guide to further reading

Hamel, P. and Keil, R. (2015) *Suburban Governance: A Global View.* Toronto: University of Toronto Press.
Keil, R. (2017) *Suburban Planet: Making the World Urban from the Outside In.* Cambridge, MA: Polity Press.

References

Addie, J.-P.D. (2016) "Theorizing Suburban Infrastructure: A Framework for Critical and Comparative Analysis." *Transactions of the Institute of British Geographers*, 41(3), pp. 273–285.
Addie, J.-P.D. and Keil, R. (2015) "Real Existing Regionalism: The Region between Talk, Territory and Technology." *International Journal of Urban And Regional Research*, 39, pp. 407–417.

Amin, A. (2007) "Re-Thinking the Urban Social." *City*, 11(1), pp. 100–114.

Amin, A. (2013) "Telescopic Urbanism and the Poor." *City: Analysis of Urban Trends, Culture, Theory, Policy, Action*, 17(4), pp. 476–492.

Belina, B. and Keil, R. (2017) "Decentralizing the Global City Region: Suburban Identities in Frankfurt and Toronto?" *MONU*, 26, pp. 34–39.

Bevir, M. and Rhodes, R.A.W. (2010) *The State as Cultural Practice*. Oxford: Oxford University Press.

Bloch, R. (2015) "Africa's New Suburbs." In: Hamel, P. and Keil, R. (eds.) *Suburban Governance: A Global View*. Toronto: University of Toronto Press, pp. 253–277.

Bowen, W.M., Dunn, R.A., and Kasdan, D.O. (2010) "What Is 'Urban Studies'? Context, Internal Structure and Content." *Journal of Urban Affairs*, 32(2), pp. 199–227.

Brenner, N. and Schmid, C. (2015) "Towards a New Epistemology of the Urban?" *City: Analysis of Urban Trends, Culture, Theory, Policy, Action*, 19(2–3), pp. 151–182.

Bunnel, T. and Maringanti, A. (2010) "Practicing Urban and Regional Research beyond Metrocentricity." *International Journal of Urban and Regional Research*, 34(2), pp. 415–420.

Caldeira, T. (2016) "Peripheral Urbanization: Autoconstruction, Transversal Logics and Politics in the Cities of the Global South." *EPD Society and Space*, 35(1), pp. 3–20.

Charmes, E. and Keil, R. (2015) "The Politics of Post-Suburban Densification in Canada and France." *International Journal of Urban and Regional Research*, 39, pp. 581–602.

Donald, J. (1999) *Imagining the Modern City*. Minneapolis, MN: University of Minnesota Press.

Dunham-Jones, E. and Williamson, J. (2011) *Retrofitting Suburbia: Urban Design Solutions for Redesigning Suburbs*. Hoboken: John Wiley and Sons, Inc.

Easterling, K. (2014) *Extrastatecraft: The Power of Infrastructure Space*. London and New York: Verso.

Ekers, M., Hamel, P., and Keil, R. (2012) "Governing Suburbia: Modalities and Mechanisms of Suburban Governance." *Regional Studies*, 46(3), pp. 405–422.

Elden, S. (2014) "Globe-World-Planet." Commentary on "Dissolving cities: Planetary Metamorphosis." *Environment and Planning D: Society and Space*, 32, pp. 199–380. Available at http://societyandspace. org/2014/04/21/globe-world-planet-stuart-elden/

Filion, P. (2013) "The Infrastructure Is the Message: Shaping the Suburban Morphology and Lifestyle." In: Keil, R. (ed.) *Suburban Constellations: Governance, Land and Infrastructure in the 21st Century*. Berlin: Jovis, pp. 39–45.

Filion, P. and Pulver, N. (eds.) (2018) *Global Suburban Infrastructure: Social Restructuring, Governance and Equity*. Toronto: University of Toronto Press.

Firman, T. (2011) "Post-Suburban Elements in an Asian Extended Metropolitan Region: The Case of Jabodetabek (Jakarta Metropolitan Area)." In: Phelps, N.A. and Wu, F. (eds.) *International Perspectives on Suburbanization: A Post-Suburban World?* London: Palgrave Macmillan, pp. 195–209.

Guney, M., Keil, R., and Ucoglu, M. (eds.) (Forthcoming) *Massive Suburbanization*. Toronto: University of Toronto Press.

Gururani, S. and Kose, B. (2015) "Shifting Terrain: Questions of Governance in India's Cities and Their Peripheries." In: Hamel, P. and Keil, R. (eds.) *Suburban Governance: A Global View*. Toronto: University of Toronto Press, pp. 278–302.

Hamel, P. and Keil, R. (2015) *Suburban Governance: A Global View*. Toronto: University of Toronto Press.

Hanlon, B., Short, J.R., and Vicino, T.J. (2010) *Cities and Suburbs: New Metropolitan Realities in the US*. New York: Routledge.

Harris, R. (1996) *Unplanned Suburbs: Toronto's American Tragedy, 1900–1950*. Baltimore, MA and London: The Johns Hopkins University Press.

Harris, R. (2004) *Creeping Conformity: How Canada Became Suburban, 1900–1960*. Toronto: University of Toronto Press.

Harris, R. and Lehrer, U. (eds.) (2018) *The Suburban Land Question: A Global Survey*. Toronto: University of Toronto Press.

Harris, R. and Vorms, C. (eds.) (2017) *What's in a Name? Talking about "Suburbs."* Toronto: University of Toronto Press.

Heinrichs, D. and Nuissl, H. (2015) "Suburbanization in Latin America: Towards New Authoritarian Modes of Governance at the Margin." In: Hamel, P. and Keil, R. (eds.) *Suburban Governance: A Global View*. Toronto: University of Toronto Press, pp. 216–238.

Hoch, C. (1985) "Municipal Contracting in California: Privatizing with Class." *Urban Affairs Review*, 20(3), pp. 303–323.

Jessen, J. and Roost, F. (eds.) (2015) *Refitting Suburbia: Erneuerung der Stadt des 20 Jahrhunderts in Deutschland und den USA*. Berlin: Jovis Verlag.

Keil, R. (2017) *Suburban Planet: Making the World Urban from the Outside In.* Cambridge, MA: Polity Press.

Keil, R. (2018) "Extended Urbanization, 'Disjunct Fragments' and Global Suburbanisms." *Environment and Planning D: Society and Space*, 36, pp. 494–511. DOI: 10.1177/0263775817749594.

Keil, R. and Addie, J.-P.D. (2016) "'It's Not Going to Be Suburban, It's Going to Be All Urban': Assembling Postsuburbia in the Toronto and Chicago Regions." *International Journal of Urban and Regional Research*, 39, pp. 892–911.

Keil, R., Hamel, P., Boudreau, J.-A., and Kipfer, S. (eds.) (2017) *Governing Cities through Regions: Canadian and European Perspectives.* Waterloo: Wilfrid Laurier University Press.

Keil, R., Hamel, P., Chou, E., and Williams, K. (2015) "Modalities of Suburban Governance in Canada." In: Hamel, P. and Keil, R. (eds.) *Suburban Governance: A Global View.* Toronto: University of Toronto Press.

Knox, P. (2017) *Metroburbia: The Anatomy of Greater London.* London: Merrell.

Kuhn, T.S. (1962) *The Structure of Scientific Revolutions.* Chicago, IL: University of Chicago Press.

Lefebvre, H. (2003 [1970]) *The Urban Revolution.* Minneapolis, MN: University of Minnesota Press.

Lefebvre, H. (2014) "Dissolving City, Planetary Metamorphosis." *Environment and Planning D: Society and Space*, 32, pp. 203–205.

Le Goix, R. (2017) *Sur le front de la métropole: Une géographie suburbaine de Los Angeles.* Paris: Publications de la Sorbonne.

Mabin, A. (2013) "Suburbanisms in Africa?" In: Keil, R. (ed.) *Suburban Constellations: Governance, Land and Infrastructure in the 21st Century.* Berlin: Jovis Verlag, pp. 154–160.

Massey, D. (2005) *For Space.* London: Sage.

Massey, D. (2013) "Doreen Massey on Space." *Social Science Bites.* Available at: www.socialsciencespace.com/2013/02/podcastdoreen-massey-on-space/.

McGee, T. (2013) "Suburbanization in a 21st Century." In: Keil, R. (ed.) *Suburban Constellations: Governance, Land and Infrastructure in the 21st Century.* Berlin: Jovis Verlag, pp. 18–25.

McGee, T. (2015) "Deconstructing the Decentralized Urban Spaces of the Mega-Urban Regions in the Global South." In: Hamel, P. and Keil, R. (eds.) *Suburban Governance: A Global View.* Toronto: University of Toronto Press, pp. 325–336.

Monte-Mor, R.L. (2014a) "Extended Urbanization and Settlement Patterns in Brazil: An Environmental Approach." In: Brenner, N. (ed.) *Implosions/Explosions.* Berlin: Jovis, pp. 109–120.

Monte-Mor, R.L. (2014b) "What Is the Urban in the Contemporary World?" In: Brenner, N. (ed.) *Implosions/Explosions.* Berlin: Jovis, pp. 260–267.

Moos, M. and Walter-Joseph, R. (eds.) (2017) *Still Detached and Subdivided.* Berlin: Jovis Verlag.

Ong, A. (2011) "Introduction: Worlding Cities, or the Art of Being Global." In: Roy, A. and Ong, A. (eds.) *Worlding Cities: Asian Experiments and the Art of Being Global.* Oxford: Blackwell Publishing, pp. 1–26.

Ortega, A.A.C. (2016) *Neoliberalizing Spaces in the Philippines: Suburbanization, Transnational Migration, and Dispossession.* Lanham, MD: Lexington Books and Rowman & Littlefield Press.

Park, R.E., Burgess, W., and McKenzie, R. (1925) *The City.* Chicago, IL: University of Chicago Press.

Perry, B. and Harding, A. (2002) "The Future of Urban Sociology: Report of Joint Sessions of the British and American Sociological Associations." *International Journal of Urban and Regional Research*, 26, pp. 844–853.

Phelps, N.A. (2015) *Sequel to Suburbia: Glimpses of America's Post-Suburban Future.* Cambridge, MA: MIT Press.

Phelps, N.A. and Wu, F. (eds.) (2011) *International Perspectives on Suburbanization: A Post-Suburban World?* London: Palgrave Macmillan.

Pinson, G. (2015) "Gouvernance et sociologie de l'action organisée. Action publique, coordination et théorie de l'État." *L'année sociologique*, 65(2), pp. 483–519.

Quinby, R. (2011) *Time and the Suburbs: The Politics of Built Environments and the Future of Dissent.* Winnipeg: Arbeiter Ring Publishing.

Robinson, J. (2006) *Ordinary Cities: Between Modernity and Development.* London: Routledge.

Robinson, J. (2016) "Thinking Cities through Elsewhere: Comparing Tactics for a More Global Suburban Studies." *Progress in Human Geography*, 40(1), pp. 3–29.

Roy, A. (2009) "The 21st Century Metropolis: New Geographies of Theory." *Regional Studies*, 43(6), pp. 819–830.

Roy, A. (2011a) "Urbanisms, Worlding Practices and the Theory of Planning." *Planning Theory*, 10(1), pp. 6–15.

Roy, A. (2011b) "Conclusion: Postcolonial Urbanism: Speed, Hysteria, Mass Dreams." In: Roy, A. and Ong, A. (eds.) *Worlding Cities: Asian Experiments and the Art of Being Global.* Oxford: Blackwell Publishing, pp. 307–335.

Roy, A. (2015) "Governing the Postcolonial Suburbs." In: Hamel, P. and Keil, R. (eds.) *Suburban Governance: A Global View.* Toronto: University of Toronto Press, pp. 337–348.

Sheppard, E., Leitner, H., and Maringanti, A. (2013) "Provincializing Global Urbanism: A Manifesto." *Urban Geography*, 34, pp. 893–900.

Shields, R. (2015) "Re-Spatializing the City as the City-Region?" In: Jones, K.E., Lord, A. and Shields, R. (eds.) *City-Regions in Prospect? Exploring Points between Place and Practice.* Montreal and Kingston: McGill-Queen's University Press, pp. 53–72.

Sieverts, T. (2003) *Cities without Cities: An Interpretation of the Zwischenstadt.* London: Routledge.

Simmel, G. (1950 [1903]) "The Metropolis and Mental Life." In: Wolff, K.H. (ed.) *The Sociology of Georg Simmel.* Glencoe: Free Press, pp. 409–424.

Soja, E.W. (2000) *Postmetropolis: Critical Studies of Cities and Regions.* Oxford: Blackwell Publishing.

Tilly, C. (2008) *Explaining Social Processes.* Boulder and London: Paradigm Publishers.

Tonkiss, F. (2013) *Cities by Design: The Social Life of Urban Form.* Cambridge, MA: Polity Press.

Walks, A. (2013) "Suburbanism as a Way of Life, Slight Return." *Urban Studies*, 50(8), pp. 1471–1488.

Weber, M. (1966 [1921]) *The City.* New York: Free Press.

Wu, F. and Phelps, N.A. (2008) "From Suburbia to Post-Suburbia in China? Aspects of the Transformation of the Beijing and Shanghai Global City Regions." *Built Environment*, 34, pp. 464–481.

Wu, F. and Shen, J. (2015) "Suburban Development and Governance in China." In: Hamel, P. and Keil, R. (eds.) *Suburban Governance: A Global View.* Toronto: University of Toronto Press, pp. 303–324.

5

Suburbanization in Latin America

Lawrence A. Herzog

Introduction

The suburbanization of Latin America, over the last three decades, can be viewed as a subset of a larger phenomenon – what one might call the *globalization* of urban sprawl. Since the end of the last century, there has been a global diffusion of the American suburban model, or the idea of an American-type suburb (rather than its literal physical form) across the border to the Americas' other nations. As this wave of suburban-style development spreads across the distant outskirts of city-regions it reproduces ecological problems that have come to define U.S. suburbs. As such, these "global suburbs" (American-style suburbs adapted globally into other urban cultures and contexts) are as unsustainable as their northern counterparts (Herzog, 2015). In fact, given the economic and social development challenges facing Mexico, Central, and South America, the long-term implications of these new suburbs are worrisome.

For the purpose of this chapter, I will argue that "suburb," as a category of urban development in the Americas, should be defined in the context of the original post-war American urban design prototype – single-family, detached homes laid out on low-density subdivisions, typically with curvilinear street patterns, surrounded by separate shopping centers, schools, offices, and other amenities generally only reachable by car. This design model evolved in the United States in the 1950s and 1960s. The suburb was typically located in a peripheral district, away from the downtown and deliberately divided into separate land use zones – residential, shopping, office – all designed principally around automobile travel (Duany et al., 2000).

Further, I consider "suburb" as an urban planning "narrative" that embodies a set of cultural values, notably – privacy, exclusivity, and security. American suburbs, above all else, place a large emphasis on privacy and the notion of safety – and protection from the perceived "ills" of the inner-city (Kunstler, 1993). Though the architectures of the emerging elite and middle-class "global suburbs" south of the border may take on different densities and visual forms, the fact is, this idea of building fortress enclaves somehow buffered from the existing city has taken root from Mexico to South America. America's "suburbia" model has spread south.

The building of suburbs can be seen as one plank in the larger project of American modernity. Marshall Berman defined modernism as "any attempt by modern men and women to become subjects as well as objects of modernization to get a grip on the modern world and make themselves at home in it" (Berman, 1988, p. 5). Thus, one must remember that as American cities grew after World War II, and as freeways and suburban communities were being laid out, the world was watching. In Brazil, architects Lucio Costa and Oscar Neimeyer were so enamored of the modernist project they designed an entire city, the new national capital of Brasília, using the principles of modern urban design. That city, constructed in only three years, quickly revealed the dangers of designing with a limited palette of modernist ideals. In Berman's words, Brasília quickly became "immense empty spaces in which the individual feels lost, as alone as a man on the moon" (ibid., p. 7).

The critical point here is that America's suburb was part of what one scholar has termed a "global project" (Beauregard, 2006). Following World War II, the United States became the center of a global economy based on free trade. However, that global dominance was not only economic, it was also cultural. The U.S. government emphasized a program of spread-out cities, in part because the federal government saw the marketing value of projecting the image of a prosperous, modern America, where families live in safe suburbs. This became the idealized American habitat model, to which nations all over the world might aspire. In this sense, the suburb became a place that exemplified American freedom, security, and the promise of prosperity. It has also been argued that homeownership, being created in the new suburbs, would be another way to protect the values of capitalism against the tide of socialism. Suburban developer Levitt, who built the iconic Levittowns on the east coast was quoted as saying, "No man who owns his own house and lot can be a communist" (ibid., p. 156).

Across the planet, after World War II, America's image as a democracy, and as world economic leader also became a cultural model for consumerism and community-building. American images and ideas central to the lifestyle of its new suburbs – supermarkets, automobiles, superhighways, and motels – were exported to other nations. The American post-war dream was solidified in the form of the single-family suburban home, the cul-de-sac community, the local shopping center, and the family car, as opposed to the crowded, high crime, industrial cities.

The rise of global suburbs in Latin America

The first major development on the periphery of most Latin American cities was not U.S.-style "suburbs," but rather, the massive construction of spontaneous "squatter settlements" or "shantytowns," which began in the 1950s. By then, millions of rural migrants in the Americas were migrating toward cities in search of work. As we now know from more than five decades of research (Perlman, 1980; Neuwirth, 2004; Davis, 2007), these migrants arrived with virtually no liquid capital and no permanent source of employment. They soon became part of a booming "informal" street and housing economy, which only provided temporary sources of income, and no guaranteed right to own land. Faced with no real prospect for renting or purchasing homes in the "formal" housing market, millions of city-ward migrants chose to squat on, or invade, land parcels, usually on the outskirts of cities, and often in floodplains, canyons, on steep sloping hillsides or other less desirable land isolated from the "formal city." Over time, these temporary encampments of shacks and cardboard or wooden homes morphed into permanent slums, which now crowd the periphery of Latin America's urban areas, with few signs of going away.

Thus, the building of "suburbs" on the outskirts of Latin American cities is now occurring on terrain that was previously the domain of *asentamiento irregulares* (irregular settlements) – spontaneous colonization and informal settlement construction by the most disadvantaged citizens of Latin American nations.

But, over the last two or three decades, globalization has begun to reshape Latin American cities and their peripheries. The construction of highways, shopping malls, and new suburbs is restructuring the outskirts, forming what has been termed an "archipelago" of gated spaces and fortified centers, all served by the automobile and all disconnected from each other. These closed off "islands" of activity include gated elite communities, malls, private schools, office complexes, and social or recreation clubs. The new developments are largely dependent on automobiles, and, in some cases, on private toll roads that allow the wealthy to get to work more quickly (Borsdorf et al., 2007). This pattern has been observed across the Americas, especially in Argentina, Chile, Mexico, and Brazil (Salcedo and Torres, 2004; Coy and Pohler, 2002; Jones and Moreno Carranco, 2007).

Today, two critical elements define the global suburbs evolving in Latin America. First, gated communities and fenced suburbs are booming, either wedged in and around the existing squatter neighborhoods, or on lands cleared of spontaneous settlements. These new gated suburbs have varying nomenclatures; they are called "*barrios privados*" (private communities) in Argentina, *condominios* in Chile, *conjuntos* or *urbanizaciones cerradas* (closed urbanizations) in Ecuador, *condominios fechados* (closed condominiums) in Brazil, and *fraccionamientos cerrados* (closed subdivisions) in Mexico. In Buenos Aires, middle- and upper-class suburban enclaves have been built; they are now surrounded by illegal settlements and garbage dumps. This emerging pattern has been termed "small-scale segregation" – wealthy and upper-middle class pockets amidst the poverty of the outskirts. The enclaves are fortified with fences and walls, but localized on plots that become available depending on topography, land markets, and other factors (Sabatini et al., 2001).

The second recent trend in the Latin American periphery is the building of "mega-projects," developments that are not merely gated communities, but, rather, "gated cities," giant complexes of 30,000 and more inhabitants – in Chile, Brazil, Argentina, Mexico, and elsewhere. Most of these complexes include residential, commercial, office, and light industrial development. One of the first mega-suburbs was Alphaville, on the edge of São Paulo, Brazil, which began construction in 1974. Scholars claim that Alphaville and other projects in São Paulo were influenced by North American ideas about suburbs and marketing (Coy, 2006). From Mexico to Brazil, we find evidence either of direct foreign investment in mega-projects, or the influence of foreign design and marketing strategies in building those projects.

The first U.S.-style suburbs

The first experiments with suburban developments occurred, not surprisingly, in Mexico's national capital, Mexico City. Many important cultural projects often begin in Mexico City, the nation's epicenter for cultural expression, economic power, and political decision-making. The first two experiments with the American-style suburb unfolded just after World War II – one in the northwestern part of the capital, in a development called Ciudad Satelite, the other in the southern part of the city, in the project known as Jardines de Pedregal. Because both were built in the late 1940s and 1950s, one could argue that if they did not outright physically copy the U.S. suburban model from that era (which was just then taking form to the north), they did embody the "suburban American imaginary," the narrative of the American suburb, of a place where one could escape the density of the city to a neighborhood closer to nature, and where one could own one's house and travel around with the modern convenience of the automobile. Both of these early Mexican suburbs borrowed heavily from references to the United States once they began to market themselves in newspapers and magazines (Capron and de Alba, 2010).

Jardines de Pedregal (Gardens of Pedregal) was planned and developed in the mid-late 1940s, on the Pedregal lava fields south of downtown where a volcano had erupted in 5000 B.C. The site was adjacent to an important pre-Colombian settlement called Cuicuilco, which was the largest city in the Valley of Mexico in 300 B.C. Luis Barragan, one of Mexico's most important modernist architects, proposed to develop the new suburb. His idea was to promote harmony between architecture and the landscape.

North of downtown Mexico City, *Ciudad Satelite* (literally translated as "Satellite City") echoes the experience to the south. *Satelite* was made possible by the development of major highways, especially the ring road or *periférico*, which linked it to the wealthy districts in the west (Lomas Chapultepec, Santa Fe) and south, and the *viaducto*, which linked it to downtown and the airport. Like the Jardines de Pedregal, Ciudad Satelite was planned as a utopian suburb by architects Mario Pani and Luis Barragan. Pani and Barragan utilized the English garden city model adapted in early U.S. utopian suburban designs in places like Radburn, New Jersey and Columbia, Maryland. It included residential superblocks with cul-de-sac designs that separated pedestrian flows from automobiles and preserved huge swaths of green space. The architects employed a street layout system of wide oval circuits or *"circuitos,"* each named for different professional careers – Scientists, Engineers, Architects, Sculptors, Economists, and Novelists. The plan was approved in 1948 and built over the next decade. Its status as a symbolic and functional center of the new modern Mexico was recognized by the 1958 completion of one of Mexico's iconic public art projects, the *Torres de Satelite* (Satelitte Towers), designed by Luis Barragan and Mathias Goerritz. These colorful high-rise, stone towers were metaphors for modern Mexico, a nation of high-rises and technology, though somewhat ironically they were located at the entrance to a mainly low-rise garden suburb.

Mega-projects and the new global suburbs

A signature example of a mega-project in Latin America is Santa Fe, a high- profile corporate and residential development planned and built in the 1980s and 1990s on the outskirts of Mexico City. Ten miles to the west of the historic center, Santa Fe was originally a toxic garbage dump where about 2,000 poor people lived in illegal cardboard shacks and earned a small wage recycling garbage. The squatters were eventually removed (and, later, from other parts of Mexico City) to make way for "mega-project" buildings, including giant corporate, residential, and commercial complexes such as those in Santa Fe (Valenzuela, 2007).

The idea was to turn Santa Fe into the new center of an emerging, increasingly global Mexico City economy, oriented more toward service than industry. Up until the 1990s, Mexico City's wealth was generated mainly by industry, especially import substitution, a strategy dating to the first half of the twentieth century that had become obsolete. Santa Fe was designed to house the free trade (NAFTA) companies that would be doing business in a globalizing Mexico by the twenty-first century, as well as all the high-tech, financial, computer, and other emerging post-industrial sectors. Planned and built in the 1980s and 1990s, Santa Fe became an enclave of multinational corporate headquarters, the new center for global finance in Mexico. Among the transnational (mainly U.S.) companies with headquarters here include Hewlett Packard, General Electric, IBM, Goodyear, Pepsi, Federal Express, and Kraft. There are five hotels with four and five-star ratings, 40 restaurants, and seven schools, including two universities. One of the city's largest and most lavish shopping malls, Centro Comercial Santa Fe, was built in the 1990s; it is one of larger malls in Latin America, with 300 stores and 14 movie theaters.

Santa Fe is symptomatic of the new globalized, elite suburbs appearing on the outskirts of cities throughout Latin America after the 1980s and 1990s, a kind of barometer of the fast pace of globalization, and the way in which it is producing a new type of urban region. It mirrors the qualities of "fortified enclaves" (Caldeira, 1996) in cities across the Americas – the emphasis on spatial segregation of the wealthy and upper-middle class, the heightened obsession with security, increasing privatization and the development of spaces of consumption (shopping malls) as central to the identity of these suburban places. Another example, outside Puebla, Mexico's fourth-largest city, is the mega-suburb of Angelopolis. It has middle- and upper-income gated residential developments, five-star hotels, two private universities, three giant shopping malls, a business park, and a government complex. Observers report that enclaves like Angelopolis evoke a sense of elite separation, glass-enclosed malls and cafes where the upper classes sip frappuccinos and stare into their smartphones, further contributing to social fragmentation and polarization in Latin American cities (Jones and Moreno-Carranco, 2007).

The social ecology of elite suburbs and giant shopping malls raises larger issues about globalization, politics, and consumerism. This is clearly illustrated by the example of the "Wal-Martization of Mexico." Wal-Mart owns nearly a thousand businesses for its franchise in Mexico, not only Wal-Mart superstores, but also several other of national big-box style supermarket franchises (Bodegas Aurrerá, Superamas, Suburbias) as well as the popular Vips restaurant chain. Overall national sales among Wal-Mart-owned establishments in Mexico is estimated to be around $20 billion per year, with some 150,000 people employed. This makes Wal-Mart Mexico's largest private employer (Walker et al., 2006). Wal-Mart super-centers are one of the driving forces that perpetuate urban sprawl, by creating one single giant destination that requires larger numbers of automobile trips, while often causing neighborhood-scale commercial enterprises to go out of business.

Mass working-class suburbs

After 1950, most Latin American cities experienced massive growth in the form of unplanned, spontaneous land invasions and the building of irregular settlements, usually on the outskirts of urban areas. Over time, faced with their own inability to provide formal public housing for this huge wave of new urban dwellers, governments often assisted in the "regularization" of these districts. This included bringing in electricity, piped water, roads, and schools. However, because residents did not necessarily receive title to their land, and because services were not always provided, "urban social movements" evolved to mobilize squatters to defend their rights to their homes and neighborhoods, and their need for services (Foweraker and Craig, 1990).

Some national governments experimented with a policy approach that sought to build subsidized housing for the poor. In Mexico, for example, the National Housing Fund (FOVI) built over 800,000 units from 1970–1992; and the National Institute for the Development of Workers' Housing (INFONAVIT) built around one million units during the same time period. Not enough housing was completed, however, to address the massive demand (in the millions) for urban housing during this period of urban expansion in Mexico. Also, these housing units were mainly available to either salaried government employees or middle-income city dwellers.

Therefore, beginning in the 1990s, the Mexican government shifted its housing policy strategy, by converting INFONAVIT from a producer of housing to a financial institution. The new policy approach was market-driven, allowing those eligible to purchase homes, mainly in the suburbs, with financing through INFONAVIT. It made possible a new era of large-scale development in the periphery and opened the door to rampant sprawl.

A massive construction boom of master-planned, suburban housing developments followed. Many of the projects built low- and medium-density complexes aimed toward the middle and working-class sectors of the nation, and even toward the lower classes in some cases (Monkkonen, 2009). For the first-time, large masses of working and low-income Mexicans were being offered an alternative to self-help housing, with its uncertain ownership, and poor access to electricity, piped water, and paved roads. However, these new suburbs often contained poorly designed homes in subdivisions that were located quite far away from shopping, schools, or other urban services.

Most of these suburban housing developments – sprawling on the edges of large cities like Mexico City, Guadalajara, Monterrey, Ciudad Juarez, Mexicali, or Tijuana – are being built by a handful of mega-home builders who have taken advantage of the change in government policy. They market these giant subdivisions by appealing to working-class desire for security and a better lifestyle. It has also been pointed out that young people in Mexico seek this kind of housing because they desire freedom and independence from their parents, a new trend away from the traditional model where children often remained with their parents even after they began to have their own families (Hiernaux and Lindon, 2002).

Some of the new developments are gated communities, while others are semi-closed. However, the vast majority of these developments lack well planned roads, any connection to public transit or other public services, like schools, hospitals, clinics, or shopping facilities. Observers have noted the poor quality of the houses themselves, the lack of a sense of community in the new districts, and the poor access to public transit. It's been argued that, rather than creating a solution to the housing and community needs of low-income families, the larger impact of these projects has been to enrich a small number of mega-construction companies (Garcia Peralta and Hofer, 2006).

Vertical suburbs: some examples from Brazil

The "vertical suburb" – a high-rise residential neighborhood – is the quintessential model for peripheral upper and middle-upper income communities in Latin America. The idea of a vertical cityscape is itself premised on a twentieth-century modernist idea in the world of architecture and city planning – the single tower isolated on a lot, and with open spaces around it. It was a product of the CIAM (International Congress of Modern Architecture), which emerged in Europe after its creation in 1928 by the Swiss architect Le Corbusier. This legacy of modern urban design had a huge impact on Latin America's modern city-building, one that remains today. For example, Corbusian high-rise urbansim was the approach adapted on a grand scale for the design and building of Brasília, the first planned national capital, with its stand-alone residential towers in the "superblocks" and high-rise ministries along the Monumental Axis.

Examples of verticle high-rise suburban growth abound in Latin American metroplitan areas, from Buenos Aires to Mexico City. Some of the best examples are found in the continent's most urbanized nation, Brazil. And within Brazil, the most illuminating case of vertical suburban growth is the São Paulo metropolis. São Paulo is one of the largest metropolitan regions on the planet, with nearly 20 million inhabitants. It is a primary center for global corporate headquarters, and thus one of the key financial megacities of the Americas. Its phenomenal expansion in the second half of the twentieth century inspired a scale of peripheral growth previously unseen in the Americas. In every sense, São Paulo represents a template for globalization and urban growth in Brazil, if not the rest of the world.

Since the 1980s urban space has become more and more "walled off," with the middle- and upper-income classes increasingly residing in their vertical high-rise enclaves, while the poor remain either in slum apartments (*corti os*) or *favelas*, the generic term in in Brazil for the poorest squatter settlement zones. For the wealthy, fear of crime led to the building of thousands of high-rise "closed condominiums," a Brazilian version of the "gated community" (Caldeira, 2000). São Paulo is a skyscraper mega-region, whose suburbs consist of complexes of luxury apartment towers and elite gated compounds. The compounds were inspired by the British garden city model, though unlike the original European garden cities, the São Paulo versions are entirely automobile-oriented. Furthermore, they are distinguished by architectures of "social policing" – walls, fences, and private security companies (Caldeira, 1996). The primary gathering spaces are shopping malls, of which there are more in São Paulo than anywhere else in Brazil (Collet Bruna and Comin Vargas, 2009).

One example of an elite high-rise suburb in São Paulo is Alphaville-Tamboré, a gated community built in the early 1970s about 15 miles southwest of downtown São Paulo. It has been compared with the scale and style of Irvine, the giant suburb in Orange County, California. The community grew to some 33 gated developments, mostly high-rise condominiums, with a few low-rise, single-family gated areas. It has more than 22,000 residential units, and a population estimated at about 90,000, making it one of Brazil's largest elite suburbs (Gotsch, 2009). This mega-suburb has 11 schools, universities, shopping malls, and office complexes. Some 150,000 people travel in and out of Alphaville each day, which eventually led the company and local government to build a toll freeway for affluent commuters. The Alphaville project was so celebrated, the developer – Alphaville Urbanismo, S.A. – successfully proceeded to market it to many cities across Brazil, including Salvador, Fortaleza, Belo Horizonte, Manaus, Natal, and Rio de Janeiro.

Another notable vertical suburb is Barra da Tijuca, a mega-suburb of over one-quarter million people, on the outskirts of Rio de Janeiro, along the southwestern coast. It was designed by Lucio Costa, the master planner who designed the new capital at Brasília. Costa's Pilot Plan for Barra displayed much of the same thinking he had employed in Brasília – but Brasília was much more a public sector-driven plan with a great deal of emphasis on civic and national pride. Its design gave priority to monumental and ceremonial spaces. Barra da Tijuca emphasized residential space, private commercial shopping malls, and security. It became a giant seaside enclave for the affluent, a Brazilianized form of the American sprawl that has produced many of the same public health and environmental problems found in U.S. suburbs (Herzog, 2015).

However, in terms of the city's physical design, the parallels with Brasília are noteworthy. They include the creation of a cross-like design, much the same as the airplane-shaped morphology of Brasília. As he had done in Brasília, Costa emphasized in Barra the modernist principle of single-use zones, specialized sectors for housing, commerce, offices, and public buildings. The plan also had a monumental formality and a sense of abstraction typical of Costa's style. However, despite Costa's creative design instincts and his dream of socioeconomic equality, Barra became an elite suburb, detached from the urbanized core of Rio de Janeiro. It lacked walkable public meeting places, a feeling of community, or street life. Its dependence on automobiles is creating traffic jams and air pollution. It is regarded as an unsustainable model for Brazil's future urban growth, yet many Brazilians continue to prefer living in closed condominium developments in U.S.-style suburbs like Barra da Tijuca.

Challenges for global suburbs in Latin America

The suburbanization of Latin America has created a number of important policy challenges. They include: segregation, ecological destruction, sprawl, and isolation.

Segregation

As governments began exploring suburban-type development on the outskirts of cities in Latin America, the value of land often increased dramatically, thus raising the cost of any housing being produced. As a result, suburban development in places like Buenos Aires, São Paulo, Rio de Janeiro, Santiago, Guadalajara, and Mexico City has tended to exacerbate social class segregation. In the mass working-class suburbs of Mexico, mentioned earlier, the new developments tend to segregate income groups within, while their lack of integration into the larger city is excluding these new suburban dwellers from urban life, creating a kind of "ruralization" of the city (Garcia Peralta and Hofer, 2006).

But in many new suburbs, land values have risen so quickly that large segments of the population are being excluded, thus enhancing social fragmentation. The example of Barra da Tijuca, the giant suburb outside Rio de Janeiro, is instructive. Once Barra's pilot land use plan was developed, real estate interests began purchasing land, driving up its value, and, as critics would note, turning it into a highly exclusive gated condominium beach suburb. From a population of 45,000 in 1980, Barra grew to nearly 250,000 by 2010, but the form of growth became even more privatized and exclusionary. Over time, observers noted that the development evolved toward more and more inward-looking gated complexes. Shopping malls began to add security and put up other real and psychological barriers to public entry. All of this was exacerbated by an urban plan that allowed large real estate interests to dominate the land market, further driving up prices at a time housing was becoming more expensive.

Ecological harm

Suburban development has caused the destruction of valuable ecosytems. Land is bulldozed, vegetation cleared, local ecological systems ignored. Returning to Rio's sprawling mega-suburb of Barra da Tijuca, the construction project has been profoundly damaging to the beach ecosystem. The Barra de Tijuca zone is called a "double barrier lagoon coast," due to its long narrow barrier island that is sited alongside a system of lagoons and waterways. This geographic formation of lowland lagoons, canals, and marshes is highly vulnerable to climate change. Storms and heavy rainfall cause strong tidal flows and shifts in waves that result in floods and the blocking of channels. This, then, causes "overwash," where waters back up into lagoons and flood-adjacent lands. Since global climate change is now understood to raise sea levels across the planet, lowland coastal zones like Barra da Tijuca are even more endangered by this phenomenon.

Urban sprawl in this kind of fragile ecological setting is problematic. Among other things, sprawl encourages fragmented development, which impedes managing coastal bioregions. This explosive growth in Barra has spread in sprawl-like fashion across the basin, invading the edges of canals, lagoons, and bays, encroaching on wildlife and interrupting the natural flows of water. Since Barra urbanized, the extent of flooding, landslides, beach erosion, and blocked waterways has exponentially increased.

Meanwhile, direct dumping of effluent from residential and industrial sites is the main source of pollution. Residential sewage comes from unregulated closed condominium complexes as well as from favelas in the lowland region. There is also an additional problem of garbage flowing into the watershed. The result is high levels of pathogenic microorganisms in the water, including coliforms and e-coli, all of which pose significant public health dangers (Zee and Sabio, 2006).

Automobile dependence/sprawl

From Argentina, Chile, and Brazil in the southern cone, to Mexico in the north, the vast majority of suburban developments being built are far from urban centers and not connected to any mass transit lines. The working-class suburbs constructed outside large cities in Mexico, for example, were built with virtually no bus or rail transit connections, few schools, shopping centers, or other amenities nearby, thus forcing millions of residents to become completely dependent on automobiles. This is a classic example of unsustainable urban sprawl. The Santa Fe mega-project outside Mexico City is also a community that is designed only for automobiles, with virtually no efficient mass transit service. It is now one of the more congested places in the metropolitan zone. In the southern cone, the suburbs outside Rio de Janeiro, São Paulo, Santiago, and Buenos Aires display the same auto-centric design pattern.

Isolation

The immense social, cultural, and functional disconnect that many of these new suburbs experience has also led to a pattern of abandonment. In one study of new suburbs in Mexico, from 2006 and 2009, a shocking 25.9 percent of new suburbs were found to be unoccupied or abandoned. This amounts to some 356,000 homes nationally. Of that total, almost half of the homes have been abandoned. In Guadalajara, the same study suggests that, in one municipality on the edge of the city, called Tlajomulco de Zuñiga, out of 251 subdivisions built in the new suburbs, and over 57,000 homes, nearly 33 percent are uninhabited (Mendiburu, 2011).

Even those who have cars find the process of living in distant suburbs difficult. In Mexico, the federal government finally realized that the working-class suburban model was not working, and vowed to end the program that had been building so many sprawling, isolated, tiny, box-like houses (some as small as between 200 and 400 square feet) in the middle of nowhere. In the words of one Mexican government planning official: "I think that, yes, the model for the future should look for a development that is less horizontal and with more density" (ibid., p. 36).

Guide to further reading

Caldeira, T. (2000) *City of Walls: Crime, Segregation and Citizenship in São Paulo.* Berkeley, CA: University of California Press.
Coy, M. and Pohler, M. (2002) "Gated Communities in the Latin American Megacities." *Environment and Planning B: Planning and Design*, 29, pp. 355–370.
Davis, M. (2007) *Planet of Slums.* London: Verso.
Herzog, L. (2015) *Global Suburbs: Urban Sprawl from the Rio Grande to Rio de Janeiro.* New York: Routledge.
Jones, G. and Moreno Carranco, M. (2007) "Mega-Projects: Beneath the Pavement, Excess." *City*, 11, pp. 144–164.
Kunstler, J.H. (1993) *The Geography of Nowhere.* New York: Touchstone.
Sabatini, F., Caceres, G., and Cerda, J. (2001) "Residential Segregation in Major Cities in Chile." *EURE*, 27, pp. 21–42.

References

Beauregard, R. (2006) *When America Became Suburban.* Minneapolis, MN: University of Minnesota Press.
Berman, M. (1988) *All That Is Solid Melts into Air: The Experience of Modernity.* New York: Penguin Press.
Borsdorf, A., Hidalgo, R., and Sanchez, R. (2007) "A New Model of Urban Development in Latin America: The Gated Communities and Fenced Cities of Santiago and Valparaiso, Chile." *Cities*, 24, pp. 365–378.
Caldeira, T. (1996) "Fortified Enclaves: The New Urban Separation." *Public Culture*, 8, pp. 303–328.
Caldeira, T. (2000) *City of Walls: Crime, Segregation and Citizenship in São Paulo.* Berkeley, CA: University of California Press.
Capron, G. and de Alba, M. (2010) "Creating the Middle Class Suburban Dream in Mexico City." *Culturales*, 6, pp. 159–183.
Collet Bruna, G. and Comin Vargas, H. (2009) "The Shopping Centers Shaping the Brazilian City." In: Del Rio, V. and Siembieda, W. (eds.) *Beyond Brasília.* Gainesville, FL: University Press of Florida, pp. 104–119.
Coy, M. (2006) "Gated Communities and Urban Fragmentation in Latin America: The Brazilian Experience." *Geojournal*, 66, pp. 121–132.
Coy, M. and Pohler, M. (2002) "Gated Communities in the Latin American Megacities." *Environment and Planning B: Planning and Design*, 29, pp. 355–370.
Davis, M. (2007) *Planet of Slums.* London: Verso.

Duany, A., Plater-Zyberk, E., and Speck, J. (2000) *Suburban Nation.* New York: North Point Press.

Foweraker, J. and Craig, A. (eds.) (1990) *Popular Movements and Political Change in Mexico.* Boulder: Lynne Rienner.

Garcia Peralta, B. and Hofer, A. (2006) "Housing for the Working Class: A New Version of Gated Communities in Mexico." *Social Justice,* 33, pp. 129–141.

Gotsch, P. (2009) *NeoTowns: Prototypes of Corporate Urbanism,* Ph.D. diss. Germany: Karlsruhe Institute of Technology.

Herzog, L. (2015) *Global Suburbs: Urban Sprawl from the Rio Grande to Rio de Janeiro.* New York: Routledge.

Hiernaux, D. and Lindon, A. (2002) "Modos de vida y utopias urbanas." *Ciudades,* 53, pp. 26–32.

Jones, G. and Moreno-Carranco, M. (2007) "Mega-Projects: Beneath the Pavement, Excess." *City,* 11, pp. 144–164.

Kunstler, J.H. (1993) *The Geography of Nowhere.* New York: Touchstone.

Mendiburu, D. (2011) "Vivir en gueto." *Emeequis,* pp. 29–36. Available at: www.m-x.com.mx/2011-07-10/vivir-en-un-gueto/.

Monkkonen, P. (2009) "The Housing Transition in Mexico: Local Impacts on National Policy." Working Paper, D 09–001. University of California, Berkeley.

Neuwirth, R. (2004) *Shadow Cities.* New York: Routledge.

Perlman, J. (1980) *The Myth of Marginality.* Berkeley, CA: University of California Press.

Sabatini, F., Caceres, G., and Cerda, J. (2001) "Residential Segregation in Major Cities in Chile." *EURE,* 27, pp. 21–42.

Salcedo, R. and Torres, A. (2004) "Gated Communities in Santiago: Wall or Frontier." *International Journal of Urban and Regional Research,* 28, pp. 27–44.

Valenzuela, A. (2007) "Santa Fe: Megaproyectos para una ciudad dividida." *Cuadernos Geograficos,* 40, pp. 53–66.

Walker, M., Walker, D., and Villagomez Velasquez, Y. (2006) "The Wal-Martification of Teotihuacan: Issues of Resistance and Cultural Heritage." In: Brunn, S. (ed.) *Wal-Mart World.* New York: Routledge, pp. 213–224.

Zee, D.M. and Sabio, C.M. (2006) "Temporal Evaluation of the Contaminated Plume Dispersion at Barra da Tijuca beach, Rio de Janeiro, Brazil." *Journal of Coastal Research,* 39, pp. 531–536.

6

Suburbanization in Australia

Robert Freestone, Bill Randolph, and Simon Pinnegar

Introduction

In *The Lucky Country*, Donald Horne (1964) characterized Australia as "the first suburban nation." The label has stuck. Australia has been a deep-seated suburban society since the European takeover in the late eighteenth century. But the suburban environment in physical, demographic, economic, cultural, and environmental terms is neither homogenous nor static. It is increasingly a realm of difference and diversity, with emergent forms and challenges distinct from the legacies of the early colonial and early twentieth-century eras. Australian suburbs have often been approached in a simplistic way. While they may remain "the crucible of Australian life," they are nonetheless "poorly understood and their dynamism is so often not appreciated" (Gleeson, 2002, p. 229). This chapter seeks to address a phenomenon at once both taken-for-granted and *terra incognita*. In so doing, we intersect with new historical and contemporary perspectives in suburban studies: approaching suburban history through a lens of longitudinal transformation and metropolitan integration (McManus and Ethington, 2007) and present day suburban forms as ongoing enterprises in place-making to optimize social requirements (Kirby and Modarres, 2010).

A suburb in the Australian context has no formal political or administrative status. The difference between city and suburb is not legally defined. Suburbs connote the great extent of suburban space outside Central Business Districts (CBDs), but many suburban local government entities are designated "cities" just to complicate things (O'Hanlon, 2005). There is an archetypal definition of the suburb as peripherally located, country as much as town, low-density, socially segregated, dominated by single-family dwellings and commuting. But this is increasingly contested when set within a dynamic city of shifting spatial, functional, and morphological characteristics (McManus and Ethington, 2007). Predominantly high value inner suburbs begin at the edge of the CBD; the outer suburbs define the periphery of the metropolitan footprint and accommodate an upwardly mobile suburban population; and in between, come the aging low-density middle ring or "third city" suburbs far from static and facing distinctive challenges (Randolph and Freestone, 2012). All three zones – albeit declining outward – are now punctuated by high-density and/or high-rise development zones, which have revalorized discarded brownfield land and in so doing complexified simple geometric models of socioeconomic and life cycle change. But "a minority urban tradition has long existed beside the suburban majority" back to the flats, boarding houses, and hostels of the late nineteenth century (O'Hanlon, 2005, p. 187). That tradition has intensified, as discussed later, but the main focus in this chapter is on the traditional lower-density areas, enscribed decisively from the late nineteenth century.

Our analysis blends a review of recent literature[1] and original research. The first half of the chapter is contextual and synoptic. It examines the cultural and ideological notions of Australian suburbanism, the basic demographic parameters of suburbia, and its physical evolution. The second half selects four contemporary issues for closer examination: densification, social polarization, suburban regeneration, and metropolitan governance. The conclusion reflects on the distinctiveness of the Australian experience.

Suburban hopes and fears

Leaving aside the romance of the outback, suburbia has always been a more realistically quintessential encapsulation of Australian living standards:

> A portrait of the typical twentieth century Australian male would show not a heavily muscled sheep shearer, but a more rotund figure in t-shirt and stubby shorts, a can of beer in one hand and a garden hose in the other, maintaining the suburbanite's long summer vigil to keep his front lawn green, or at least alive.
>
> *(Davison, 1995, p. 40)*

At various junctures suburbia has stood positively for different ideals: egalitarianism, populism, independence, security, privacy, healthy living, and family values. Well into the 1970s, metropolitan planning policies enthusiastically privileged suburban development.

A countervailing tradition of anti-suburbanism has an equally long history (Gilbert, 1988). Cultivated by the intelligentsia and mirroring overseas critiques, suburbs have been portrayed by a succession of social critics as dull, dreary, unfulfilling, homogenous aesthetic wastelands. Before World War I, playwright Louis Esson wrote, "the suburban home must be destroyed . . . the suburban home is a blasphemy. It denies life . . . in the suburbs all is repression, stagnation, a moral morgue" (Esson, 1973, p. 73). After World War II, comedian Barry Humphries mocked the banality of suburbanism through his archetypally drab middle-class personae, Edna Everage. Architect Robin Boyd spoke for many peers in declaring "the Australian suburb has been consistent in its ignorance and emotional immaturity for nearly a century" (McKay et al., 1971, p. 8). There has thus been a deep ambivalence towards suburban development (Davison, 2006) and tension between the majority of the population and cultural elites (Flew, 2011).

But the nature of the Australian suburb has fundamentally changed since the 1960s, "as the form and density of many new housing developments has changed, as accommodation has become less affordable for younger people, and as Australian cities have become more cosmopolitan and diverse" (Hamnett and Maginn, 2015, p. 7). Multiple positionalities are evident. If anything, anti-suburban strains have become stronger; suburbs continue to be "mined for comic effect" (Turnbull, 2008, p. 15), and have latterly become the realm of crass, commercial, and conformist "bogans" (Nichols, 2011). Suburban design standards are seen as inferior to urban hot spots, and in many other ways the terms of reference have changed for the worse. Suburban life is critiqued for its unsustainable dependence on automobiles, profligate consumption of land, distancing of people from social and economic opportunities, cultural conservatism, and high energy consumption.

Nevertheless, suburban life endures as a populist dream for many latter-day advocates, politicians, and commentators. Hugh Stretton in *Ideas for Australian Cities* (1970, p. 20) made an articulate and influential defense of suburbanism: "You don't have to be a mindless conformist to choose suburban life." Patrick Troy's (2004) suburbs represent a deeply ingrained and sensible housing preference that have become a target for neoliberal hegemony, as governments engage in a "race for the bottom" to minimize public expenditures on maintaining quality urban environments and infrastructure. Brendan Gleeson's (2006a) endorsement is expressed through similar fears and warns that despite their innate resilience, the "heartlands" of Australian cities are under siege from powerful forces intent on impoverishing the public realm.

The battleground has been scaled up and become more complex. According to Davison (2012, p. 27), the suburb is now "as wasteful and dangerous as it once seemed safe and boring." Dodson (2016, p. 26) captures the challenges posed by broader cultural and political attitudes to planning policies:

> Australian suburbia now finds itself in a place of ambiguity. No longer is Australian suburbia assured of its position as the promised land of the post-World War II Australian imagination where opportunity beckons for households holding aspirations of a better life through homeownership. The contemporary social, cultural, and political complexity, the interwoven contradictions and combinations of opportunity and vulnerability, the environmental costs and the infrastructure needs of Australian suburbia now come together as a 'problem' in national life.

Suburbs – far from being a classic "place apart" – find themselves at the epicenter of debates about the future of Australian cities, which are seen as increasingly divided "between young and old, rich and poor, homeowners and renters, the outer suburbs and the inner-city" (Kelly and Donegan, 2015, p. 184). They face pressures from all fronts – population growth, changing household size and composition, densification and renewal, social and economic polarization, and changing mobilities (Randolph, 2004).

Suburban cities

Most Australian city dwellers continue to live at relatively low suburban densities. The scale of Australian suburbanization was evident within a century of white settlement. In 1899, American statistician Adna Weber not only observed "the most remarkable concentration of population" in Australian cities (p. 138) but that its "movement towards the suburbs" was outpacing the rest of the world, including the United States. He included figures for Sydney in the 1880s showing that, while the growth of the central city had stalled, the suburbs more than doubled their population to a ratio of 2.5:1. This was all the more noteworthy given the relatively low population density of the center. He endorsed the "rise of the suburbs" as furnishing "the solid basis of a hope that the evils of city life, so far as they result from overcrowding, may be in large part removed" (Weber, 1899, p. 475).

The suburban condition has persisted, a compelling illustration of powerful centrifugal forces of path dependency. Recent demographic research by Gordon et al. (2015) confirms Australia's status as "a suburban nation" based on a transit modal share analysis. State and territory capital cities are on average 85 percent suburban (see Table 6.1). Moreover, this conclusion is reinforced by the distribution of recent population growth, with a substantial majority in suburban neighborhoods, mostly in automobile-oriented commuter suburbs.

Table 6.1 Population structure of Australian capital cities

City	Population (2011, '000s)	Active cores	Transit suburbs	Auto Suburbs	Exurbs	Total suburban
Sydney	4,392	16%	14%	66%	4%	84%
Melbourne	4,000	14%	6%	73%	7%	86%
Brisbane	2,066	10%	8%	71%	1%	90%
Perth	1,729	13%	9%	71%	7%	87%
Adelaide	1,225	9%	13%	72%	7%	91%
Canberra	392	18%	7%	74%	0%	82%
Hobart	212	19%	13%	39%	29%	71%
Darwin	121	8%	0%	74%	18%	92%

Source: Based on Gordon et al. (2015)

Table 6.2 Comparative population densities of Australian cities

City	Weighted population density (persons per hectare)
Brisbane	25
Melbourne	32
Sydney	52
Vancouver	65
Montreal	71
London	97

Source: Spencer et al. (2015)

A complementary analysis by Spencer et al. (2015) shows that, benchmarked against three representative global cities, Australian cities have relatively lower proportions of their total populations living in mid-density areas (between 60 and 200 persons per hectare) and instead feature pockets of high-density against a pervasive background of low-density (see Table 6.2). Moreover, related work suggests that the average population densities of Australian cities have in fact been falling at about 0.8 percent per annum since the early 1990s (Roberts, 2007).

Evolution

Davison (1995, p. 52) wrote that, "European Australia was born urban and quickly grew suburban." Indeed, suburbia has reigned supreme across successive phases of capitalist economic development from the early colonial era. The predominant urban form has been promoted by numerous interlocking factors. Australia was constructed willfully but wrongly as a *terra nullius* of wide-open spaces, and the social and geographic marginalization of Indigenous Australians into fringe camps or inner-city ghettoes was institutionalized early (Johnson, 2015). European settlement came when the suburb was emerging as a solution to urban ills, with public health professionals concerned by the link between density and ill health; in these terms Australia was "the farthest suburb of Britain" (Davison, 1995, p. 52). British imperialism promoted the bungalow abroad while the limitations of early building technology impeded higher rise development. Housing development was fueled by the desire for outright homeownership of detached dwellings. After early privations, the trajectory of development was inexorably toward a relatively high standard of living, and investment in suburban housing became a critical economic development driver (Butlin, 1964). Provision of community infrastructure by colonial then state (from 1901) governments significantly lowered the threshold of development costs. In time, rules and regulations around minimum space and other building standards locked in a distinctly suburban morphology, making departures at least difficult and at worst illegal.

The first suburbs were created in Sydney as the state granted large allotments for the colonial elite. As the town established itself, the immediately surrounding residential quarters were developed from the 1830s as closely packed neighborhoods of attached terrace and row housing. Frost (1991) differentiates Sydney (along with Brisbane and Hobart) from later capitals as exemplifying a land intensive urbanism more typical of the older new world cities of the Atlantic coast and distinct from the expansive metropolitan forms that would follow in, for example, the later settled Adelaide and Perth. Terraces were the dominant housing form in Sydney and Melbourne for the inner suburbs from the 1850s to 1890s, becoming increasingly ornate into the late Victorian era. They were built incrementally by owner-builder landlords and the subdivision of lands further out was similarly spawned by small-scale speculative behavior.

The cities began to become truly suburban in the 1870s, with the physical separation of home and work for the majority of workers bridged only by train and tram commuting. The first services were provided by private enterprise before soon falling into public ownership. The Government Statistician of Victoria – himself a land speculator – provides a picture of greater Melbourne in the 1890s with a radius of 10 miles with the "suburban portions . . . scattered – gardens, grounds, and paddocks, some of large size, being attached to most residences" and with "room for much more building, without crowding" and "extensive suburban villages . . . springing up outside the metropolitan limits" (Hayter, 1892, p. 546). The enduring norm was the "quarter acre block" (over 1000 square meters), more rhetorical than reality, with 800–900 square meters more typical of suburban blocks to the mid-twentieth century.

From the early twentieth century the garden city movement stood against inner-city apartment and tenement development, endorsing a more manicured, pre-planned version of the suburban idyll, but it was the ordinary, do-it-yourself suburb that proliferated. Extensions and electrification of metropolitan railway systems further promoted suburban expansion, but from the 1950s road building and increased car ownership saw growth "eventually outrun" public transport systems to create far-flung cities (Frost and Dingle, 1995, p. 20). The "national hobby" of land speculation continued to fuel broad acre subdivisions (Sandercock, 1979), but this form of growth was reined in during the post-war period with the implementation of statutory regional and local planning.

State governments assisted with low-cost loan schemes from the early 1900s, while the federal government launched a War Service Homes program in the late 1910s. From the 1930s, states also progressively introduced public housing schemes, and an historic financial agreement with the federal government in 1945 formed the basis for development of extensive suburban estates into the 1970s (Troy, 2012). Houses were usually constructed of cheap materials and their public tenants, predominantly working-class families, were precariously dependent upon manufacturing employment. A distinctive suburban form emerged, dubbed "Holdenist suburbia" after the locally designed car mass-produced by General Motors that was the lifeline for these fringe communities from the 1950s (Winter and Bryson, 1998). Many suburban pioneers roughed it in "do-it-yourself" dwellings alongside the growing diffusion of the "cream brick frontier" (Davison et al., 1995). The bleak, often treeless fringe suburbs sparked corrective aesthetic, environmental, and social responses from the 1960s, with calls for more innovative site planning as cluster or courtyard homes, environmentally sensitive design, more compact and denser neighborhoods, and better integration with transport and employment planning. The public housing estates were the subjects of revitalization programs in the 1990s involving exterior refurbishment, internal upgrades, the "de-radburnization" of layouts, and also privatization.

At least three broad macro physical trends are evident into the twenty-first century. First is the shrinking size of average suburban blocks. Older, more generous allotments promoted self-sufficiency in food production, water and waste management, and during World War II, an estimated 48 percent of suburban households produced some sort of food (Gaynor, 2006). In the mid-1990s, the average site area of new houses in Australian capital cities was around 800 square meters; by the early 2000s, it was down to about 700 square meters; in the 2010s, it has fallen further. In Melbourne, the most common block size now is approximately 650 square meters with even smaller blocks of 180–200 square meters common (Charting Transport, 2016). The backyard as an iconic feature of traditional Australian suburban spaces, supporting biodiversity, food production, and social interaction has been shrunk or effectively eliminated in new suburban estates (Hall, 2010).

Second, and not unrelatedly, the average new Australian house has increased from a cottage of around 100 square meters in the early 1900s to a 245 square meters house just over a century later, despite average household size steadily decreasing. Most of the increase is attributable to larger living spaces and in the number and size of bedrooms (McMullan and Fuller, 2015). The "gigantic proportions" of such houses are said to make them some of the biggest new suburban houses in the world (Johnson, 2015, p. 124); this heralded the phenomenon of the "McMansion."

Third, the rough and ready owner-built suburb gave way almost entirely to master-planned communities as the dominant suburban form. These estates were very different from post-war "cookie-cutter" and "Holdenist" subdivisions of suburban "battlers." Here, community is a commodity spruiking a range of sales features: lifestyle, safety, security, infrastructure, and amenity. These Australian adaptations of American-style "privatopias" have attracted a significant niche academic literature (Dowling et al., 2010), with assessments ranging across the spectrum from pariah to panacea (Johnson, 2010). While influenced by the new urbanism, these are neither neo-traditional in design nor physically gated, the latter a less pervasive phenomenon mostly found in lifestyle and tourism regions such as the Gold Coast.

Contemporary issues

Is there an average Australian suburb? One recent attempt to measure that against national norms identified Oak Park in Melbourne's northwestern suburbs as the most statistically typical, with an average deviation of less than 4 percent across a range of social indicators: with a household size of 2.54 persons, median age of 36, 29.4 percent born overseas, and mortgages and tertiary qualifications almost identical to the Australian averages of 35 percent and 22 percent, respectively (Id, 2013). The wider reality is heterogeneity. Studies of residential differentiation abound since the 1970s. Working with a basket of indicators, Healy and Birrell's (2006) cross-sectional analysis of the residential population of Melbourne produced five suburban types: high amenity near-city suburbs dating from the pre-1940s; transitional near-city suburbs undergoing gentrification (see Table 6.1); Holdenist low-amenity suburbs built predominantly in the 1950s and 1960s with increasing concentrations of social disadvantage; middle-class suburbia as legacy of a dynamic suburban fringe; and outer suburbia defining the current suburban frontier. Even further out could be added peri-urban suburbs, the exurban rural-residential zone latterly coming under increasing growth and environmental pressures (McKenzie, 1997).

This typology is evident in other large Australian cities, emphasizing a range of housing submarkets with their own distinctive social ecology and policy implications. Forces of physical, social, economic, and environmental change since the 1970s have produced uneven outcomes and challenges. These include economic restructuring and loss of jobs; social isolation; lack of public transport and other community services; exposure to rising costs of both mortgages and fuel (Dodson and Sipe, 2008); places now defined by minority, ethnic or racial identities (Turner, 2008); loss of biodiversity and urban heat islands; and obesity and climate change concerns. At the same time residents in the more vulnerable communities resist stigmatization and can display considerable resourcefulness in improving their quality of life (Kirby and Modarres, 2010; Johnson et al., 2016).

In the following sections, we focus on four trends and challenges confronting many Australian suburbs: densification through medium and high-rise development, social polarization intensifying the unevenness of life chances, the renewal of suburban fabric through spontaneous owner interventions, and the problem of effective governance.

Densification

Attempts to densify Australia's suburbs have been ongoing since at least the 1980s, with metropolitan planning policies increasingly promoting market-driven higher density development (Bunker, 2014). Growing pressure on land and the mounting costs of providing infrastructure for low-density development challenged economic and environmental sustainability. Early attempts by Australian land use planners to encourage greater suburban densities under the mantle of "urban consolidation" took the form of lowering block sizes in fringe development, splitting large existing blocks into two titles ("dual occupancy"), and permitting secondary dwellings – so-called "granny flats."

But the real driver of greater densification in Australian suburbs is the transformation wrought by the passage of world-leading strata title legislation in the early 1960s that allowed sale of individual apartments (Randolph and Easthope, 2014). This sparked a rapid process of redevelopment, especially in Sydney, as local councils, pushed by local development interests, rezoned land with the capacity for higher density with a focus on the edges of suburban town centers near railway stations. The suburban apartment market mushroomed as detached suburban home sites were redeveloped into three or four story apartments, typified as "gun-barrel" blocks because of their layout, stretching back from narrow street frontages. Elsewhere, suburban densification has been more gradual, mainly reflecting differential pressures of land value. Melbourne has rapidly caught up in the early twenty-first century (Buxton and Tieman, 2005), although medium-density development had been a feature of suburban renewal since the 1950s (Lewis, 1999). Densification in other cities has been less pronounced until more recently. In Perth, for example, suburban densification has been modest and undertaken primarily by redevelopment of separate house blocks into single-story attached home units and villa developments, typically with six new homes on the one lot (Bunker and Troy, 2015).

However, the promotion of higher urban densities has now reached new levels. Stimulated in large part by investors from both home and overseas, the transition of the development industry from one that delivered a predominantly low-density single house product on the suburban fringe to one where multi-unit housing provides 50 percent of all housing supply has been transformative. Figure 6.1 charts this change, which has overtaken many suburbs in the last decade, particularly in Sydney and Melbourne. The question is how far and quickly this process will progress, and exactly what impact it will have on community structures. Large-scale developments in established communities are frequently contested. Critically, they also test the capacity of existing and new infrastructure – hard and soft – to support rapidly growing suburban populations.

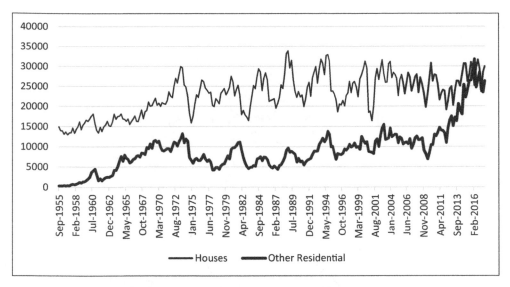

Figure 6.1 Australia quarterly private sector dwelling units commenced, 1955–2016

Source: Australian Bureau of Statistics (2016)
Series 8731.0 Building Approvals Table 10

Socioeconomic disadvantage

Systemic spatial variations in service provision, social dysfunction, and poverty have been recognized by urban academic researchers and policymakers for decades (Stimson, 1982). These have been exacerbated through major changes over the last 30 years. The collapse of manufacturing industry and the decline of transport infrastructure hubs (docks and rail) from the 1970s led to widespread job losses in inner-city industrial cores (Stimson, 1982; Troy, 1981). The corollary was the outwards shift in new jobs, often based on servicing the growing suburban population, as well as out-of-town office employment and administration, and road-based warehousing and logistics. Urban development, driven by land speculation and heavy public investment in road transport, accelerated the process. However, within a generation, this picture has radically changed. In line with trends elsewhere (Katz and Bradley, 2013), the inner cities have been revitalized as new knowledge-based industries have centralized in parallel with widespread gentrification. Much of this has been spurred by government planning policies and major public investment, especially in large-scale revitalization projects that have pump-primed renewal in older industrial and dockside locations.

The resulting spatial "inversion" of urban socioeconomic fortunes has been profound (Ehrenhalt, 2012; Randolph, 2017). While there has been long-standing concern over the social position of Australian suburban development, in large part a result of major post-war "Holdenist" building led by state housing commissions (Winter and Bryson, 1998), the outcome of 30 years of socio-spatial restructuring has meant that Australian inner-city suburbs no longer feature as places of disadvantage, beyond a few residual public housing estates. Instead, the weight of disadvantage has shifted to the middle and outer suburbs (Randolph and Tice, 2016). Figure 6.2 illustrates the changing location of the 15 percent most disadvantaged census tracts in Melbourne between 1986 and 2011, as measured by the Australian Bureau of Statistics Index of Relative Socioeconomic Disadvantage. The resulting shift in the locus of disadvantage, in part driven by the arrival

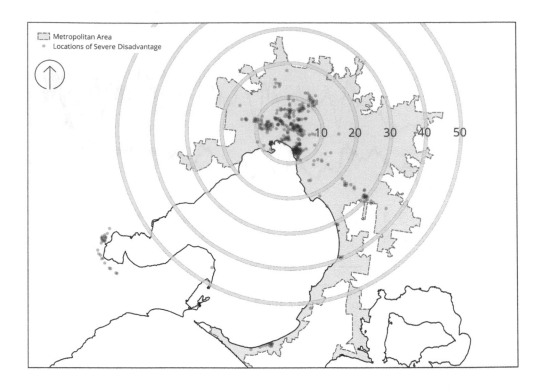

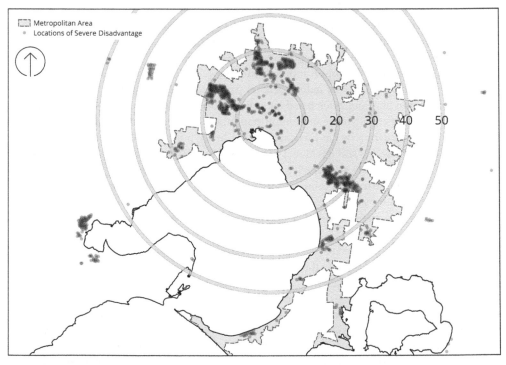

Figure 6.2 Spatial distribution of disadvantage in Melbourne, 1986 (top) and 2011 (below)

Source: Randolph and Tice (2016)

of new immigrant communities, has significant implications for the delivery of urban services, infrastructure, and broader urban policy (Dodson, 2016).

Badcock (1997) was one of the first to point out that the 1980s was a pivotal moment when peaks of urban social disadvantage had begun to relocate away from the inner-city. Randolph (2004) and Baum et al. (2005) both echoed this observation, noting the emergence of a distinct band of social disadvantage in what had become middle suburbia, focused on concentrations of lower value privately rented flats and apartments around suburban town centers. More recently, Pawson et al. (2015) have documented this shift more extensively and identified both housing and labor market drivers. But it is not all bad news. Gleeson's (2006a) perceptive commentary on the changing fortunes of the Australian suburban "heartlands" acknowledges shifting fortunes but robust communities working toward a better life.

Suburban renewal

With Australia's cities continuing to experience strong population growth over the last two decades, strategic planning aspirations to promote the compact city through infill and urban renewal have demanded a reworked engagement with suburbs. Urban consolidation can be tracked through the relative split of new development provided through land release in greenfield fringe locations and through additions in existing built-up areas. Housing market and policy settings have colluded to encourage residential development into the latter, with Sydney's infill ratio rarely falling below 70 (existing areas): 30 (new release) since the early 1980s. Australia's other cities are further behind in their drive to higher densities; in Melbourne's most recent plan, the target split moving forward to 2050 is 65:35.

However, "infill" is an inadequate term for capturing the myriad drivers and built form outcomes reshaping existing suburbs, and it is instructive to break down current suburban renewal activity into three parallel strands. The first, and of primary interest in a strategic planning context, relates either to brownfield redevelopment around activity centers, typically transit nodes, or opportunities for large-scale, multi-unit development in rezoned priority precincts. In recent instances, residents themselves have taken a combined and coordinated approach into negotiations with developers to maximize their sales prices. The second strand captures the relatively few, but high profile, public housing estate renewal projects delivered through complex public-private partnership arrangements. Such projects also evince a systematic approach to suburban renewal underpinned by extracting the benefits of densification through the reworking of government-owned land.

Suburban renewal is also taking place to an equal if not greater degree through a third category: more diffuse and less strategic spontaneous, incremental activity. The distinctive characteristics of the post-World War II build-out of Australian suburbs, where independent owner-builders contributed a significant proportion of housing stock, are still conducive to change through individually led, small-scale activity. These inherent characteristics are reinforced by dwelling obsolescence as those neighborhoods age and upward shifting land values. Such conditions are highly favorable to more incremental forms of renewal, such as "knockdown rebuild," the one-for-one replacement of single-family homes on an individual lot (Pinnegar et al., 2015; Pinnegar et al., 2018). A desire to foster a more coordinated policy focus for these incremental processes is captured in the concept of "greyfield" regeneration to enable benefits of scale through more community-oriented locally driven arrangements (Newton and Glackin, 2014).

Research into housing choice and constraint in the face of affordability concerns has also thrown an increased spotlight on the "missing middle" in terms of the limited diversity of housing types (Kelly, 2011). Addressing a lack of mid-sized options between single detached dwelling and high-density units – whether in the form of townhouses, terraces or innovative low-rise schemes – has been pitched as one means of improving affordable housing supply and choice. In late 2016, the New South Wales (NSW) Government ran a *Missing Middle Design Competition*, encouraging architects to find innovative solutions for low-rise, medium-density housing.

Metropolitan governance

Integral to the above three issues shaping Australia's contemporary suburban landscapes are questions of urban governance, and the ongoing challenges posed by reconciling complex drivers and outcomes in cities experiencing high levels of growth. While Australian cities can demonstrate a long trajectory of strategic planning at the metropolitan scale (Bunker and Searle, 2009; Freestone, 2010), implementation of those plans has been more challenging (Dodson, 2009; Hamnett and Freestone, 2018). In part, this reflects the fiscal imbalance between the three tiers of government (commonwealth, state/territory, local) and in particular reconciling state government attempts to offer citywide direction across a fragmented jigsaw of under-resourced local authorities.

While there have been continually evolving formal, informal, and voluntary attempts to institute stronger spatial arrangements at the appropriate scale (for example, through council amalgamations or entities such as the Regional Organizations of Councils [ROCs] in NSW), effective governance arrangements at the subregional scale have been elusive. Thus, outside the strategic gaze focused upon the CBD and major knowledge-based centers, much of suburbia has played a rather subservient role in recent decades. Councils are frequently treated as receptacles for top-down strategic pronouncements – merely to accept their fair share of intensified housing supply targets, for example – rather than as genuine partners in the "global, liveable" narratives that infuse current discourse. The result is impaired capacity to implement major infrastructure initiatives and tackle fundamental issues such as social polarization (Tomlinson and Spiller, 2018).

The clarion call for a rethink on Australian metropolitan governance has been heard, and recent developments in both Sydney and Melbourne are instructive, not least in terms of holding out the prospect for communities to gain a stronger "voice for their collective will" (Spiller, 2014, p. 361). The establishment of the Greater Sydney Commission in 2015 through an Act of Parliament hands over responsibilities for strategic planning at the metropolitan and subregional level to an independent statutory authority. The *Greater Sydney Region Plan* rearticulates Sydney's global city narrative beyond the CBD and global economic corridor and envisages a metropolis of three cities (Greater Sydney Commission, 2018). The "Western Parkland City" marks a significant re-engagement with large tracts of the city's outer ring suburbs, as well as a more networked understanding of future growth around a new airport. It is instructive that one of Australia's first "city deals" (a policy import in name at least from the UK) is for Western Sydney, and it signals the most recent foray of Commonwealth Government interest in encouraging an enhanced and collaborative commitment to infrastructure provision and productivity in the major cities. The development of five *District Plans* – subregional documents with the task of linking citywide strategic issues to more local implementation – as one of the first deliverables of the new authority also holds out the prospect for better strategic engagement with suburban Sydney.

Melbourne's restructured metropolitan governance arrangements have taken a different tack. The Victorian capital shares Sydney's need to break down the preeminence of its historic city center, particularly in terms of jobs density, and encourage a more polycentric urban form. However, rather than establishing a metropolitan body, overall responsibility for the city's new strategic framework, *Plan Melbourne 2017–2050*, is retained in a restructured Department of Environment, Land, Water, and Planning (DELWP), but makes it clear the importance and close synergies with other agencies (Goodman, 2018). Within DELWP, a new ministerial portfolio for suburban development has been established. This will be responsible for creating five-year plans focused on jobs, services, and infrastructure for each of Melbourne's six metropolitan regions.

Conclusion

Many of the themes and trends explored here will resonate internationally but there remains a distinctive Australian suburbanism. Despite low urban densities, state and territory planning systems progressively put in place after World War II have ensured that unbridled scatteration has not resulted. Land release, growth management, development control, and infrastructure planning inject a significant public oversight of land development. The legacy of the traditional twentieth-century suburb is "relative design diversity," as owners have customized their living quarters to changing needs and contemporary organic renewal processes perpetuate the variegated landscape (Gleeson, 2006b, p. 12). Influences from abroad abound but their impacts have been adapted through cultural filters. While master-planned estates are an increasingly mainstream business model, few have adopted "the picket-fence purity of North American New Urbanism, or the theme park whimsy of its niche spin-offs" (Gleeson, 2006b, p. 12).

This chapter has reviewed philosophical, physical, policy, and related attributes of Australian suburbanisation, highlighting different historical and contemporary attitudes to suburban culture, the increasing heterogeneity of suburban development, and a number of issues posing future challenges, particularly for policymakers. Kellett (2011, p. 266) argues that "the morphology of Australian cities appears to be at a crucial point in its history" with increasing densification. Yet, "despite continuing attempts to promote a more compact and cosmopolitan form, detached living in suburban settings remains the preferred housing type and setting of most Australians" (Burton, 2015, p. 504). In this sense, the future trajectory of Australian cities still appears to be "decidedly suburban" (Gordon et al., 2015, p. 3).

Note

1 Our thanks to Caitlin Buckle for her initial assistance in pulling together some key recent works.

Guide to further reading

Bryson, L. and Winter, I. (1999) *Social Change, Suburban Lives*. Sydney: Allen and Unwin.
Davison, G. (2016) *City Dreamers: The Urban Imagination in Australia*. Sydney: New South.
Dingle, T. (1999) "'Gloria Soame' the Spread of Suburbia in Post-War Australia." In: Harris, R. and Larkham, P. J. (eds.) *Changing Suburbs: Foundation Form and Function*. London: E & F N Spon, pp. 187–201.
Forster, C. (2004) *Australian Cities: Continuity and change*. 3rd ed. Melbourne: Oxford University Press.
Frost, L. (2015) "Urbanisation." In: Ville, S. and Withers, G. (eds.) *The Cambridge Economic History of Australia*. Melbourne: Cambridge University Press, pp. 245–264.
Tomlinson, R. (ed.) (2012) *Australia's Unintended Cities: The Impact of Housing on Urban Development*. Melbourne: CSIRO Publishing.

References

Australian Bureau of Statistics (2016) *The Census of Population and Housing*. Canberra: Australian Bureau of Statistics.
Badcock, B. (1997) "Recently Observed Polarising Tendencies and Australian Cities." *Australian Geographical Studies*, 35, pp. 243–259.
Baum, S., O'Connor, K., and Stimson, R. (2005) *Fault Lines Exposed: Advantage and Disadvantage across Australia's Settlement System*. Melbourne: Monash University Press.

Bunker, R. (2014) "How Is the Compact City Faring in Australia?" *Planning Practice & Research*, 29, pp. 449–460.

Bunker, R. and Searle, G. (2009) "Theory and Practice in Metropolitan Strategy: Situating Recent Australian Planning." *Urban Policy and Research*, 27, pp. 101–116.

Bunker, R. and Troy, L. J. (2015) *The Changing Political Economy of the Compact City and Higher Density Urban Renewal in Perth*. Sydney: City Futures Research Centre and University of New South Wales.

Burton, P. (2015) "The Australian Good Life: The Fraying of a Suburban Template." *Built Environment*, 41, pp. 504–518.

Butlin, N. (1964) *Investment in Australian Economic Development, 1861–1900*. Cambridge, MA: Cambridge University Press.

Buxton, M. and Tieman, G. (2005) "Patterns of Urban Consolidation in Melbourne: Planning Policy and the Growth of Medium Density Housing." *Urban Policy and Research*, 2, pp. 137–157.

Charting Transport (2016) Available at: https://chartingtransport.com/2016/05/22/are-melbourne-suburbs-full-of-quarter-acre-blocks/.

Davison, A. (2006) "Stuck in a Cul-de-Sac? Suburban History and Urban Sustainability in Australia." *Urban Policy and Research*, 24, pp. 201–216.

Davison, G. (1995) "Australia: The First Suburban Nation." *Urban History*, 22, pp. 40–74.

Davison, G. (2012) "Suburbs." In: Beilharz, P. and Hogan, T. (eds.) *Sociology: Antipodean Perspectives*. Sydney: Oxford University Press, pp. 26–31.

Davison, G., Dingle, T., and O'Hanlon, S. (eds.) (1995) *The Cream Brick Frontier: Histories of Australian Suburbia*. Melbourne: Monash University Press.

Dodson, J. (2009) "The 'Infrastructure Turn' in Australian Metropolitan Spatial Planning." *International Planning Studies*, 14, pp. 109–123.

Dodson, J. (2016) "Suburbia in Australian Urban Policy." *Built Environment*, 42(1), pp. 23–36.

Dodson, J. and Sipe, N. (2008) *Unsettling Suburbia: The New Landscape of Oil and Mortgage Vulnerability in Australian Cities*. Urban Research Program and Griffith University.

Dowling, R., Atkinson, R., and McGuirk, P. (2010) "Privatism, Privatisation and Social Distinction in Master-Planned Residential Estates." *Urban Policy and Research*, 28, pp. 391–410.

Ehrenhalt, A. (2012) *The Great Inversion and the Future of the American City*. New York: Knopf Doubleday.

Esson, L. (1973 [1912]) *The Time is Not Yet Ripe*. Sydney: Currency Press.

Flew, T. (2011) "Right to the City, Desire for the Suburb?" *M/C Journal*, 14(4). Available at: http://journal.media-culture.org.au/index.php/mcjournal/article/view/368.

Freestone, R. (2010) *Urban Nation: Australia's Planning Heritage*. Melbourne: CSIRO Publishing.

Frost, L. (1991) *The New Urban Frontier: Urbanisation and City-Building in Australasia and the American West*. Sydney: University of New South Wales University Press.

Frost, L. and Dingle, T. (1995) "Sustaining Suburbia: An Historical Perspective on Australia's Growth." In: Troy, P. (ed.) *Australian Cities: Issues, Strategies and Policies for Urban Australia in the 1990s*. Cambridge, MA and Melbourne: Cambridge University Press, pp. 20–38.

Gaynor, A. (2006) *Harvest of the Suburbs: An Environmental History of Growing Food in Australian Cities*. Perth: University of Western Australia Press.

Gilbert, A. (1988) "The Roots of Australian Anti-Suburbanism." In: S. Goldberg and F. Smith (eds.) *Australian Cultural History*. Melbourne: Cambridge University Press, pp. 33–49.

Gleeson, B. (2002) "Australia's Suburbs: Aspiration and Exclusion." *Urban Policy and Research*, 20, pp. 229–232.

Gleeson, B. (2006a) *Australian Heartlands: Making Space for Hope in the Suburbs*. Sydney: Allen and Unwin.

Gleeson, B. (2006b) "Towards a New Australian Suburbanism." *Australian Planner*, 43(1), pp. 10–13.

Goodman, R. (2018) "Melbourne: Growing Pains for the Liveable City." In: Hamnett, S. and Freestone, R. (eds.) *Planning Metropolitan Australia*. London: Routledge, pp. 51–75.

Gordon, D., Maginn, P. J., Biermann, S., Sisson, A., and Huston, I. (2015) *Estimating the Size of Australia's Suburban Population*. Perth: Patrec Perspectives.

Greater Sydney Commission (2018) *Greater Sydney Region Plan: A Metropolis of Three Cities*. Sydney: NSW State Government.

Hall, T. (2010) *The Life and Death of the Australian Backyard*. Melbourne: CSIRO Publishing.

Hamnett, S. and Freestone, R. (eds.) (2018) *Planning Metropolitan Australia*. London: Routledge.

Hamnett, S. and Maginn, P. J. (2015) "Australian Cities in the 21st Century: Suburbs and Beyond." *Built Environment*, 42(1), pp. 5–22.

Hayter, H.H. (1892) "The Concentration of Population in Australian Capital Cities." *Australasian Association for the Advancement of Science*, 4, pp. 541–546.

Healy, E. and Birrell, B. (2006) *Housing and Community in the Compact City*, AHURI Final Report No. 89. Melbourne: Swinburne-Monash Research Centre, Australian Housing and Urban Research Institute.

Horne, D. (1964) *The Lucky Country*. Melbourne: Penguin Books Australia.

Id (2013) "What Is Australia's Most Typical Suburb?" Available at: http://blog.id.com.au/2013/population/ australian-demographic-trends/what-is-australias-most-typical-suburb/ [Accessed 2 May 2013].

Johnson, L.C. (2010) "Master Planned Estates: Pariah or Panacea?" *Urban Policy and Research*, 28, pp. 375–390.

Johnson, L.C. (2015) "Governing Suburban Australia." In: Hamel, P. and Keil, R. (eds.) *Suburban Governance*. Toronto: University of Toronto Press, pp. 110–129.

Johnson, L.C., Andrews, F., and Warner, E. (2016) "The Centrality of the Australian Suburb: Mobility Challenges and Responses by Outer Suburban Residents in Melbourne." *Urban Policy and Research*, published online. http://dx.doi.org/10.1080/08111146.2016.1221813.

Katz, B. and Bradley, J. (2013) *The Metropolitan Revolution: How Cities and Metros Are Fixing Our Broken Politics and Fragile Economy*. Washington, DC: The Brookings Institute.

Kellett, J. (2011) "The Australian Quarter Acre Block: The Death of a Dream?" *Town Planning Review*, 82, pp. 263–284.

Kelly, J.-F. (2011) *The Housing We'd Choose*. Melbourne: Grattan Institute.

Kelly, J.-F. and Donegan, P. (2015) *City Limits: Why Australia's Cities Are Broken and How We Can Fix Them*. Melbourne: Grattan Institute and Melbourne University Press.

Kirby, A. and Modarres, A. (2010) "The Suburban Question: Notes for a Research Program." *Cities*, 27, pp. 114–121.

Lewis, M. (1999) *Suburban Backlash: The Battle for the World's Most Liveable City*. Melbourne: Bloomings Books.

McKay, I., Boyd, R., Stretton, H., and Mant, J. (1971) *Living and Partly Living*. Melbourne: Nelson.

McKenzie, F. (1997) "Growth Management or Encouragement? A Critical Review of Land Use Policies Affecting Australia's Major Exurban Regions." *Urban Policy and Research*, 15, pp. 83–99.

McManus, R. and Ethington, P.J. (2007) "Suburbs in Transition: New Approaches to Suburban History." *Urban History*, 34, pp. 317–337.

McMullan, M. and Fuller, R. (2015) "Spatial Growth in Australian Homes (1960–2010)." *Australian Planner*, 52, pp. 314–325.

Newton, P. and Glackin, S. (2014) "Understanding Infill: Towards New Policy and Practice for Urban Regeneration in the Established Suburbs of Australia's Cities." *Urban Policy and Research*, 32, pp. 121–143.

Nichols, D. (2011) *The Bogan Delusion*. Melbourne: Affirm Press.

O'Hanlon, S. (2005) "Cities, Suburbs and Communities." In: Lyons, M. and Russell, P. (eds.) *Australia's History: Themes and Debates*. Sydney: University of New South Wales University Press, pp. 172–189.

Pawson, H., Hulse, K., and Cheshire, L. (2015) *Addressing Concentrations of Disadvantage in Urban Australia*. AHURI Final Report No. 247. Melbourne: Australian Housing and Urban Research Institute.

Pinnegar, S., Freestone, R., and Wiesel, I. (2018) "Incremental Change with Significant Outcomes: Regenerating the Suburbs through Knockdown Rebuild." In: Ruming, K. (ed.) *Urban Regeneration in Australia: Policies, Processes and Projects of Contemporary Urban Change*. London: Routledge, pp. 295–310.

Pinnegar, S., Randolph, B., and Freestone, R. (2015) "Incremental Urbanism: Characteristics and Implications of Residential Renewal through Owner-Driven Demolition and Rebuilding." *Town Planning Review*, 86, pp. 279–301.

Randolph, B. (2004) "The Changing Australian City: New Patterns, New Policies and New Research Needs." *Urban Policy and Research*, 22, pp. 481–493.

Randolph, B. (2017) "Emerging Geographies of Suburban Disadvantage." In: Hannigan, J. and Richards, G. (eds.) *The Handbook of New Urban Studies*. London: Sage, pp. 159–178.

Randolph, B. and Easthope, H. (2014) "The Rise of Micro-Government: Strata Title, Reluctant Democrats and the New Urban Vertical Polity." In: Gleeson, B. and Beza, B. (eds.) *The Public City: Essays in Honour of Paul Mees*. Melbourne: Melbourne University Press, pp. 210–224.

Randolph, B. and Freestone, R. (2012) "Housing Differentiation and Renewal in Middle-Ring Suburbs: The Experience of Sydney, Australia." *Urban Studies*, 49, pp. 2557–2575.

Randolph, B. and Tice, A. (2016) "Relocating Disadvantage in Australian Cities: Socio-Spatial Polarisation Under Neo-Liberalism." *Urban Policy and Research*, published online. http://dx.doi.org/10.1080/0811 1146.2016.1221337.

Roberts, B.H. (2007) "Changes in Urban Density: Its Implications on the Sustainable Development of Australian Cities." Proceedings of the State of Australian Cities National Conference, Adelaide, Australia, 28–30 November.

Sandercock, L. (1979) *The Land Racket: The Real Costs of Property Speculation*. Canberra: Silverfish.

Spencer, A., Gill, J., and Schmahmann, L. (2015) "Urban or Suburban? Examining the Density of Australian Cities in a Global Context." Proceedings of the State of Australian Cities National Conference, Gold Coast, Australia, 9–11 December.

Spiller, M. (2014) "Social Justice and the Centralisation of Governance in the Australian Metropolis: A Case Study of Melbourne." *Urban Policy and Research*, 32, pp. 361–380.

Stimson, R. J. (1982) *The Australian City: A Welfare Geography*. Sydney: Longman Cheshire.

Stretton, H. (1970) *Ideas for Australian Cities*. Melbourne: Georgian House.

Tomlinson, R. and Spiller, M. (2018) *The Metropolitan Imperative: An Agenda for Governance Reform*. Melbourne: CSIRO Publishing.

Troy, P. (ed.) (1981) *Equity in the City*. Sydney: Allen and Unwin.

Troy, P. (2004) "Saving Our Cities with Suburbs." *Griffith Review*, Summer 2003–4, pp. 80–90.

Troy, P. (2012) *Accommodating Australians: Commonwealth Government Involvement in Housing*. Sydney: Federation Press.

Turnbull, S. (2008) "Mapping the Vast Suburban Tundra: Australian Comedy from Dame Edna to Kath and Kim." *International Journal of Cultural Studies*, 11, pp. 15–32.

Turner, G. (2008) "The Cosmopolitan City and Its Other: The Ethnicizing of the Australian Suburb." *Inter-Asia Cultural Studies*, 9, pp. 568–582.

Weber, A. F. (1899) *The Growth of Cities in the Nineteenth Century: A Study in Statistics*. New York: Columbia University and Palgrave Macmillan.

Winter, I. and Bryson, L. (1998) "Economic Restructuring and State Intervention in Holdenist Suburbia: Understanding Urban Poverty in Australia." *International Journal of Urban and Regional Research*, 22, pp. 60–75.

7

Suburbanization in Europe

A focus on Dublin

Ruth McManus

Dublin: setting the scene

The ancient city of Dublin, set against the stunning natural backdrop of Dublin Bay to the east, rich farmland to the north, and the dramatic skyline of the mountains to the south, is blessed with an attractive hinterland. Although the authorities celebrated the city's millennium in 1988, its origins go back much farther. The early Christian ecclesiastical settlement at the River Liffey predates the establishment of a semi-permanent Viking encampment in 841 (Simms, 2001). With the arrival of the Anglo-Normans in 1171, Dublin became capital of the English Lordship in Ireland, and its development since that date can be seen within this colonial context.

Medieval Dublin was walled and, like most similar towns, was extensively developed beyond these walls (Simms, 2001, p. 50). Clarke (1998) has suggested that some 80 percent of Dubliners in the early-fourteenth century lived outside the town walls. The archbishop and major monastic houses held large tracts of suburban land, known as "liberties," with ecclesiastical and municipal authorities competing at times. Such competition, though with different players, would be a recurrent theme in Dublin's suburban development.

Waves of outward movement continued in subsequent centuries as the city evolved from a walled medieval city to an elegant Georgian one. Economic prosperity and political change, notably the 1661 reinstatement of an Irish parliament in Dublin and the arrival of the Duke of Ormonde as viceroy five years later, set the stage for a development boom. By the 1680s, Dublin Corporation, the municipal authority, expressed concern that the area outside the walls was greater than that within them, potentially leaving the city exposed to attack, although unlike many continental cities it did not fortify its suburbs (Sheridan, 2001, p. 75). New speculative development shifted northwards and eastwards to private estates centered on Henrietta Street and Sackville Mall (now O'Connell Street) and later to the southeastern sector of the city (see Figure 7.1). The resultant fragmentation of the urban landscape, with poor connectivity between the various estates and the medieval core, ultimately required a radical solution. That came with the 1757 creation of Europe's first formal planning authority, the Wide Streets Commission, which brought order and cohesion to Dublin through a process of street widening and new street creation (Sheridan, 2001, p. 69). The need for formal intervention to curb or repair the excesses of uncoordinated and unregulated speculative development is another recurring feature of Dublin's suburban

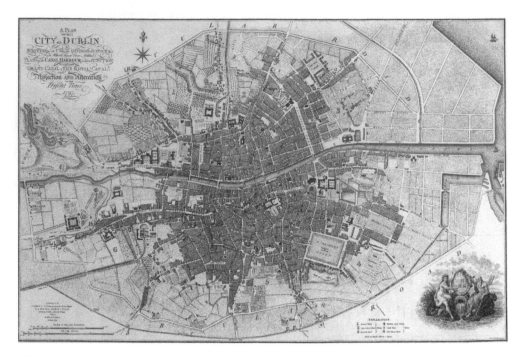

Figure 7.1   Faden map of Dublin

Source: Library of Congress, Geography, and Map Division

development. Concern at the scale of suburban expansion is not a recent phenomenon. In 1786, an anonymous pamphlet sent to parliament argued for the improvement of existing areas such as Temple Bar rather than promoting suburban expansion towards Ringsend, Donnybrook or Ballybough (Ó Gráda, 2015, p. 102).

The city's relationship with its surroundings evolved during the Georgian era, a period of affluence associated with the successive reigns of Kings George I to IV (1714 to 1830), as new landscapes of leisure were created in the urban hinterland. Seaside areas such as Howth, Blackrock, and Clontarf were developed, initially as areas of resort and summer lodging, for bathing and taking the sea air, but increasingly for permanent residence (Lennon, 2014; Ó Gráda, 2015). Suburban villas for better-off residents began to proliferate at a further remove from the city, often within reach of the sea breeze and with distant views of the bay. This development has been associated with a range of factors, including an increasing desire for single-family houses and social exclusivity, a growing ability and willingness to commute, the Romantic movement and taste for the picturesque, and perceptions of the health benefits of seaside resorts and rural air. Such villas could combine these many benefits with access to the city for association (Ó Gráda, 2015). Improving communications, including better roads and an expanded postal service, attested to existing suburban growth and acted as stimuli for further development. Nevertheless, this leading edge of development still generally took the form of "country" villas whose occupants also enjoyed the benefits of a "town" house; for example, Lord Charlemont kept a town house at Rutland Square and country residence at Marino. Therefore, the extent to which these dwellings for the elite in the urban hinterland were truly "suburban" is debatable.

These early phases of Dublin's suburban development echo Thompson's observations (1982, p. 2) that "there were suburbs long before the nineteenth century in the sense of places beyond city limits, the outskirts of towns hanging on to the central area physically and economically." He has pointed to the increasing desirability of suburban residences in London from the middle of the eighteenth century. Nevertheless, in Dublin as in London, it was not until the early-nineteenth century that modern suburban development got underway on a significant scale.

Dublin's Victorian suburban experience

By the end of the eighteenth century, Dublin was one of the premier European cities, frequently claimed as the second city of the British Empire. Its economic success related to its position both within Ireland and the colonial context. Dublin was Ireland's leading port, a national center of distribution, location of the higher law courts, financial services and higher education, as well as a political and social center due to the presence of a vice-regal court and national legislature (Dickson, 2001). Political, economic, and social change following the Act of Union (1801), which transferred power from the Dublin-based parliament to Westminster and thereby reduced the city's status, was to have ongoing repercussions for both city and suburbs, discussed later. The nineteenth century saw major expansion, as the city's population increased by over 43 percent and new middle-class residential suburbs were established (Daly, 1984). From about mid-century, the city began to experience the same impetus towards suburbanization that characterized most European cities at this time. The suburban trend was generally associated with the increasing separation of home and workplace, the expansion of the middle classes, the desire to move from the increasingly polluted and impoverished industrial city, advances in transport technology that made commuting more feasible, and the ongoing speculative development process, facilitated by mass production in the building industry. In the case of Dublin, these common processes played out against a very particular socio-religious and political backdrop, which gave Dublin's suburbs a unique character.

Dublin remained a small, relatively compact city well into the nineteenth century. The city was almost entirely enclosed by the North and South Circular Roads, which were bordered for much of their 9-mile length by two canals (the Royal, to the north, and the Grand, to the south). This enclosed an area slightly over three miles long by two miles wide, comprising 3,800 acres (O'Brien, 1982). Early-nineteenth-century maps reveal ribbon development along the main routes out of the city, but even by mid-century Dublin's exclusive suburbs were no more than one or two miles from the city center. The opening of the railway from Dublin to Kingstown (now Dún Laoghaire) in 1834 facilitated the creation of coastal suburbs at picturesque locations in Dalkey, Killiney, Blackrock, and exclusive Monkstown, although many of the fine villas erected there were occupied only seasonally (see Figure 7.2).

Dublin's suburban exodus resulted in increasing social and spatial segregation, as single-class residential areas emerged, particularly to the south of the Grand Canal in Ballsbridge, Donnybrook, Rathmines, and Rathgar. The available research suggests that growth in the city area reversed in the 1830s, while the suburban population grew by one-fifth between 1821 and 1841. This suburban growth was most marked along the coastline to the southeast, with a population increase of just under 44 percent in the area that embraced Pembroke, Blackrock, and Kingstown (Dickson, 2001). Indeed, bourgeois Dublin continued to relocate outside the municipal boundaries, with a particular flight from the older parts of the city leading to dereliction in the Liberties area. As the better-off residents moved to the suburbs, vacating their single-family residences, impoverished rural to urban migrants who continued to stream into the city re-occupied these now-subdivided houses, known as tenements (Prunty, 2001). Over time the city became increasingly associated with poverty and slum conditions, while the upper echelons of society who resided elsewhere could, in the words of Engels (1845), avoid the "misery and grime which form the complement of their wealth."

Figure 7.2   Seaside villas of Dublin

Source: Ruth McManus

Dublin's new suburbs in the Victorian era were constituted as independently governed entities, known as "townships," through the enactment of private legislation in Westminster. Thus, middle-class suburbanization brought about municipal independence and had financial as well as territorial implications. Suburban residents could avoid the disrepair, pollution, and increasing dereliction of the city, and enjoy healthier surroundings away from its unhealthy population, while also avoiding the financial costs of municipal taxation that supported workhouses, hospitals, and police. Although the nature and style of development, and the avoidance of what was perceived as an increasingly unsavory central city, were familiar to other cities, the continued political independence of Dublin's suburbs was unique, for reasons described as follows.

Social class, religion, and political persuasion

The interaction of class, religion, and politics had a very direct bearing on the development of Dublin's suburbs. After Catholic Emancipation (1829) and legislation in 1840 that allowed a wider franchise, power shifted in Dublin Corporation. Protestant big business members were becoming a minority, while the increase in the Catholic middle class was reflected in the composition of the Corporation, which became more liberal and nationalist (Barry, 2011). After 1841, when Nationalist leader Daniel O'Connell became the first Roman Catholic lord mayor since the 1600s, there was an additional impetus to suburbanization, with many Protestant and Unionist citizens choosing to find refuge in the independently governed Tory suburban townships to the south of the city (Daly, 1984; Dickson, 2001; Ó Maitiú, 2003). The city was thereby deprived of their financial backing (through the system of rates) as well as their political influence. In some ways, the decision to avoid the jurisdiction of Dublin

Corporation resulted in a self-selecting system of "apartheid" among the Unionist and Protestant middle classes. There was a perception that the suburbs would offer a "well-ordered genteel environment away from the slums and the perceived unresponsive and high-taxing Corporation" (Barry, 2011, p. 50).

Ó Maitiú (2003, p. 15) highlighted the "complex socio-religious gulf between city and suburbs." While middle-class flight to the suburbs was common across cities of the United Kingdom during the nineteenth century, he points to the religious and political complexion of the social gulf between those left behind in the neglected city center and the businessmen, officials, and clerks who moved to Dublin's suburbs. In particular, the two southern "townships" of Rathmines and Pembroke provided fine residential environments where the "Protestant and unionist Dublin middle-class could evade the unpleasant reality that they were a minority which was increasingly losing political control in both Ireland and in the city of Dublin" (Daly, 1984, p. 123). In Dublin, suburban disconnection from the city was, significantly, not simply a response to changing residential fashions, but also a deliberate choice to disengage socially and politically from a hostile city. "Within a ten mile radius, eight town halls represented their different communities" (Wallace, 2012, p. 934).

The speculative development process

The landscape impact of suburbanization was variously interpreted positively, as progress, or negatively, as despoilization of the countryside. In 1859, the newly established *Dublin Builder* trade newspaper described the rapid transformation occurring at the city's edges where

> we find the same steady march in the path of improvement . . . where villas, single, and semi-detached, terraces &c., are springing up with an almost fairy-like rapidity, and the green sward speedily gives way to macadamized roads and populated thoroughfares, justifying the supposition that there is a universal move in that direction.
>
> *(Dublin Builder, 1 Nov 1859)*

The suburban development process, as outlined by a witness to an 1885 housing inquiry, involved small-scale speculative operators, many of them new to the building trade (see Figure 7.3).

> It is not large capitalists that build these houses, but small builders and enterprising industrious artisans. . . . When they have a little money, they come to me and get a plot of ground, and I get them a loan of money; and they build houses; and then, when they are built, they are sold to small capitalists, also men with savings of £400 or £500, grocers and butchers, and that class of people.
>
> *(Comment by Edward McMahon in Third Report of Her Majesty's Commissioners for Inquiring into the Housing of the Working Classes Ireland, 1885, p. 97, para. 24,603; see Great Britain, 1885)*

The development of Dublin's Victorian suburbs was broadly similar to that described by Dyos (1961) for London's Camberwell. While smaller suburban dwellings retained the urban terraced form, a new suburban-style of house form also emerged during this era. Where wider plots and greater setbacks were available, houses were separated into pairs, resulting in a semi-detached building model with gardens front and rear (Galavan, 2013). This new house style was to become the standard preferred model for Irish suburban residents to the present day (McManus, 2011).

Figure 7.3   House variety on Hollybank Road, Drumcondra, built in the late nineteenth century

Source: Ruth McManus

Specific suburbs: the various township experiences

The eight independent suburbs of Dublin each had a unique character. To the south, the neighboring suburbs of Rathmines (founded 1847) and Pembroke (founded 1863) between them had about half the total suburban population, whereas the northern townships of Drumcondra (founded 1878) and Clontarf (founded 1869) never reached critical mass. New Kilmainham was the only working-class township. The railway-dependent townships of Blackrock, Kingstown, Dalkey, and Killiney were far beyond the city and remained relatively secluded with low population growth.

Pembroke township, to the southeast of Dublin city center, comprised almost 1600 acres to the coast, including existing fishing villages at Ringsend and Irishtown, and south to Ballsbridge, Donnybrook, and Merrion. Much of the land (seven-ninths according to the Exham Commission in 1881) was owned by the paternalistic Fitzwilliam/Pembroke family, whose agent was an ex-officio commissioner. Development was strictly controlled through the leasehold system, ensuring that the area generally retained a high-status profile. The estate subsidized the costs of the town hall, main drainage, technical school, and parks. Pembroke attracted professional and wealthier classes, and it had a substantial Protestant population. It maintained its independence until 1930 when it was incorporated into the city along with neighboring Rathmines. The high quality of the built fabric with its substantial houses on tree-lined roads has ensured that the area has retained its social cachet into the present day.

Rathmines became a township in 1847 and was extended on three subsequent occasions, so that by 1880 it included Rathgar and Milltown, with a land area of over 1,700 acres. Development was organized along purely speculative lines, as detailed by Daly (1984), and catered to the wealthier section of society. Individuals laid out roads and prepared sites that were then leased in small units. This small-scale and piecemeal development process resulted in a varied landscape with a wide assortment of house styles and streets that took years or even decades to complete (Brady, 2001). The township commissioners were members of the building trades or professional men who favored low property taxes (rates) to encourage development at the expense of services. The city's first tramline in 1872 ran to Rathmines from the city center and seems certain to have spurred development. Toward the end of the century, development began to focus on smaller houses at higher densities, in recognition of the growing demand for suburban dwellings among the clerical population of the city, who were also being catered to by the Drumcondra township.

The relatively high-status areas established at Pembroke and Rathmines gradually expanded outwards over time as the built-up area grew in size. However, they retained their status as they grew, so that in the present day, the legacy of the township era remains in the social geography of the higher status "south side" (predominantly Dublin's southeastern suburbs). To the north side of the city, middle-class enclaves did not develop to the same extent during the nineteenth century. A combination of reasons has been enumerated (Ó Gráda, 2015; Brady, 2001; Lennon, 2014) including the prevalence of toll roads, changing fashions favoring the south side, the rundown nature of the city's northern side, the prevalence of institutional land use and the insalubrious sloblands at Fairview. Both Drumcondra and Clontarf, to the north and northeast of the city, respectively, developed later and failed to achieve the degree of critical mass evident in the southern townships. Slow, incremental development in a piecemeal fashion was once again the norm, as seen in the work of particular builders such as Alexander Strain (McManus, 2002). Coastal Clontarf, which was largely controlled by the Vernon family, similar to the role played by the Pembroke estate in the township of the same name, grew slowly. Amenities were limited and the challenges to provide drainage to this extensive area, given its low base of rate-payers, were ultimately used as the definitive argument in bringing the township under city control in 1900.

Independence or annexation?

By 1899, it was being argued that Dublin was the only major city in Britain and Ireland that had not had its area extended during the previous half-century (O'Brien, 1982). The local government system confined an increasingly impoverished city within outdated boundaries, while most of the better-off, with their higher rate-paying capacity, lived in independently governed locations yet within easy reach of the city (Brady, 2001). The inequity arose from the fact that these suburbanites, over 45,000 of whom came to the city daily for work, shopping, and business, could also enjoy the many services and facilities of the city without paying for them. The valuation of the city would increase by 50 percent (£350,000) if the townships were incorporated. Increasingly, arguments of efficiency and economy were used to demand a centralization of municipal services in the greater Dublin area. Already in the early 1880s the Exham Commission (the Municipal Boundaries Commission [Ireland]) had endorsed the Corporation's proposal to annex the townships to the city proper, rejecting the claims of those "who enjoy all the advantages and escape many of the disadvantages of the citizens" (quoted in O'Brien, 1982, p. 15). The persistence of a fragmented suburban governance system denied Dublin Corporation badly needed revenue to address poverty and poor housing in the city area, while also preventing holistic management of development. Nevertheless, opposition to annexation was fierce, as seen in the evidence given to subsequent inquiries in 1899 and 1926 (Brady, 2014). The less powerful townships of Drumcondra, Clontarf, and Kilmainham were absorbed along with part of the county area in 1900, doubling the spatial extent of the city from 3,733 to 7,911 acres and adding 30,000 people, but securing a far smaller income potential than that of the surviving independent entities of Rathmines and Pembroke. While there was a general distrust of Dublin Corporation and its inefficiencies, as Wallace (2012) demonstrated, resistance to annexation in Clontarf, Rathmines, and Pembroke rested on a principle of Unionist self-government. In preserving their independence, the latter two suburbs continued to express their residents' British and middle-class identity. Whereas the Free State government dissolved the long-time Nationalist Dublin Municipal Council in 1925, Rathmines and Pembroke survived until 1930, perhaps offering some comfort and reassurance to suburban Unionists at a time of unprecedented social and political upheaval (Wallace, 2012).

Twentieth-century mass suburbanization and its implications

Dublin evolved from a still-compact city at the start of the twentieth century to a hugely expanded suburbanized and globalized city-region by its end (Horner, 1999). In the nineteenth century, Dublin's suburbs had generally been the preserve of the better-off and their servants, but during the twentieth century suburbs were provided for everyone (McManus, 2002, 2003). Suburban rehousing of the working classes was largely a response to the processes that had played out in the previous century, whereby the middle classes had left the city and its worsening slums. It can also be understood in the context, from 1922, of a newly independent state that promoted home ownership and improved housing quality. Dublin's suburbs had an uneasy relationship with a planning system largely adopted from the UK and never fully accepted.

The seeds for Dublin's mass suburbanization in the twentieth century had been sown in the nineteenth century. By the early decades of the twentieth century, there was an acknowledged need to address the appalling conditions of the working classes. Many of these residents resided in tenements and suffered from overcrowded, subdivided decaying homes of the former aristocracy. This process became associated with the modern town planning movement (Bannon, 1985). The garden city influence was strong, with Patrick Abercrombie's competition-winning plan for the city (1914, published 1922) demonstrating the potential for low-density housing (at a suburban norm of 12 houses per acre) to radically alter the population's living arrangements. The competition took place against the backdrop of a housing inquiry undertaken in October 1913, in response to the public outcry following a tenement collapse in which seven people died. The Inquiry's findings left no doubt as to the dire situation. Almost 14,000 houses were urgently required and, given the limited availability of sites and perceived desirability of a fresh start in healthy surroundings, the commissioners recommended that these be located in suburban areas. Opponents, including some members of the Dublin Corporation Housing Committee, argued strongly that many city center residents would not wish to leave their communities or afford the additional transport costs to their work. Ultimately, however, the construction of suburban housing for the working classes became a pillar of the city's housing policy. Indeed, central area flat (apartment) complexes were provided only for those who could not afford the additional costs involved in living in the suburbs. One result of the policy was increased segregation among the working classes. In the longer term, the decentralization of the population was followed by suburbanization of business, retailing, and industry by the second half of the century, with negative effects for the inner-city.

Dublin suburbs after national independence (1922)

At the beginning of the twentieth century, despite the developments of the nineteenth century and the social, political, and psychological importance of the suburbs to Dubliners, the city remained relatively compact. In 1911, almost 88 percent of Dublin county's urban population lived within four miles of the city center; this only changed significantly after 1936 (Horner, 1999). The shift toward lower-density housing for all classes was linked to the perceived importance of the modern town planning movement, itself largely associated with the garden city and garden suburb idea, in providing a solution to the slum problem.

The decision to build garden suburbs from the 1920s reflected the spirit of the age and Dublin's involvement in broader international trends (Kincaid, 2006). A further impetus to construct high quality suburbs for Dublin's working classes came with national independence in 1922. The Marino garden suburb was clearly intended as a model for others to follow, but also as a signal that the new Irish Free State would do better than the previous administration in tackling its citizens' housing problems (McManus, 2002). New housing schemes had the added benefit of providing much-needed employment in construction. As low-density suburban housing was favored, it is not surprising that the built-up area of the city began to grow significantly (Horner, 1999). In the working-class suburbs built by Dublin Corporation in the 1920s, home ownership was being promoted for the first-time through a tenant-purchase system. Kincaid (2006) suggests that both politicians and planners of the new state were middle class in their outlook and saw suburbs as capable of promoting values such as property ownership, self-reliance, and financial frugality among the working-class population. Certainly, the government preference for suburban housing and owner occupation promoted a stable society and responsible citizens (McManus, 2003).

Social segregation in twentieth-century suburbs

By mid-century, the decanting of the population from central areas to newly emerging suburbs was an established practice, not just for the middle classes but also for all members of society. This mass suburbanization did not necessarily imply increased social mixing. In general, the private speculatively developed middle-class suburbs avoided the location of new social housing suburbs provided by the city authorities, although some limited social mixing arose through the application of a "reserved areas" policy typically at the edges of the new Corporation "schemes" (McManus, 2002). To the city's south, where the pattern of middle-class respectability had already been largely laid down in the nineteenth century, there was limited scope for the city authorities to build large estates, although social housing was built at Crumlin and Kimmage to the southwest of the city in the 1930s and 1940s. By contrast, the picture to the city's north was less clearcut (Brady, 2016). Here there had been more limited development and therefore greater tracts of land remained available for construction. The social geography became much more mixed, with smaller pockets of middle-class housing interspersed with local authority (Dublin Corporation) housing estates, a pattern that has persisted to the present.

By the 1960s, suburban housing dominated the capital. The residual city center population was increasingly elderly, socially excluded, unemployed, and impoverished, residing in poor quality, privately rented accommodation or in the local authority flat complexes that had been a feature of the 1930s as well as the 1950s and 1960s. Like many Western cities, economic forces conspired with the existing centrifugal forces to cause a hollowing out of the central population.

Fifty years from Dublin's planned "new towns" to exurban sprawl

Despite the early-twentieth-century interest in town planning associated with the garden suburb ideal and limited application of planning legislation in the 1930s, urban planning and development control were relatively weak in Ireland until the introduction of the 1963 Planning Act. The newly independent state remained relatively rural (the tipping point where over 50 percent of the population was urbanized was not reached until the 1970s), with a weak economy, where construction was generally viewed positively because of its job creation potential. Furthermore, the Irish constitution of 1937 generally promoted property rights, going so far as to guarantee "to pass no law attempting to abolish the right of private ownership or the general right to transfer, bequeath, and inherit property" (Article 43.1.2). The overwhelming support for private property, at the expense of the common good, can perhaps be attributed to Ireland's post-colonial sensibilities. Whatever the cause, it had negative implications for development control.

While emigration remained a feature of Irish life until the 1990s, Dublin's population continued to grow throughout the twentieth century because of rural to urban migration. Many new suburbanites therefore had rural roots and both desired and expected to live in a house with its own garden. A 1967 study remarked on the "obsession with the land" that saw a preference for single-family houses with gardens (see McManus, 2012). The attractions of a pastoral ideology remain important in the constitution of new suburbs into the twenty-first century, helping to inculcate a sense of place (Corcoran, 2010). The expectation of home ownership was also a factor, associated with the land struggle and thus another legacy of Ireland's colonial experience, which probably contributed to the typical suburban-style. Until the 1970s, the sale of private flats (apartments) to tenants or availability of mortgages for the same was problematic (McManus, 2012).

By the time that effective town plans were being introduced in the late 1960s, the suburban housing preference was well-established. Under the Myles Wright advisory report and development plan for the Dublin region published in 1967, growth would now be directed into four (later three) "new towns" to the west of the city, Tallaght, Blanchardstown, Lucan, and Clondalkin. However, given their scale and distance from the capital, these "new towns" (based around existing settlements) never became truly independent of the city, as was the case with their British antecedents. Instead, as implemented in the 1970 Dublin City Development Plan, the outcome can be seen as an acceleration of suburbanizing tendencies. Again, the characteristic suburban-style housing with its low densities, single-family houses of a standard type, with gardens front and rear, remained the favored type. This preference also hampered the provision of services, particularly public transport, contributing to sprawl and social isolation for non-car owners (MacLaran, 1993).

The impact of the Wright plan was effectively the intensification of the existing suburbanization processes. Inner-city residents were rehoused in single-class local authority housing schemes in the "new towns" (effectively suburbs) of Blanchardstown, Tallaght, and Lucan/Clondalkin. As a result, the socio-demographic structure of the inner area changed dramatically. Its population was already in decline, but now there was a reduction in families with children, a disproportionate concentration of the elderly and those in the lowest socioeconomic groups, and relatively few owner-occupiers. Overall, the population of the area within the canals/circular roads dropped by more than half between 1961 and 1991, to just 76,558 (see Census 1991, Vol. 1, Table 13 in Central Statistics Office, 1993).

A range of studies have charted Irish population growth and demonstrated Dublin's ongoing magnetic effect as a primate city from the 1960s onward, despite some government attempts at regional dispersal (Horner, 1999; Lutz, 2001). Although Dublin remains a major focus of population (about one-third of the national population) as well as economic opportunity, increasingly both have been suburbanized within an ever-expanding greater Dublin region. The 1970s witnessed five- and six-fold population increases in many places beyond the main built-up area, for example at Portmarnock, Malahide, Leixlip, and Celbridge, where suburban housing estates were built for commuters. Relatively slow change due to economic recession and high emigration in the 1980s was followed by major transformations in the 1990s as the population grew and there was a significant shift to smaller household sizes. For the first-time in the 1990s, urban renewal policies targeting the inner-city led to an increase in population of the city center, reversing the trend of decline evident since 1936. At the same, private developers began to construct city center apartment blocks aimed at the middle classes. Nevertheless, research has shown that most apartment dwellers in Dublin aspire to future suburban residence, while the scale of urban renewal-fueled central development was not sufficient to counter ever-increasing suburban centrifugal movement. Indeed, by the turn of the millennium, a series of studies and reports demonstrated that Dublin's position as a primate city within Ireland had strengthened, with significant impacts on its rapidly suburbanizing and ever-enlarging hinterland (Department of the Environment and Local Government, 2000, Horner, 1999). Although the city's innovative urban renewal policies were viewed internationally as being relatively successful, inadequate management of peripheral development was resulting in an acceleration of suburban sprawl (Williams and Shiels, 2000).

During the 1990s, it became evident that Dublin-related suburban-type development was incorporating even more distant locations into a functional relationship with the city, requiring ever-longer commutes for workers. New suburban-style housing estates generally replicated the building type found in Dublin's suburbs, but were located at a greater remove from the capital, including the outskirts of smaller villages that had not previously experienced large amounts of construction. Examples of the former include "Dublin suburbs" in Rochfortbridge, Co. Westmeath and Gorey, Co. Wexford, while the latter can be found at Clonard and Robinstown (Meath), Prosperous (Kildare), Baltinglass (Wicklow) and Collon (Louth). Corcoran (2010) has explained some of this development on the "leading edge of the suburbs" as relating to the new residents' aspirations for an idealized rural lifestyle (see also McGrath, 2013).

The most recent phase of suburbanization and exurban development around Dublin has, perhaps, cemented the city's position as a typical Western globalized metropolis, with all the challenges that entails. In their 2007 report for the Society of Chartered Surveyors, Williams et al. demonstrated the increasing outward growth of the commuter belt, now stretching over 60 miles (100 kilometers) from the capital, contrary to official national and regional planning policy objectives, creating sprawl-type settlement with a near-total dependency on the private motorcar. They argued, "the absence of integration of housing, land use and transportation policies, poses major problems for accessibility, sustainability, and quality of life" (Williams et al., 2007, p. 4). Relatively modern suburbs with an aging profile have been experiencing population losses, resulting in potential under-utilization of social infrastructure, although some existing suburbs served by public transport, such as Dundrum, have begun to experience a population resurgence associated with new apartment development. However, the most recent suburban housing constructed during Ireland's economic boom (c. 1995–2008) is inherently unsustainable, with the designs of newer developments encouraging car dependency (Caulfield and Ahern, 2014).

Despite the challenges outlined earlier, and the frequently negative portrayal of suburban life in the media and in academia, an important comparative study of Dublin's recently constructed suburbs has revealed the resilience of new suburban communities and the degree to which new suburban affiliations become established (Corcoran et al., 2010). In the same way that the mass suburbanization of Dublin in the twentieth century can be seen as an almost inevitable response to the slum problems of the nineteenth, so too can the present day issues of sprawl and long-distance commuting trace their lineage back to decisions made almost a century before. The provision of single-family suburban dwellings for all social classes inevitably led to rapid expansion of the city, increasing outward pressure and demand for land. Although contemporary commentators generally portray suburban sprawl in a negative light, the majority of Dublin's citizens still choose to replicate their own childhood experience by bringing up their families in suburban surroundings.

Guide to further reading

Brady, J. (2016) *Dublin 1950–1970, Houses, Flats and High-Rise*. Dublin: Four Courts Press.
Brady, J. and Simms, A. (eds.) (2001) *Dublin through Space and Time, c. 900–1900*. Dublin: Four Courts Press.
Corcoran, M.P., Gray, J., and Peillon, M. (2010) *Suburban Affiliations: Social Relations in the Greater Dublin Area*. New York: Syracuse University Press.
Daly, M.E. (1984) *Dublin, the Deposed Capital*. Cork: Cork University Press.
McManus, R. (2002) *Dublin 1910–1940: Shaping the City and Suburbs*. Dublin: Four Courts Press.
Ó Maitiú, S. (2003) *Dublin's Suburban Towns 1834–1930*. Dublin: Four Courts Press.
Wallace, C. (2012) "Fighting for Unionist Home Rule: Competing Identities in Dublin 1880–1929." *Journal of Urban History*, 38(5), pp. 932–949.

References

Bannon, M.J. (ed.) (1985) *The Emergence of Irish Planning, 1880–1920*. Dublin: Turoe Press.
Barry, M. (2011) *Victorian Dublin Revealed, the Remarkable Legacy of Nineteenth-Century Dublin*. Dublin: Andalus Press.
Brady, J. (2001) "Dublin at the Turn of the Century." In: Brady, J. and Simms, A. (eds.) *Dublin through Space and Time*. Dublin: Four Courts Press, pp. 221–281.
Brady, J. (2014) *Dublin 1930–1950, the Emergence of the Modern City*. Dublin: Four Courts Press.
Brady, J. (2016) *Dublin 1950–1970, Houses, Flats and High-Rise*. Dublin: Four Courts Press.
Caulfield, B. and Ahern, A. (2014) "The Green Fields of Ireland: The Legacy of Dublin's Housing Boom and the Impact on Commuting." *Case Studies on Transport Policy*, 2(1), pp. 20–27.
Central Statistics Office (1993) *Census 1991 Volume 1 Population Classified by Area*. Dublin: Government of Ireland.
Clarke, H. (1998) "*Urbs et suburbium*: Beyond the Walls of Medieval Dublin." In: Manning, C. (ed.) *Dublin and beyond the Pale: Studies in Honour of Patrick Healy*. Bray: Wordwell, pp. 45–58.
Corcoran, M.P. (2010) "'God's Golden Acre for Children': Pastoralism and Sense of Place in New Suburban Communities." *Urban Studies*, 47(2), pp. 2537–2554.
Corcoran, M.P., Gray, J., and Peillon, M. (2010) *Suburban Affiliations: Social Relations in the Greater Dublin Area*. New York: Syracuse University Press.
Daly, M.E. (1984) *Dublin, the Deposed Capital*. Cork: Cork University Press.
Department of the Environment and Local Government, Spatial Planning Unit, National Spatial Strategy (2000) "The Irish Urban System and Its Dynamics." (PDF). Available at: http://nss.ie/docs/report8.pdf [Accessed 28 April 2017].
Dickson, D. (2001) "Death of a Capital? Dublin and the Consequences of Union." In: Clark, P. and Gillespie, R. (eds.) *Two Capitals: London and Dublin, 1500–1840*. Oxford: Oxford University Press, pp. 111–131.
Dyos, H.J. (1961) *Victorian Suburb: A Study of the Growth of Camberwell*. Leicester: Leicester University Press.
Galavan, S. (2013) "Building Victorian Dublin: Meade & Son and the Expansion of the City." In: O'Neill, C. (ed.) *Irish Elites in the Nineteenth Century*. Dublin: Four Courts Press, pp. 51–67.
Great Britain (1885) *Third Report of Her Majesty's Commissioners for Inquiring into the Housing of the Working Classes Ireland*. London: H.M. Stationary Office.
Horner, A. (1999) "Population Dispersion and Development in a Changing City-Region." In: Killen, J. and MacLaran, A. (eds.) *Dublin: Contemporary Trends and Issues for the Twenty-First Century*. Vol. 11. Dublin: Geographical Society of Ireland, Special Publication, pp. 55–68.
Kincaid, A. (2006) *Postcolonial Dublin, Imperial Legacies and the Built Environment*. Minneapolis, MN: University of Minnesota Press.
Lennon, C. (2014) *That Field of Glory: The Story of Clontarf from Battlefield to Garden Suburb*. Bray: Wordwell.
Lutz, J. (2001) "Determinants of Population Growth in Urban Centers in the Republic of Ireland." *Urban Studies*, 38(8), pp. 1329–1340.
MacLaran, A. (1993) *Dublin: The Shaping of a Capital*. London: Belhaven Press.
McGrath, B. (2013) *Landscape and Society in Contemporary Ireland*. Cork: Cork University Press.
McManus, R. (2002) *Dublin 1910–1940: Shaping the City and Suburbs*. Dublin: Four Courts Press.
McManus, R. (2003) "Blue Collars, 'Red Forts', and Green Fields: Working-Class Housing in Ireland in the Twentieth Century." *International Labor and Working Class History*, 64(Fall), pp. 38–54.
McManus, R. (2011) "Suburban and Urban Housing in the Twentieth Century." *Proceedings of the Royal Irish Academy, Section C*, 3, pp. 253–286.

McManus, R. (2012) "'Decent and Artistic Homes': Housing Dublin's Middle Classes in the 20th Century." *Dublin Historical Record*, 65(1), pp. 96–109.

O'Brien, J.V. (1982) *Dear, Dirty Dublin: A City in Distress, 1899–1916*. Berkeley, CA: University of California Press.

Ó Gráda, D. (2015) *Georgian Dublin: The Forces that Shaped the City*. Cork: Cork University Press.

Ó Maitiú, S. (2003) *Dublin's Suburban Towns 1834–1930*. Dublin: Four Courts Press.

Prunty, J. (2001) "Improving the Urban Environment." In: Brady, J. and Simms, A. (eds.) *Dublin through Space and Time*. Dublin: Four Courts Press.

Sheridan, E. (2001) "Designing the Capital." In: Brady, J. and Simms, A. (eds.) *Dublin through Space and Time*. Dublin: Four Courts Press.

Simms, A. (2001) "Origins and Growth." In: Brady, J. and Simms, A. (eds.) *Dublin through Space and Time*. Dublin: Four Courts Press, pp. 15–65.

Thompson, F.M.L. (ed.) (1982) *The Rise of Suburbia*. Leicester: Leicester University Press and St Martin's Press.

Wallace, C. (2012) "Fighting for Unionist Home Rule: Competing Identities in Dublin 1880–1929." *Journal of Urban History*, 38(5), pp. 932–949.

Williams, B., Hughes, B., and Shiels, P. (2007) *Urban Sprawl and Market Fragmentation in the Greater Dublin Area*. Dublin: Society of Chartered Surveyors.

Williams, B. and Shiels, P. (2000) "Acceleration into Sprawl: Causes and Potential Policy Responses." *Quarterly Economic Commentary*, ESRI, June, pp. 37–62.

8

Suburbanization in Asia

A focus on India

Annapurna Shaw

Introduction

For a meaningful discussion of suburbanization in India, it is necessary to first clarify the meaning of "suburbanization" as a process and of "suburbs" as socio-spatial entities in the context of this developing country. Until the 1950s, suburbanization as a nationwide process of the city's outward growth, engulfing surrounding towns and villages, was not a common phenomenon. With the exception of the largest colonial port cities of Mumbai, Kolkata, and Chennai, the city's growth remained largely confined within municipal boundaries (Brush, 1968, 1977). However, "suburbs" as distinct and exclusionary socio-spatial entities have a much longer history. Some of the earliest mentions of the suburbs' existence come from historians of medieval and colonial India and offer important perspectives. In fact, the term "suburb," meaning an outlying residential area, occurs more frequently in the historical and architectural literature rather than in the geographical and planning literature. Geographers and planners, studying more contemporary times, have preferred the terminology "rural-urban fringe" or "peri-urban area" to "suburbs," highlighting some intrinsic differences between the way these regions outside the central city are formed and governed as compared to the United States (Jargowsky, 2005). Suburbs in the United States, as initial developments, are planned residential areas of single-family homes, generally for middle- and upper-income classes, located outside the central city with political autonomy and taxation powers. Over time, they can attract office parks and "clean" industries and become alternate employment hubs as well. In contrast, the peri-urban area in India is more heterogeneous in structure, density, and socioeconomic status. Stretches of it could be planned with single-family homes of the middle and upper classes, with gated residential complexes and multi-story housing. In contrast, other stretches could house the poor in squatter settlements and informal housing interspersed with commercial and industrial activities of both the formal and the informal economy. Politically, older suburbs could be governed by the central city, as in the case of Greater Mumbai, or there could be a large central city with suburbs as separate political entities in its surrounding areas. The Kolkata Metropolitan Area, with the Kolkata Municipal Corporation as its central city and 42 other smaller cities located within the metropolitan area, exemplifies the second type of political arrangement. Some of the newer cities, such as Gurgaon, originally started as suburbs of a central city, but they have been given separate city status because of their dynamic growth. Thus, a variety of political arrangements exist along with complex and continually evolving spatial, economic, and social structures in India's peri-urban areas.

While this discussion clearly indicates that the type of spread in Indian cities is different from that of the United States, "how periurbanization differs from conventional notions of suburbanization, for one thing, is unresolved" (Leaf, 2016, p. 130). Whether it is a difference in the suburbs of the Global North and Global South, characterizing the developed world and periurbanization a characteristic of the still-developing world, or whether suburbanization is a subset of the larger and more common process of periurbanization, remains to be explored further. In this chapter, I am using the terms interchangeably.

History of suburbanization in India

Suburbanization as a nationwide process affecting large parts of the urban system began after the 1950s; however, its roots are quite old and traceable to the overspill of population experienced by northern India's larger cities in the Mughal era, or the period 1500 to 1750 (Hambly, 1984). It was a period of urban prosperity and expansion, with the larger cities tending to outgrow their walls, often with a ring of encircling suburbs around the original core. Such suburbs usually grew around a local shrine or a sacred site, a cluster of noblemen's houses and gardens, or a military encampment. These early suburbs did not leave a lasting imprint on the built environment. After the second half of the eighteenth century, the increasing presence of European traders, and the gradual dominance of the East India Company hastened the formation of new kinds of suburbs. Thus, in the city of Madras, today known as Chennai, where the British founded a fort in 1640, according to Brown (2003, p. 156):

> the city itself expanded beyond the initial kernel of the black town and the fort, forming suburban areas fairly early in its existence. These suburbs served as a site for the construction of large bungalows and garden estates for newly wealthy European merchants.

This "eighteenth century suburban flight" manifested in the other port cities of Bombay (now Mumbai), Calcutta (now Kolkata), as well as in provincial cities such as Patna.

In the second half of the nineteenth century, two developments led to the formation of new kinds of suburbs. The first was the Mutiny or Revolt of 1857, which made security an overriding concern for the British. As a result, residential areas became segregated from the local population (Metcalf, 1998). The second was the introduction of the railways from the 1860s onward, which resulted in the addition of a new railway suburb to the towns through which it passed (Grewal, 1991). However, these railway suburbs remained small and did not significantly change the pre-existing layout of towns, unlike the larger suburbs or "civil stations" that grew outside indigenous settlements as separate colonies to house British officials and professional people. Likewise, cantonments developed as small self-contained suburban settlements catering to the military. Both the civil stations and the cantonments had a low population density and were located at some distance from the indigenous urban settlement. With their gridiron pattern, well laid out bungalows and roads, and a high level of civic amenities, they provided a stark contrast to the congested and crowded older urban core. A dualistic urban structure, as characterizing many ex-colonial countries in Asia and Africa, of a black town along with its planned white settlement, was to become a common feature across India by the end of the nineteenth century.

During the late colonial period of the twentieth century, suburban development continued to occur, but for different reasons than the imperial need for separation from the local population. After decades of stagnation, as the rate of urban growth in India began to increase in the 1920s,

> inner cities became more crowded and congested while there was also considerable suburbanization, particularly between 1920 and 1940, as middle-class Indians started to move out to the suburbs in large numbers. Cooperative housing societies, independent bungalows, and mansions of business people and industrialists located on plots of several hectares, began to be built on the outskirts of cities.
>
> *(Lang et al., 1997, p. 155)*

However, this occurred primarily in the largest cities. Elsewhere, even in large cities, infilling was the first response to rising population pressure. Even with the influx of around seven million refugees after independence and the partition of the sub-continent in 1947, infilling of existing urban cores occurred first as noted by Pachauri (1987, p. 28) in the case of Delhi: "The peripheral expansion of Delhi city up to 1951 was limited though the immigration was high, especially after the partition of India in 1947. The population accumulated within the city limits and the density of population increased."

Infilling led to the merger of the civil station or cantonment with the indigenous city, thereby gradually obliterating the duality in urban structure of black town-white town created during colonial times. It also increased population density, causing a population shift to the city's periphery and the start of full-scale suburbanization. Between 1950 and 1980, suburbanization in large cities such as Pune became well entrenched with the expansion of boundaries by the addition of fringe villages and small urban settlements, and the demarcation of metropolitan areas for the better planning and governance of an increasingly heterogeneous urban region (Shekhar and Jaymala, 2015). The liberalization of India's economy and its opening up in 1991 further encouraged the city's outward expansion, as new economy, high-tech, and service industries found peripheral areas more suitable sites than the central cities, with their old infrastructure and costly land (Shaw and Satish, 2007; World Bank, 2013).

Major policy and planning initiatives to influence suburbanization in India

The Indian state has undertaken a number of policy and planning initiatives to manage urbanization that has, often in unintended ways, impacted city growth. In the early years after independence, the most immediate need was the housing of over seven million refugees. It necessitated the creation of new residential layouts in cities such as Kolkata, Delhi, and Ambala, which led to extensions of the city and the formation of new suburbs. In addition, new towns, which were often satellites to older urban areas, were also built by the state to accommodate the newcomers (Shaw, 2009). Both these methods ultimately resulted in the extension of urban growth from the central city to surrounding areas (Shaw, 1995). A second major policy initiative, influenced by thinking in the rest of Asia in the 1960s and 1970s, was centered on slowing down the growth rate of the largest cities, as they were becoming too congested and growth needed to be redistributed towards other cities in the system (Rondinelli, 1991). Following this logic, new factories were encouraged to locate outside such cities and wherever possible, some pre-existing industries and commercial activities were to be relocated. The planning and development of the mainland across from Mumbai harbor and the creation of Navi Mumbai as a countermagnet to Greater Mumbai were direct outcomes of such thinking, and this included the relocation of the fruit and vegetable wholesale market and its associated truck terminal from central Mumbai to Vashi, in Navi Mumbai (Shaw, 2004).

After 1991, India, like the rest of Asia, gave up metropolitan city growth control measures for measures to strengthen the infrastructure of its largest central cities, recognizing their importance as "engines of growth" (Chakravorty, 1996; Shaw, 2012). The policy change was an acknowledgment that the largest cities' lateral spread continued unabated. A major driver of this was land unavailability in the central city, which forced businesses and new housing to establish locations outside city boundaries. Land unavailability itself was the result of the Indian state's restrictive land policies, highlighted in 1988 by the government's own expert commission, the National Commission for Urbanization, in its discussion of the difficulties of recycling or redeveloping land (GoI, 1988). This was compounded further by restrictions on the buildings' vertical heights via the low Floor Area Ratios (FAR)/Floor Space Index (FSI) permitted in Indian cities (Sridhar, 2010; World Bank, 2013). Comparing the Floor Space Index (FSI) in the central business districts of cities in Asia, United States, and Europe with those prevailing in India, the World Bank report (2013) noted that unlike much of East Asia, urban India's FSIs were very low. However, this has recently changed, as the latest Five-Year Plan endorsed the World Bank's recommendation of "strategic densification" or allowing higher vertical height along urban corridors or nodes where land value escalation is likely to take place (GoI, 2012, p. 329).

Another set of policies that have had an impact on suburbanization are those relating to housing finance. In the early decades following independence in 1947 until the end of the 1970s, India's housing market was weakly developed with no private banks disbursing long-term housing loans. The setting up of HUDCO (Housing and Urban Development Corporation) in 1971 and HDFC (Housing Development and Finance Corporation) in 1977 improved housing finance availability for the middle class. While HUDCO provided housing finance to public building agencies, HDFC provided long-term housing loans directly to homebuyers. After 1991, the housing finance market deepened considerably, with numerous private housing finance companies, microfinance companies, and commercial banks, both public and private, providing long-term housing loans. This led to a rush of home building on the peripheries of Indian cities where land is cheaper and more easily available.

Improvements in transportation infrastructure and services have also enabled the Indian city to spread. According to a multi-organizational comparative study of the peri-urban in India and Western countries financed by the European Commission, in India, "transport infrastructure is the primal factor influencing the process of PU (peri-urban) development" and "the speed of change is most explicit in corridors which combine high level roads and railway infrastructure" (Anna et al., 2006, p. 30). Outward population spread via such transport corridors has been followed by the shift of economic activities, for instance, manufacturing and retail to the peripheries of cities, followed by civic and transport improvements. The hectic widening and extension of roads, building of flyovers, and the laying of metro rail and BRT lines in major cities have been triggered by the rising population and increased demand for travel from both the central city and the periphery for the last two decades.

People who live in suburbs of India

Today, a heterogeneous mix of people lives on the peripheries of Indian cities, but this has not always been the case. Suburbs that evolved during the colonial period were primarily meant for the British, which later included local elites and were characterized by larger lots and bigger homes set at a distance from the older, indigenous urban settlements. In the early post-independence period, suburbs often originated as private housing developments for the upper-middle class. For example, this was the case of Valmiki Nagar in southern Chennai, where in the late 1950s and early 1960s, a group of elite bureaucrats working for the state government bought vacant land along the southern shoreline to set up their retirement homes (Arabindoo, 2009). Suburbs have also had their origins in a pre-existing village absorbed by the expansion of city boundaries. The demographic and social mix of these two kinds of suburbs is likely to be different. In the latter, the local population of farmers or fisher folk reside along with an incoming group of middle- and lower middle-class residents. The former, or Valmiki Nagar type of suburb, is likely to be more homogenous and upper-class, although mixed housing built by the state could have reduced such exclusivity in recent years (Arabindoo, 2009). Urban villages, on the other hand, with much poorer levels of basic amenities, can remain areas of largely informal building activities and cheap housing for the poor. This, together with the presence of environmentally degrading land uses such as landfills, sewage pumping stations and treatment plants, poorly built public housing for the resettlement of central city slum dwellers or other evicted populations, and new squatter clusters have led some scholars to regard the peri-urban area in Indian cities as a "degenerated periphery" (Kundu et al., 2002), that is, largely given to low-end and low-status uses.

Current and ongoing research, however, points to a more complex situation shaped by the forces of demographic growth, institutional practices, rising incomes, and globalization. A key point to keep in mind is that in a densely populated country such as India, there is no unoccupied land, even on the periphery, and the rapid outward spread of India's cities since independence has been made possible in a variety of ways. One of these is as planned colonies where, as per plan documents, the state builds new housing on land it owns or acquires on the urban periphery or the private sector buys a large tract of land, builds housing on it, and sells it on the market with legal titles. Another way is as unauthorized colonies, via land or housing provided by agents who did not get the requisite permits for land conversion and have illegally subdivided agricultural land and sold it to buyers for nonagricultural use. A third way is via squatting, or occupying land, that belongs to either the state or private owners. While the provision of housing via planned colonies accounts for a small proportion of total transactions in the housing sector, particularly for low-income groups (Sivan and Karuppannan, 2002), a common situation in much of peri-urban India is the existence of a "peripheralized middle-class" (Ranganathan, 2011). This term describes middle-class homeowners with tenuous legal tenure of the land on which their houses have been built. Due to this irregularity, these residents live in a state of uncertainty, making constant efforts to legitimize their property via recognition by the state. This "informal peri-urban middle-class" encompasses a significant portion of the urban population, as much as 40 to 70 percent of urban Indians falling outside the purview of formal planning institutions and mostly in the urban periphery (see Ranganathan, 2011). This class must be contrasted with the poor, or those squatting on land and whose tenure security is much worse, squatter settlements being regarded as "encroachments" that could be demolished by the authorities. In this matter, the state has historically acted in a biased way, demolishing the hutments of the poor but gradually regularizing middle- and lower middle-class illegal settlements, as seen in the case of Delhi's growth from 1947 to 2011 (Bhan, 2013).

A mix of social classes thus can be found in the outer fringes of the Indian city characterized by planned residential layouts as well as unplanned layouts and squatter settlements. These distinct kinds of residential uses can occur in close proximity, highlighting their economic interdependencies in terms of cheap labor for household work and related services for the middle- and upper middle-classes on the one hand, and employment in a variety of household tasks such as child care, driving the family, newspaper delivery, laundry, and security services, on the other, for the poorer inhabitants. They, and particularly the migrants among them, are most often renters, underlining the fact that unplanned layouts in the urban periphery are not only comprised of individual plots with single-family homes but increasingly also rental premises, as seen in

Hyderabad's outer areas (Sinha, 2018). These are created through individual homeowners extending their property to make additional rooms, which are rented out as well as through the building of single and multi-story rental housing. The former constitutes a class of "subsistence landlords" supplementing their income through rentals and living along with their renters, while the latter are a growing class of more capitalist landlords who run the rentals as a business and do not live in them (Naik, 2015). Naik's (2015) study of informal rental housing in the suburban city of Gurgaon reveals the wide range in rental housing typologies. These vary from *juggis*, or semi-permanent shacks, to *pukka*, permanent rooms with a separate toilet that cater to a range of needs: of those at the bottom of the economic ladder, such as waste pickers and vegetable vendors, to semi-skilled and skilled factory workers with higher job security and earnings. Poor light and ventilation, and multiple families sharing one toilet, are common characteristics across all types. Thus, urban villages and unauthorized colonies have become the leading suppliers of cheap rental housing to the low-income and low-income migrants, and the housing stock is of poor quality.

In stark contrast are the housing and amenities associated with the planned layouts of peripheral areas. Here both private capital and local state agencies have engaged in real estate development for the middle- and upper middle-classes, as well as for the upper classes. Such development has taken off in a big way since the liberalization of the Indian economy in 1991 and the gradual entry of foreign direct investment in real estate. As noted by Brosius (2010, p. 73), "a whole range of new pleasures and spaces have emerged since the 1990s." They are most visible in the satellite cities or suburbs of megacities such as Kolkata, Delhi, or Bangalore, and increasingly in smaller cities such as Lucknow, Jaipur, and Amritsar as well. Interestingly, along with increasing globalization of the real estate market, religion, and tradition are important markers at its upper end, reflecting "glocal influences" on the lifestyles of the affluent (Brosius, 2010).

Major conflicts, challenges, and problems with suburbanization in India

India's suburbanization or periurbanization is beset with a number of conflicts and challenges that emanate from the way the process is occurring, particularly its rapid pace since the 1980s, and its imbrication in larger global and national processes of change. State governments have attempted to control land use and the growth of built-up areas outside the largest cities through metropolitan planning and metropolitan regulatory authorities dating back to the 1960s. Despite these efforts, the peripheral areas of India's largest cities today are a complex mix of the planned and the unplanned co-existing in separate spaces but with myriad "entanglements" in everyday life (Srivastava, 2015). The modes of the creation of both the planned and the unplanned areas have been problematic; in the case of the former because it has generally involved acquiring land from farmers through forced acquisition by the state, and in the case of the latter, it has been through illegal subdivision of agricultural land or squatting on available land. "Entangled urbanism" (Srivastava, 2015) or the interconnections between shopping mall, slum, or "basti," and gated communities has caused both types of conflicts as well as, though less often, cooperation. The conflicts are of several kinds: first, those arising from activism by the "new" middle class to exclude others, namely, the poor, to preserve their living space and respond to the concerns of the "consumer citizen" (Fernandes, 2004; Harris, 2007; Coelho and Venkat, 2009). This often takes the form of a "bourgeois environmentalism" (Bhaviskar, 2003) that pits the needs of the poor against middle-class aspirations (Zerah, 2007). However, while middle-class associational life has generally been depicted as thriving on "politics of exclusion" and the need to maintain distance from the poor, Ranganathan (2011) distinguishes between lower middle-class groups living in Bangalore's periphery who "do not distance themselves from the working-class poor" and who are different from the English-speaking middle class. Her fieldwork in Bangalore's unplanned revenue layouts revealed little antagonism directed at the poor or slums. In fact, like the poor, the lower middle class also makes claims on the state for regularization of their settlements through legal and extra-legal means.

The second type of conflict is related to the expanding city's impact on natural resources, local residents' livelihood and quality of life, a major source of conflict being land acquisition by the state (Shaw, 1994). These two kinds of problems are interrelated, as faster economic growth since 1991 and India's opening up to the global economy have led to new demands on land in the periphery by the state and private corporations. State industrial policies have promoted the development of "global growth centers" in the urban periphery where land is cheaper and can be acquired by the government in "the public interest" for the creation of technology parks and related services, as in the case of Hyderabad (Kennedy, 2007). Opening up to global markets along with rising incomes in the middle class has also increased the demand for larger apartments/homes, leading to a higher land footprint and more use of natural resources such as water (Gajendran, 2016; Ranganathan, 2014). Ruet et al. (2007) show how ground water in the periphery of Chennai has been diverted away from farming for the city's use, creating a shift from agriculture to commercialization of water and a growing income gap between farmers who own bore wells and sell water, and those who do not.

Given the conflicting stakes, one of the most important challenges with regard to periurbanization in India is the governance of these areas (Dupont, 2007). Many parts of the peri-urban belt, even in the largest metropolitan cities, lie beyond the administrative boundaries of cities and any planning entities (Shaw, 2005). They are often still a part of rural local government bodies called *panchayats*, where building regulations are weak and the conversion of agricultural land to nonagricultural uses through illegal revenue layouts is easier. Records of land ownership are fuzzier and can be manipulated by dominant interest groups, both public and private. Roy's (2008) detailed study of Kolkata's southeastern periphery shows how the state government uses such ambiguity to its advantage in vesting land, resettling evicted slum dwellers from the inner-city, and then removing them again as the city expands and such land is required for infrastructure and other projects. Controlling peri-urban development in these ways enhances state power and increases the arbitrariness of state decisions, which inevitably go against the interests of the poor and strengthen corporate sector players' gains. Effective governance of the peri-urban will have to address the opaque issues of land use rights, land management, and popular participation in local governance institutions.

What the future holds

As India completes the second decade of the twenty-first century, all the demographic and economic signals point to a continued outward spread of its cities and an increasing proportion of its total urban population living in peri-urban areas. In fact, the central zones of several of the largest metropolitan cities have slowed down in the last two decades while areas outside them continue to grow. A good example is the Kolkata Municipal Corporation, which in 2011 accounted for around 30 percent of its metropolitan population, a significant decline from 1960s when it accounted for nearly half of it. How do we understand what the future holds for these emerging areas? Will it be an optimistic one, with increasing economic opportunities through the opening up of new spaces and an overall improvement in standards of living? Or will it be one of increasing inequalities in lifestyles and incomes, and further marginalization of the poor? Will there be greater political inclusion in the decisions that impact the lives of those most affected by periurbanization? These questions cannot be answered in isolation of the state's broader political agenda, its role and position vis-à-vis other powerful interest groups, and its own commitment, or lack of it, towards the poor. Given India's size and complexity, it is possible that all three kinds of processes, namely, economic dynamism and consequent prosperity, increasing inequalities, and political inclusion, could happen in the peri-urban areas of different parts of the country concurrently. Also, given the country's federal structure and the powers of the subnational state in regard to land and its legislation, and big differences in the level of urbanization across the states, numbers of their large cities, and size of urban population, one uniform outcome is not likely.

While the need for land and the profits to be made from its rising value would continue to make the peri-urban an attractive location for new economic activities, civil society's checks and balances along with negotiated solutions on the part of the poor and lower middle class could see a continuance of the peri-urban as an important habitat for the poorer groups. Given the shortage of serviced land and the growing population, governance mechanisms must be in place to ensure that the peri-urban continues to provide space for the "use value" activities of the poor and the lower middle class, as well as the "market value" or commercial activities of larger actors. So long as a set of governance agendas are in place that protect the land use rights of individuals and communities, provide for sustainable land management and popular participation in local governance issues, an optimistic outcome of city spread is possible. But without such safeguards, the contribution of suburbanization to making Indian cities more livable for ordinary persons will be greatly circumscribed.

Guide to further reading

Nair, J. (2005) *Bangalore: Promise of the Metropolis*. New Delhi: Oxford University Press.
Roy, A. (2009) "Why India Cannot Plan Its Cities." *Planning Theory*, 8(1), pp. 76–87.
Shaw, A. (2012) *Indian Cities*. Oxford Short Introductions. New Delhi: Oxford University Press.

References

Anna University (Chennai), IRMA (Gujarat), TUW-IVV Vienna, ITS-Leeds, Stockholm Environment Institute, and TERI (Delhi) (2006) "A Comparison between the Periurban Developments in the EU, the US and India." 51. Available at: Http://ivv.tuwien.ac.at/fileadmin/mediapoolverkehrsplanung/Bilder/Forschung/Projekte/Periurban/WP7_del_final_01.pdfpage.
Arabindoo, P. (2009) "Falling Apart at the Margins? Neighbourhood Transformations in Peri-Urban Chennai." *Development and Change*, 40(5), pp. 879–901.
Bhan, G. (2013) "Planned Illegalities: Housing and the 'Failure' of Planning in Delhi 1947-2010." *Economic and Political Weekly*, 48(24), pp. 58–70.
Bhaviskar, A. (2003) "Between Violence and Desire: Space, Power and Identity in the Making of Metropolitan Delhi." UNESCO.
Brosius, C. (2010) *India's Middle Class: New Forms of Urban Leisure, Consumption and Prosperity*. New Delhi: Routledge.
Brown, R.M. (2003) "The Cemeteries and the Suburbs: Patna's Challenges to the Colonial City in South Asia." *Journal of Urban History*, 29(2), pp. 151–172.
Brush, J.E. (1968) "Spatial Patterns of Population in Indian Cities." *Geographical Review*, 58, pp. 362–391.
Brush, J.E. (1977) "Growth and Spatial Structure of Indian Cities." In: Noble, A.G. and Dutt, A.K. (eds.) *Indian Urbanization and Planning: Vehicles of Modernization*. New Delhi: Tata McGraw-Hill, pp. 64–92.
Chakravorty, S. (1996) "Too Little, in the Wrong Places? Mega City Programme and Efficiency and Equity in Indian Urbanization." *Economic and Political Weekly*, 31(935/937), pp. 2565–2567 and 2569–2572.
Coelho, K. and Venkat, T. (2009) "The Politics of Civil Society: Neighborhood Associationism in Chennai." *Economic and Political Weekly*, 44, pp. 26–27.
Dupont, V. (2007) "Conflicting Stakes and Governance in the Peripheries of Large Indian Metropolises – An Introduction." *Cities*, 24(2), pp. 89–94.
Fernandes, L. (2004) "The Politics of Forgetting: Class Politics, State Power and the Restructuring of Urban Space in India." *Urban Studies*, 41(12), pp. 2415–2430.

Gajendran, V. (2016) "Chennai's Peri-Urban: Accumulation of Capital and Environmental Exploitation." *Environment and Urbanization Asia*, 7(1), pp. 113–131.

Government of India (GoI) Planning Commission (1988) *Report of the National Commission on Urbanization.* Vol. 2, Part 3. New Delhi: Ministry of Urban Development.

Government of India (GoI) Planning Commission (2012) *Draft of Twelfth Five Year Plan (2012–2017).* Vol. 2. Available at: http://12thplan.gov.in.

Grewal, R. (1991) "Urban Morphology under Colonial Rule." In: Banga, I. (ed.) *The City in Indian History: Urban Demography, Society and Politics.* New Delhi: Manohar Publications and Urban History Association of India, pp. 176–190.

Hambly, G.R.G. (1984) "Towns and Cities: Mughal India." In: Raychaudhuri, T. and Habib, I. (eds.) *Cambridge Economic History of India, Vol. 1, c. 1200-c. 1750.* Hyderabad: Orient Longman, pp. 434–451.

Harris, J. (2007) "Antinomies of Empowerment: Observations on Civil Society, Politics and Urban Governance in India." *Economic and Political Weekly*, June 30, pp. 2716–2724.

Jargowsky, P.A. (2005) "Comparative Metropolitan Development." In: Dupont, V. (ed.) *Peri-Urban Dynamics: Population, Habitat and Environment on the Peripheries of Large Indian Metropolises: A Review of Concepts and General Issues*, CSH Occasional Paper, No. 14. New Delhi: Centre de Sciences Humaines.

Kennedy, L. (2007) "Regional Industrial Policies Driving Peri-Urban Dynamics in Hyderabad, India." *Cities*, 24(2), pp. 95–100.

Kundu, A., Pradhan, B.K., and Subramanian, A. (2002) "Dichotomy or Continuum: An Analysis of Impact of Urban Centers on Their Periphery." *Economic and Political Weekly*, December 14, pp. 5039–5046.

Lang, J., Desai, M., and Desai, M. (1997) *Architecture and Independence: The Search for Identity-India 1880–1980.* New Delhi: Oxford University Press.

Leaf, M. (2016) "The Politics of Periurbanization in Asia." *Cities*, 53, pp. 130–133.

Metcalf, T.R. (1998) *Ideologies of the Raj.* New Delhi: Cambridge University Press.

Naik, M. (2015) "Informal Rental Housing Typologies and Experiences of Low-Income Migrant Renters in Gurgaon, India." *Environment and Urbanization Asia*, 6(2), pp. 154–175.

Pachauri, M.K. (1987) *Suburbanization Process in Delhi Metropolitan Region: A Temporal and Spatial Analysis.* Unpublished PhD Thesis. Center for the Study of Regional Development, Jawaharlal Nehru University, New Delhi.

Ranganathan, M. (2011) "The Embeddedness of Cost Recovery: Water Reforms and Associationism at Bangalore's Fringes." In: Anjaria, J.S. and McFarlane, C. (eds.) *Urban Navigations: Politics, Space and the City in South Asia.* New Delhi: Routledge, pp. 165–190.

Ranganathan, M. (2014) "Paying for Pipes, Claiming Citizenship: Political Agency and Water Reforms at the Urban Periphery." *International Journal of Urban and Regional Research*, 38(2), pp. 590–608.

Rondinelli, D.A. (1991) "Asian Urban Development Policies in the 1990s: From Growth Control to Urban Diffusion." *World Development*, 19, pp. 791–803.

Roy, A. (2008) *Calcutta Requiem: Gender and the Politics of Poverty.* New Delhi: Pearson Longman.

Ruet, J., Gambiez, M. and Lacour, E. (2007). "Private Appropriation of Resource: Impact of Peri-Urban Farmers Selling Water to Chennai Metropolitan Water Board." *Cities*, 24, pp. 110–121.

Shaw, A. (1994) "Urban Growth and Landuse Conflicts: The Case of New Bombay." *Bulletin of Concerned Asian Scholars*, 26(3), pp. 33–44.

Shaw, A. (1995) "Satellite Town Development in Asia: The Case of New Bombay, India." *Urban Geography*, 16(3), pp. 254–271.

Shaw, A. (2004) *The Making of Navi Mumbai.* Hyderabad: Orient Longman.

Shaw, A. (2005) "The Peri-Urban Interface of Indian Cities: Growth, Governance and Local Initiatives." *Economic and Political Weekly*, 40(2), pp. 129–136.

Shaw, A. (2009) "Town Planning in Postcolonial India, 1947–67: Chandigarh Re-Examined." *Urban Geography*, 30(8), pp. 857–878.

Shaw, A. (2012) "Metropolitan City Growth and Management in Post-Liberalized India." *Eurasian Geography and Economics*, 53(1), pp. 44–62.

Shaw, A. and Satish, M.K. (2007) "Metropolitan Restructuring in Post-Liberalized India: Separating the Global and the Local." *Cities*, 24(2), pp. 148–163.

Shekhar, S. and Jaymala, D. (2015) "Spatial-Temporal Changes in an Urban Fringe: A Case Study of Pune." *Annals of the National Association of Geographers, India*, 35(2), pp. 25–34.

Sinha, N. (2018) "A Comparative Institutional Analysis of Rental Houng Sub-Markets in Low-Income Settlements in Hyderabad, India." In: Sengupta, U. and Shaw, A. (eds.) *Issues and Trends in Housing in Asia: Coming of an Age.* New Delhi: Routledge, pp. 250–277.

Sivan, A. and Karuppannan, S. (2002) "Role of State and Market in Housing Delivery for Low-Income Groups in India." *Journal of Housing and Built Environment*, 17, pp. 69–88.

Sridhar, S.K. (2010) "Impact of Landuse Regulations: Evidence from India's Cities." *Urban Studies*, 47(7), pp. 1541–1569.

Srivastava, S. (2015) *Entangled Urbanism: Slum Gated Community and Shopping Mall in Delhi and Gurgaon*. Oxford: Oxford University Press.

World Bank (2013) *Urbanization beyond Municipal Boundaries: Nurturing Metropolitan Economies and Connecting Peri-Urban Areas in India*. Washington, DC: World Bank.

Zerah, M.H. (2007) "Conflict between Green Space Preservation and Housing Needs: The Case of the Sanjay Gandhi National Park in Mumbai." *Cities*, 24(2), pp. 122–132.

9

Suburbanization in Asia

A focus on Jakarta

Deden Rukmana, Fikri Zul Fahmi, and Tommy Firman

Introduction

Jakarta is the capital of Indonesia and the largest metropolitan area in Southeast Asia, with tremendous population growth, land use change, and new town and industrial estate development. In the twentieth century, the overall population of the Jakarta region grew from approximately 150,000 people to 30 million. Jakarta's metropolitan region is also called *Jabodetabek*, taken from the initial letters of the administrative units of Jakarta, Bogor, Depok, Tangerang, and Bekasi. The center of Jabodetabek is Jakarta, also called the Special Capital Region of Jakarta (*Daerah Khusus Ibukota* Jakarta), which covers a total area of 664 square kilometers. The inner peripheries of Jakarta's metropolitan region include four municipalities (City of Tangerang, City of South Tangerang, City of Depok, City of Bekasi), and the outer peripheries of Jabodetabek include the City of Bogor, Tangerang Regency, and Bekasi Regency. This metropolitan region covers a total area of 5,897 square kilometers (Hudalah and Firman, 2012).

Jakarta, or the Special Capital Region of Jakarta, has "provincial government level" status. The peripheries of Jabodetabek are within the jurisdiction of two provinces. The City of Bogor, City of Depok, City of Bekasi, and Bekasi Regency are within the jurisdiction of West Java Province, whereas the City of Tangerang, City of South Tangerang, and Tangerang Regency are within the jurisdiction of Banten Province. The four municipalities within the inner peripheries of Jabodetabek are newer municipalities founded in the 1990s and 2000s. The City of Tangerang, City of Bekasi, City of Depok, and City of South Tangerang were founded in 1993, 1996, 1999, and 2008, respectively. The City of Tangerang and City of South Tangerang seceded from Tangerang Regency. Meanwhile, the City of Depok was part of Bogor Regency and the City of Bekasi seceded from Bekasi Regency.

Population growth of the metropolitan area of Jakarta

Jakarta has been the capital of Indonesia since the Dutch colonial era. The population of Jakarta in 1900 was about 150,000. In the first nationwide census of the Dutch colonial administration (1930), Jakarta's population increased to 409,475. Over the next 10 years, the population increased to 544,823, with an annual growth rate of 3.30 percent. After independence, Jakarta's population increased by nearly three times to 1.43 million by 1950, to 2.91 million in 1960, and

Table 9.1 Population of the metropolitan region of Jakarta in 1980–2010 (in millions)

Area	1980	1990	2000	2010
Core	6.50	8.26	8.39	9.60
Jakarta	6.50	8.26	8.39	9.60
Inner peripheries	n.a	n.a	4.93	7.22
City of Tangerang	n.a	n.a	1.33	1.80
City of South Tangerang	n.a	n.a	0.80	1.29
City of Depok	n.a	n.a	1.14	1.75
City of Bekasi	n.a	n.a	1.66	2.38
Outer peripheries	5.41	8.88	7.31	11.20
City of Bogor	0.25	0.27	0.75	0.95
Tangerang Regency	1.53	2.77	2.02	2.84
Bekasi Regency	1.14	2.10	1.62	2.63
Bogor Regency	2.49	3.74	2.92	4.78
Megacity of Jakarta	11.91	17.14	20.63	28.02

Source: Rukmana (2014)

4.47 million in 1970. The annual growth rates are 10.35 percent and 5.36 percent (1950–1960 and 1960–1970, respectively).

Table 9.1 shows Jakarta's metropolitan region's population, Jakarta, the inner and outer peripheries of Jakarta, from 1980 to 2010. The megacity of Jakarta's population increased from 11.91 million in 1980 to 17.14 million in 1990 and from 20.63 million in 2000 to 28.01 million in 2010. The population in 2010 was 11.79 percent of Indonesia's total population despite residing in less than 0.3 percent of Indonesia's total area. The proportions of Jabodetabek's population to the total population of Indonesia have steadily increased from 8.07 percent to 9.56 percent, and again to 10 percent (in 1980, 1990, and 2000, respectively).

Transformation of Jakarta

The modern city of Jakarta was initiated by President Soekarno's strong vision to build Jakarta into the greatest city possible (Cybriwsky and Ford, 2001). He gave the city Monas, his most symbolic new structure. The national monument stands at 132 meters and included spacious new government buildings, department stores, shopping plazas, hotels, the sport facilities of Senayan that were used for the 1962 Asian Games, the glorious mosque of Istiqlal, new Parliament buildings, and the waterfront recreation area at Ancol.

Such construction continued under the New Order regime that began in 1967. Under this regime, Indonesia enjoyed steady economic growth, along with a reduction in the percentage of the population living under the poverty line. Investment in the property sector, including offices, commercial buildings, new town development, high-rise apartments, and hotels, grew substantially. Jakarta contained the largest concentration of foreign and domestic investment in Indonesia, receiving U.S.$32.5 billion and Rp. 68,500 billion from foreign and domestic investments, respectively, from January 1967 through March 1998 (Firman, 1999). Jakarta, by the mid-1990s, was heading toward global city status.

In the New Order regime's early administration, some projects completed included the Ismail Marzuki Arts Center, the industrial zones at Tanjung Priok and Pulo Gadung meant to attract further foreign investments, plus the unique theme park Taman Mini Indonesia Indah. During the 32 years of the New Order regime, Jakarta changed considerably, with generally rapid economic growth allowing Jakarta to expand and develop into a modern city. Hundreds of new office towers, hotels, and high-rise condominiums were built in many parts of the city.

The Golden Triangle – a new style commercial zone – was built in Thamrin-Sudirman corridor to push the urban skyline upward in response to high land costs in key areas and the convenience of the automobile (Cybriwsky and Ford, 2001). This zone aimed to accommodate internationally invested high-rise mega-blocks, a result of the regional competition among "global cities" (Firman, 1998; Goldblum and Wong, 2000). Jakarta is linked with other "global cities" in a functional system built around telecommunications, transportation, services, and finance. A parade of tall buildings, one after the other, rose above both sides of the major streets, housing the offices of Indonesian and multinational corporations.

The monstrous economy crisis that hit Indonesia in 1998 resulted in major disruptions of Jakarta's urban development, shifting Jakarta from "global city" to "city in crisis." The crisis – commonly known in Indonesia as *krismon* – squeezed Jakarta's economy and in order to survive, a large number of workers became food traders or engaged in other informal sector jobs. The number of street vendors – commonly known in Indonesia as *pedagang kaki lima* – increased rapidly, from about 95,000 in 1997 to 270,000 in 1999 (Firman, 1999).

This shrinkage of economic activities meant less demand for office space, which dropped from 300,000 square meters in 1997 to 85,000 square meters in 1999. Similarly, the demand for high-class apartments dropped from 49,000 in December 1997 to 16,000 in February 1998. The housing market in the megacity nearly collapsed due to increasing costs of building materials and higher interest rates for mortgages. Most construction projects in Jakarta's periphery slowed down or even stopped completely (Firman, 2004).

In order to mitigate the impact of the *krismon*, in July 1998, the government, with the assistance of the IMF, launched a variety of social safety net programs, in addition to other political and economic reforms implemented during the recovery process. Civil unrest and political uncertainty, which had heightened during the *krismon*, gradually decreased during the recovery process.

As of early 2005, Indonesia's economic performance was more robust. Indonesia's rate of economic growth was 5.73 percent per year from 2004 through 2008. The economic growth resulted in an increased number of construction projects in Jakarta, including malls, apartments, and office buildings. Winarso (2010) reported 12 malls and shopping centers built in Jakarta between 2004 and 2006, and another seven between 2013 and 2016. The land area of malls in Jakarta increased from 1.7 million square meters in 2000 to 4.8 million square meters in 2009 (Suryadjaja, 2012).

Jakarta has been a stronghold of Indonesia's economy since the colonial era (Salim and Kombaitan, 2009), and Indonesia's most attractive area for both domestic and foreign investments. Nearly one-fourth of total approved foreign investments in Indonesia over the period of 2000 through 2005 were in Jakarta, due to the city's high concentration of skilled labor and entrepreneurs (Firman, 2008).

Jakarta's contribution to Indonesia's GDP in 2010 increased to 16.7 percent from 14.9 percent in 2000, primarily caused by the city's dominance in the financial and business sectors. The high economic growth also attracted new residents. Kenichiro (2015) identified returning to the city as a trend after the *krismon*, indicated by higher population growth from 2000 to 2010 than the growth Jakarta experienced from 1990 to 2000.

Beginning in 2005, luxury high-rise apartments were constructed in many parts of Jakarta, the investors of which came from several Asian countries including China, Singapore, Hong

Kong, and Japan (Colliers International, 2017). The cumulative supply of luxury apartments reached more than 100,000 units by 2012 (Kenichiro, 2015), and the market has remained strong in the last decade. In the first four months of 2017 alone, 2,790 units were completed in three different projects (Colliers International, 2017).

According to the Council on Tall Buildings and Urban Habitat, as of 2017, Jakarta had a total of 377 tall buildings that satisfy the minimum height of a hundred meters, ranking twelfth among cities in the world for the number of tall buildings (CTBUH, 2017). This strong trend for vertical urbanism, as demonstrated by the construction of numerous high-rise buildings (Alexander et al., 2016), shows no signs of abating. A total of 66 high-rise buildings are currently under construction or in development, including the Signature Tower that will become Jakarta's tallest building in 2022.

Along with such tremendous population growth, Jakarta has faced a wide range of urban problems in the last few decades. Two major problems are traffic congestion and flooding. Jakarta's urbanization and suburbanization are strongly associated with its traffic congestion. The city loses an estimated U.S.$3.5 billion every year because of traffic, which can't be separated from the high growth rate of vehicle ownership (Wismadi et al., 2013). Residents rely heavily on road transportation and about 80 percent of trips are made by private vehicles (Sugiarto et al., 2015).

According to the Jakarta's Bureau of Statistics (2016), in 2014, nearly three-quarters (74.66 percent) of the city's vehicles were motorcycles. The number of motorcycles increased at a rate of 13.35 percent per year, from 6.76 million in 2008 to 13.08 million in 2014, while the number of passenger cars increased at a rate of 8.65 percent per year, from 2.03 million in 2008 to 3.27 million in 2014. Over the same period of time, the total road length in Jakarta increased at a rate of 0.90 percent per year.

Several programs have been implemented to alleviate the acute traffic congestion in Jakarta, including the expansion of the inner-city toll roads and the development of Bus Rapid Transit (BRT) and Mass Rapid Transit (MRT). The total length of inner-city toll roads increased from 112.9 kilometers in 2008 to 123.73 kilometers in 2014. The BRT, popularly known as TransJakarta, was introduced in 2004, and its service expanded to 12 corridors with a total of 669 buses and 111.6 million passengers by 2014.

For at least 20 years, the proposed MRT has been in discussion by the Jakarta administration and the government of Indonesia. Activists and nongovernmental watchdogs have seen the MRT proposal as a possible bonanza for corrupt politicians and contractors. Eventually, in 2009, the government secured a $1.6 billion loan agreement with the Japanese International Cooperation Agency (JICA) for funding, and construction of the MRT began on October 10, 2013. The first MRT track will serve 173,000 passengers per day (Rukmana, 2014). By June 30, 2017, the first MRT track was nearly 75 percent complete. The city administration expects to launch the MRT for trial purposes in August 2018.

Jakarta lies in a lowland area in the vicinity of 13 rivers. All tributaries and basin areas of these rivers are located in the megacity's peripheries and strongly associated with city's flooding problems. Many of the industrial parks and new towns built in the peripheries have converted water catchment areas, green areas, and wetlands; however, such land conversions have affected the severity of flooding in Jakarta, increasing the threat posed by the floods and the residents' resulting yearly woes.

Flooding has had a critical impact on the infrastructure and population. In 2008, floods inundated most parts of the city. Nearly a thousand flights were delayed or diverted while 259 were canceled. In 2012, floods submerged hundreds of homes along major Jakarta waterways and displaced 2,430 people. In January 2013, many parts of Jakarta were inundated following heavy rains, and, as reported by the National Disaster Mitigation Agency (BNPB), the ensuing floods killed a minimum of 20 people and forced at least 33,502 to flee their homes (Rukmana, 2014)

In the aftermath of these annual floods, the government normally attempts to dredge the rivers and release floodwater as quickly as possible into the sea via the East Flood Canal. Construction of the East Flood Canal began in the aftermath of 2002's major floods and finally reached the sea on December 31, 2009, after a lengthy and protracted process due to complicated land acquisitions. The East Flood Canal has been considered the most viable means of preventing future flooding in Jakarta, but one canal clearly cannot prevent flooding entirely.

New towns and industrial estates in the suburbs of Jakarta

In order to understand the suburbanization of Jakarta's metropolitan region, it is essential to recognize the socioeconomic dualism pervading urban Indonesian society. This dualism manifests in the presence of the modern city and the kampung city in urban areas. The *kampong* – village in Indonesian – is associated with informality, poverty, and the retention of rural traditions within an urban setting. Firman (1999) argues the existence of *kampungs* and modern cities reflect spatial segregation and socioeconomic disparities.

The growing numbers of migrants and poor Jakarta natives have produced new squatter *kampungs* on the periphery of Jakarta (Cybriwsky and Ford, 2001). Construction in the central city has also resulted in the eviction of some *kampung* residents, who are then relocated to the periphery (Silver, 2007). The periphery also attracts migrants because of its improved infrastructures and facilities (Goldblum and Wong, 2000).

Beginning in the early 1980s, agricultural areas and forests in Jakarta's suburbs were converted into large-scale subdivisions and new towns (Silver, 2007). Between 1990 and 2010, more than 30 large new towns were built in the suburbs, ranging in size from 500 to 30,000 hectares (Firman, 2014; Winarso and Firman, 2002).

The suburbs' massive development was the result of a series of deregulation and de-bureaucratization measures enacted by the Suharto government in the 1980s (Winarso and Firman, 2002). The subsidized housing finance program and municipal permit system for land development also contributed to this development. Developers strongly linked with the New Order regime have benefited from these policies the most (Leaf, 1994; Arai, 2015).

The residential enclave for narrowly targeted moderate- and high-income families characterized Jakarta's suburban area (Firman, 1998; Leaf, 1994). Located on the city's periphery, these settlements were built in automobile-accessible areas with various high quality amenities. High-income families also moved away from the city in search of better quality of living (Goldblum and Wong, 2000). The high cost of houses and the need for automobiles prevented low-income families from exploring the suburban housing market. One in five families in Jakarta's suburbs owned an automobile (Leaf, 1994).

The first new town in the suburbs of Jakarta was Bintaro Jaya in Pondok Aren, South Tangerang in 1979. Planned by the developer of PT. Jaya Raya Property, Tbk., the town has a total area of 2,321 hectares and includes schools, recreational facilities, and health facilities. The second new town was a collaborative project of Bumi Serpong Damai in Serpong, South Tangerang in 1984. This new town was created for an eventual population of 600,000 in a total area of 6,000 hectares; the project was developed by several private developers and led by the largest – the Ciputra Group. A list of the Jakarta suburbs' new towns is presented in Table 9.2.

New towns in the Jakarta suburbs are aimed at middle- and upper-middle income groups (Goldblum and Wong, 2000; Firman, 2004). They are mostly furnished with golf courses, shopping malls, cinemas, hospitals, and hotels. They are heavily influenced by American design

Table 9.2 New towns in the suburbs of Jakarta in 2017 (> 500 ha)

Name of New Towns	Developer	Planned total size (ha)	District	Municipality
Bintaro Jaya	PT. Jaya Real Property, Tbk	2,321	Pondok Aren	South Tangerang
Bumi Serpong Damai	PT. Sinarmas Land	6,000	Serpong	South Tangerang
Alam Sutera	PT Alam Sutera Realty, Tbk	700	North Serpong	South Tangerang
Kota Modern	PT. Modernland Realty	770	Cipondoh	Tangerang
Gading Serpong	PT. Summarecon Agung and PT. Paramount Land	975	Kelapa Dua	Tangerang Regency
Lippo Karawaci	PT. Lippo Karawaci, Tbk	2,266	Kelapa Dua	Tangerang Regency
Citra Raya	PT. Ciputra Group	2,760	Cikupa	Tangerang Regency
Kota Tigaraksa	Tangerang Regency	3,000	Tigaraksa	Tangerang Regency
Kota Legenda	PT. Sinarmas Land	1,100	South Tambun	Bekasi Regency
Lippo Cikarang	PT. Lippo Karawaci, Tbk	2,216	South Cikarang	Bekasi Regency
Cikarang Baru	PT. Jababeka, Tbk	1,400	South Cikarang	Bekasi Regency
Jababeka	PT. Jababeka, Tbk	2,140	North Cikarang	Bekasi Regency
Kota Deltamas	PT. Pembangunan Deltamas	3,000	Central Cikarang	Bekasi Regency
Kota Harapan Indah	PT. Hasana Damai Putra	3,000	Medan Satria	Bekasi
Kota Wisata	PT Sinarmas Land	750	Cileungsi	Bogor Regency
Kota Taman Metropolitan	PT. Metropolitan Land	600	Cileungsi	Bogor Regency
Harvest City	PT. Dwikarya Langgeng Sukses	1,050	Cileungsi	Bogor Regency
Sentul City	PT. Sentul City, Tbk	3,100	Babakan Madang	Bogor Regency
Sentul Nirwana	PT. Bakrieland Development	12,000	Jonggol	Bogor Regency
Citra Indah	PT. Ciputra Group	1,200	Jonggol	Bogor Regency
Telaga Kahuripan	PT. Kahuripan Raya	750	Parung	Bogor Regency
Rancamaya	PT. Suryamas Dutamakmur, Tbk	500	Ciawi	Bogor Regency
Bogor Nirwana	PT. Bakrieland Development	810	South Bogor	Bogor

Sources: Kenichiro (2015); Firman (2014)

concepts to offer luxury, security, self-sufficient neighborhoods, and improved lifestyles. Many new towns also led to large-scale displacement of farmers and existing residents, such as Tigaraksa, which evicted about 1,400 farmers (Firman, 2004).

In a number of these new towns, the State Housing Provider Agency (Perumnas) joined with private developers to assure that some housing was targeted for low- and moderate-income families (Cybriwsky and Ford, 2001). Most of the new towns offered relatively few employment opportunities. Their initial concept was to create self-contained communities, but this was rarely implemented. Instead, the new towns became "bedroom suburbs for city-bound commuters" (Cybriwsky and Ford, 2001) and were still heavily dependent on the central city (Firman, 1999; Silver, 2007). The development of large-scale housing projects intensified the daily interaction between the fringe areas and the central city, worsening the traffic problems in metropolitan

Jakarta. People who live on the outskirts of Jakarta can save as much as 30 percent of their transportation costs using motorcycles to commute to work rather than public transportation.

Winarso and Firman (2002) revealed almost all large developers were connected to the President Suharto's family and inner circle, and this connection to the First Family helped the developers expand their businesses. Interlinking also occurred among the large developers through cross-shareholding, shared directorships and joint ventures, a process that turned potential competitors into collaborators and created oligopolistic types of land and housing markets.

In addition to residential zones, the periphery of Jakarta is also made up of specialized zones of commercial and industrial enterprises. These areas complement Jakarta's other districts: the central business districts on the Thamrin-Sudirman corridor, the government offices around Medan Merdeka, the international seaport of Tanjung Priok, and the growing network of freeways. The development of industrial zones on the peripheries also indicates a spatial restructuring that shifted manufacturing from the central city to the periphery. Firman (1998) reported that the central city attracted disproportionate investment in service industries, trade and hotels, and restaurant construction.

The peripheries attracted most of the industrial construction, including textiles, apparel, footwear, plastics, chemicals, electronics, metal products, and foods (Cybriwsky and Ford, 2001). The total area of industrial estates in the suburbs increased from 11,000 hectares in 2005 to 18,000 hectares in 2010 (Firman, 2014; Hudalah et al., 2013). About 40 percent of the industrial estates in the region were located in the district of Bekasi, including seven large industrial estates.

Three industrial estates in the district of Bekasi also integrated their industrial areas with residential and other urban activities. They created towns rather than estates (Hudalah and Firman, 2012). Jababeka also built an inland port named Cikarang Dry Port and opened it in 2010. The Cikarang Dry Port offers one-stop service for cargo handling for international export and import and domestic distribution. The seven large industrial estates in the district of Bekasi are Indonesia's largest concentration of industry; in 2005, they produced about 46 percent of the national non-oil and gas exports, at a value of USD$66.428 billion (Hudalah and Firman, 2012). The industrial activities in Bekasi also generate taxes for the central and local governments, as much as 3.4–6 trillion rupiahs in 2005. Nearly 10,000 expatriates also lived in Bekasi in 2005 due to the industrial activities.

The development of private industrial parks in the peripheries followed the development of the three highways stretching from Jakarta to the peripheries – the Jagorawi toll road, the Jakarta-Cikampek toll road, and the Jakarta-Merak toll road highways (Henderson and Kuncoro, 1996; Hudalah et al., 2013). Private industrial parks in the peripheries range from 50 to 1,800 hectares, although the average size is about 500 hectares (Hudalah et al., 2013); major industrial centers are located in Cikupa-Balaraja of Tangerang Regency and Cikarang of Bekasi Regency. The industrial center of Cikarang, with a total industrial land area of nearly 6,000 hectares, is the largest planned industrial center in Southeast Asia (Hudalah and Firman, 2012).

The industrial estates in the suburbs are becoming increasingly specialized and intensifying the region's polycentrism (Firman, 2014; Hudalah et al., 2013). Each industrial estate built its own facilities and infrastructure including roads, waste treatment plants, and communication networks, resulting in a fragmented industrial complex (Hudalah et al., 2013).

Post-suburbanization of Jakarta

Urban development in metropolitan Jakarta has continued and expanded beyond the suburbs. Jabodetabek fringe areas that used to be "traditional" dormitory towns have transformed into more independent areas with a strong economic base. Agricultural land in these areas has been converted for various urban land purposes, including new town and large-scale residential areas, industrial estates, and shopping centers. The core of the metropolitan region, in contrast, is experiencing low population growth due to considerable population spillover to fringe areas. While population growth in Jakarta City was 3.1 percent between 1980 and 1990, it was only 1.5 percent between 2000 and 2010 (see also Table 9.1). As a result of new town and industrial development in fringe areas, commuting is evident in Jabodetabek, where millions of people commute between the city and the peripheral areas daily by trains, buses, motorcycles, and cars. Likewise, many inhabitants commute between the city and the smaller, newer towns on the outskirts, as they work there but still live in Jakarta (Firman, 2014).

As Firman and Fahmi (2017) explain, recent Jabodetabek development reflects some signs of the early stages of post-suburbanization. Post-suburban development in Jabodetabek is, however, less likely to fully resemble that of Western cities (Feng et al., 2008) "because so many people choose to continue to live in the traditional core and commute out to suburban developments for work, as well as other activities" (Firman and Fahmi, 2017, p. 77). Post-suburbanization in Jabodetabek is triggered by privatization of land development and management, particularly in fringe areas. The private sector has gained stronger control over land in that it can aggressively acquire, develop, and manage land in fringe areas, most notably for residential and industrial activities. The prominent role of the private sector in land development has indeed materialized for a long time. Currently, the private sector plays a more significant role: it is able to direct land development and manage the areas "exclusively" by providing municipal services traditionally delivered by local governments in the areas.

The shift of power in land development from the public to the private sector is strongly driven by decentralization and its associated reforms in Indonesia. For the Jabodetabek case, the central government still plays a strong role in suburban development, in that many industrial activities in fringe areas are made possible by foreign direct investments, which are subject to the central government's approval. On the other hand, local governments now have the authority to direct spatial plans and development in their areas, as well as grant building permits to private developers.

Industrial centers in Jabodetabek are becoming increasingly diversified, so that fringe areas are becoming more polycentric and fragmented industrial regions (Hudalah et al., 2013). This development can be associated with the behavior of private developers, both foreign and domestic, seeking economic benefits from the ongoing industrialization process as well as the pro-growth economic policies of both central and local governments. The central government has stimulated the development of industrial estates in fringe areas by subsidizing the provisions of infrastructure and other facilities built and managed by "licensed companies" (Hudalah et al., 2013). According to Government Regulation 142/2015, the licensed companies, those holding permits from either central or local government, have the exclusive right to develop and manage specific industrial areas, provide and manage ongoing utilities and facilities exclusively for the firms located in these areas. The license to develop and manage industrial parks is to be granted by the local government, where the potential estates are located, and by the provincial government if the potential location extends into two or more municipalities/districts. If the potential area extends over two or more bordering provinces, or if it is to be operated by a foreign company, the developer must acquire additional permits from the central government. After a private developer obtains the license to manage an industrial park, it has the exclusive authority to sell land units to other companies that wish to start businesses inside the industrial estates.

The shift of power from the public to the private sector is also reflected by new town development in fringe areas. Private developers build new towns and large-scale residential projects in response to the local needs driven by economic growth and diversification in fringe areas. They gain permits from the local governments to design the new towns as gated suburban developments surrounded by walls and separated from nearby communities (Leisch, 2002). Private developers not only provide infrastructure exclusive to the inhabitants within the gated communities, but also administer municipal services as if they were these communities' "government." In so doing, they appoint their own "city" managers to ensure service delivery and security of the area. Local governments enable this development by granting building permits to private developers, although these sometimes do not comply with the legalized spatial plans. For example, new town projects are built on land that is supposed to serve as catchment areas. The local autonomy rights given to the local governments have cultivated a competitive climate, so they are now eager to promote economic development in their regions and exploit regional resources more intensively. In many cases, economic growth is preferred over enforcing spatial plans (Rukmana, 2015). Decentralization has also intensified the practice of "clientelism," or patronage relationships, between the local government and the private sector (Rukmana, 2015). Spatial plans are often prepared and easily altered to accommodate the developers' interests rather than plan for more sustainable regions (Firman, 2004; Rukmana, 2015). Driven by political pressures and interests in placing what are perceived to be profitable economic activities, spatial plans are often negotiated and violated. This condition actually illustrates contradictory facts. On the one hand, local governments have strong power to direct local development and also to empower developers to perform their profit-seeking behaviors, although this violates the spatial plans (Cowherd, 2005; Kenichiro, 2015). On the other hand, this reflects the local government's inability to enforce the legalized plans, as if they are "powerless" when they have to face the developers.

The fact that the private sector takes over some governmental tasks, on the one hand, can be seen as an opportunity to improve upon the local governments' limited ability to provide basic services. On the other hand, the private sector focuses mainly on profit and often pays less attention to the spatial plans that aim to create sustainable cities and regions. As local governments have the authority to direct local development and the central government has less power to intervene in it, the making and enforcement of spatial plans in Greater Jakarta has been fragmented (Kusno, 2014). As such, recent post-suburbanization of Jakarta reveals new, significant challenges in managing urban development and enforcing spatial plans, which require innovative governance solutions.

Conclusion

This chapter has presented Jakarta's transformation from a concentric and radial-patterned urban structure to a city in the early stages of post-suburbanization. Jakarta has been the nation's capital and the largest city in Indonesia since the Dutch colonial era, although before independence Jakarta was far smaller. After independence, Jakarta began to grow beyond the city's boundaries and formed a metropolitan region consisting of several administrative districts and municipalities (i.e., Jabodetabek). During the New Order's regime (1967–1998), as the country enjoyed rapid economic growth, Jakarta had a chance to expand its construction and develop into a modern city. The central government's pro-growth economic policy at that time supported large-scale industrial activities on the peripheries. Although the monetary crisis forced a pause in development at the beginning of the New Millennium, it resumed after the crisis passed and has continued since. Current development indicates some signs of the early stages of post-suburbanization, in which the traditional core remains preeminent, but the peripheral areas have become more independent satellite cities with strong economic bases and diversified activities.

This development was triggered by the privatization of land development and management, particularly in fringe areas (Firman and Fahmi, 2017). The private sector has indeed played a crucial role in developing industrial and large residential activities in fringe areas. However, it now plays an even more significant role, as it can direct land development and manage the areas "exclusively" by providing municipal services traditionally delivered by local governments. As a result, regional development of greater Jakarta, which consists of several districts and municipalities, is potentially even more fragmented and unsustainable. A forum, namely the Coordinating Body of Jabodetabek Development, is designed to integrate local government actions in managing the region's development. However, this body seems ineffective, as under the Indonesian New Decentralization law the real authority of local development is owned by the local government. This condition suggests that it is now crucial to designate a form of Metropolitan Authority that works above the local government and is authorized to coordinate the region's development.

Guide to further reading

Cybriwsky, R. and Ford, L.R. (2001) "City Profile: Jakarta." *Cities*, 18(3), pp. 199–210.

Firman, T. (2004) "New Town Development in Jakarta Metropolitan Region: A Perspective of Spatial Segregation." *Habitat International*, 28(3), pp. 349–368.

Kusno, A. (2014) *After the New Order: Space, politics and Jakarta*. Honolulu, HI: University of Hawaii Press.

References

Alexander, N., Sukamta, D., and Handoko, S. (2016) "State-of-the-Practice in Design and Construction of Deep Basement in Jakarta." Conference Proceeding Cities to Megacities: Shaping Dense Vertical Urbanism.

Arai, K. (2015) "Jakarta since Yesterday: The Making of the Post-New Order Regime in an Indonesian Metropolis." *Southeast Asian Studies*, 4(3), pp. 445–486.

Colliers International. (2017) *Jakarta Apartment Colliers Quarterly*, April 25.

Cowherd, R. (2005) "Does Planning Culture Matter? Dutch and American Models in Indonesian Urban Transformations." In: Bishwapriya, S. (ed.) *Comparative Planning Culture*. New York and London: Routledge.

CTBUH. (2017) "Council on Tall Buildings and Urban Habitat Height and Statistics." Available at: www.ctbuh.org/TallBuildings/HeightStatistics/tabid/1735/language/en-US/Default.aspx.

Cybriwsky, R. and Ford, L.R. (2001) "City Profile: Jakarta." *Cities*, 18(3), pp. 199–210.

Feng, J., Zhou Y.X., and Wu, F. (2008) "New Trends of Suburbanization in Beijing Since 1990: From Government-Led to Market-Oriented." *Regional Studies*, 42, pp. 83–99.

Firman, T. (1998) "The Restructuring of Jakarta Metropolitan Area: A 'Global City' in Asia." *Cities*, 15(4), pp. 229–243.

Firman, T. (1999) "From 'Global City' to 'City of Crisis': Jakarta Metropolitan Region under Economic Turmoil." *Habitat International*, 23(4), pp. 447–466.

Firman, T. (2004) "New Town Development in Jakarta Metropolitan Region: A Perspective of Spatial Segregation." *Habitat International*, 28(3), pp. 349–368.

Firman, T. (2008) "In Search of a Governance Institution Model for Jakarta Metropolitan Area (JMA) under Indonesia's New Decentralisation Policy: Old Problems, New Challenges." *Public Administration and Development*, 28(4), pp. 280–290.

Firman, T. (2014) "The Dynamics of Jabodetabek Development: The Challenging of Urban Governance." In: Hill, H. (ed.) *Regional Dynamics in a Decentralized Indonesia*. Singapore: Institute of Southeast Asian Studies.

Firman, T. and Fahmi, F.Z. (2017) "The Privatization of Metropolitan Jakarta's (Jabodetabek) Urban Fringes: The Early Stages of 'Post-Suburbanization' in Indonesia." *Journal of the American Planning Association*, 83(1), pp. 68–79.

Goldblum, C. and Wong, T.-C. (2000) "Growth, Crisis and Spatial Change: A Study of Haphazard Urbanization in Jakarta, Indonesia." *Land Use Policy*, 17, pp. 29–37.

Henderson, J.V. and Kuncoro, A. (1996) "The Dynamics of Jabotabek Development." *Bulletin of Indonesian Economic Studies*, 32(1), pp. 71–95.

Hudalah, D. and Firman, T. (2012) "Beyond Property: Industrial Estates and Post-Suburban Transformation in Jakarta Metropolitan Region." *Cities*, 29, pp. 40–48.

Hudalah, D., Viantari, D., Firman, T., and Johan Woltjer, J. (2013) "Industrial Land Development and Manufacturing Deconcentration in Greater Jakarta." *Urban Geography*, 34, pp. 1–22.

Jakarta's Bureau of Statistics (2016) *Transportation Statistics of DKI Jakarta*. Jakarta: BPS DKI Jakarta.

Kenichiro, A. (2015) "Jakarta 'since Yesterday': Making of the Post-New Order Regime in an Indonesian Metropolis." *Southeast Asian Studies*, 4(3), pp. 445–486.

Kusno, A. (2014) *After the New Order: Space, Politics and Jakarta*. Honolulu, HI: University of Hawaii Press.

Leaf, M. (1994) "The Suburbanization of Jakarta: A Concurrence of Economics and Ideology." *Third World Planning Review*, 16(4), pp. 341–356.

Leisch, H. (2002) "Gated Communities in Indonesia." *Cities*, 19, pp. 341–350.

Rukmana, D. (2014) "Peripheral Pressures: Jakarta." In: Connah, R. (ed.) *Archeology of the Periphery of Megacities*. Moscow: Strelka Press, pp. 158–167.

Rukmana, D. (2015) "The Change and Transformation of Indonesian Spatial Planning after Suharto's New Order Regime: The Case of the Jakarta Metropolitan Area." *International Planning Studies*, 20(4), pp. 350–370.

Salim, W. and Kombaitan, B. (2009) "Jakarta: The Rise and Challenge of a Capital." *City*, 13(1), pp. 120–128.

Silver, C. (2007) *Planning the Megacity: Jakarta in the Twentieth Century*. London and New York: Routledge.

Sugiarto, S., Miwa, T., Sato, H., and Morikawa, T. (2015) "Use of Latent Variables Representing Psychological Motivation to Explore Psychological Motivation Citizens' Intentions with Respect to Congestion Charging Reform in Jakarta." *Urban Planning and Transport Research*, 3, pp. 46–67.

Suryadjaja, R. (2012) "Jakarta's Tourism Evolution: Shopping Center as Urban Tourism." A Presentation to the Fifth International Forum on Urbanism, Barcelona, 25–27 February.

Winarso, H. (2010) "Urban Dualism in the Jakarta Metropolitan Area." In: Sorensen, A. and Okata, J. (eds.) *Megacities: Urban Form, Governance, and Sustainability*. New York: Springer, pp. 163–191.

Winarso, H. and Firman, T. (2002) "Residential Land Development in Jabotabek, Indonesia: Triggering Economic Crisis?" *Habitat International*, 26, pp. 487–506.

Wismadi, A., Soemardjito, J., and Sutomo, H. (2013) "Transport Situation in Jakarta." In: Kutani, I. (ed.) *Study on Energy Efficiency Improvement in the Transport Sector through Transport Improvement and Smart Community Development in the Urban area*. ERIA Research Project Report 2012–29, pp. 29–58.

10

Suburbanization in Asia

A focus on Seoul

Chang Gyu Choi and Sugie Lee

Urbanization of Seoul City in the twentieth century

Seoul has been the capital city of Korea since 1394. For over 500 years, the small city's size – the radius approximately 4 kilometers – did not change much. In the twentieth century, railway construction during the Japanese colonial period and the development of the suburbs progressed gradually to the southwest, but the city's scale was not changed dramatically. After the 1953 Korean War, a large number of refugees and members of the rural population were pushed into the shantytowns surrounding Seoul.

Since 1960, Korea has experienced accelerated economic development centered around manufacturing industries. Seoul became a center of manufacturing and industry, and a major destination, reducing the rural areas' populations and producing explosive urban population growth. As shown in Figure 10.1, the population of Seoul in 1970 was about 5.4 million, which increased to 10.6 million in 1990 and decreased to 9.9 million in 2015. The population of the entire Seoul Metropolitan Area (SMA) was 9.4 million in 1970 and 25.3 million in 2015. While the population in Seoul has been stable since 1990, the population in the SMA has increased gradually over time.

During the 1960s, a large number of shantytowns were built on near the rivers, streams, hills, and mountains surrounding the old city. In addition to the problems of slums, residential developments were needed to meet the increasing demand for middle-class housing. As shown in Figure 10.2, the southern part of the Han River basin across from the present-day Seoul area, which was used for rice paddies and rural fields, became the center of suburban development. The suburban housing supply became a top priority. The Gangnam area, one of the three urban centers in current Seoul, saw the beginning of such construction in 1967, and its first development plan was a residential complex in the suburbs.

Since the 1970s, Seoul has expanded its administrative boundaries toward the north and west in order to accommodate the soaring urban population. In addition, the Korean government designated the green belt areas between 20 and 30 kilometers from the CBD. The green belt policies were adopted not only to manage unplanned development and protect the natural environment, but also to suppress potential development and population growth near the military border with North Korea.

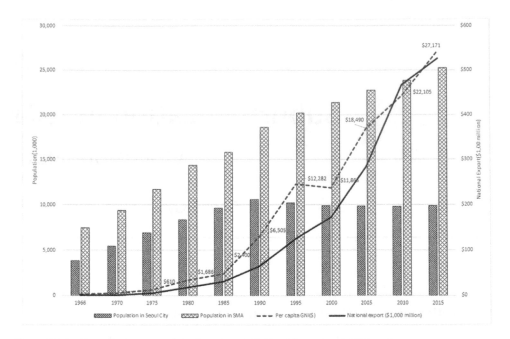

Figure 10.1 Population growth in Seoul and SMA with national GNI and export

Source: Korean Statistical Information Service (1966–2015); Korean Statistical Information Service, 2017a, 2017b, 2017c

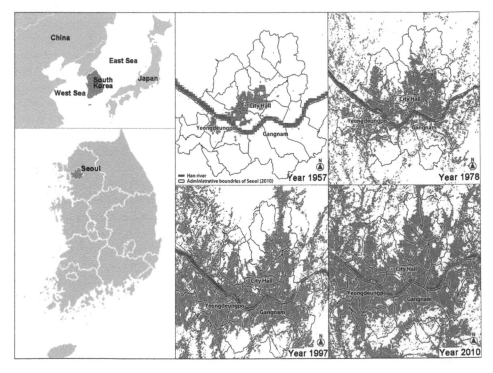

Figure 10.2 The urbanization process of Seoul City, Korea (1957–2010)

Source: Adapted from the Seoul Research Data Service (SRDS, 2017, pp. 40–41)

However, Seoul expanded rapidly due to natural population growth and in-migration from rural areas during the 1980s. With a shortage of housing and prices skyrocketing as a result, the Korean government initiated the new town developments beyond green belt areas toward Gyeonggi province, which transformed the city of Seoul into the SMA. In 1989, the government has completed the first generation of new town development in five locations, with an approximate supply of 584,000 housing units, and has almost finished the second-generation development in more than 11 locations, with an approximate total housing supply of 623,000 units in suburban areas of the SMA.

During the 2000s, the population of Seoul was relatively stable, while its suburban areas in Gyeonggi province grew to 11 million by 2010. To accommodate the soaring SMA population, the government promoted suburban developments and urban regeneration while adopting a rigid urban planning system and strong land use regulations.

Greenbelts and land readjustment projects before the mid-1980s

The exponential inflow of people to Seoul was a concern for the Korean government. The ostensible concern was that the rapid urbanization would create urban problems regarding housing, traffic, and the environment. However, the most important concern was the fear of military defense. The Korean War ended in 1953; South Korea was at the forefront of the Cold War, and military conflicts with North Korea were frequent. Seoul was only about 40 kilometers from the demilitarized zone (DMZ), the shared border with North Korea. The capital city fell in three days when North Korea invaded in 1950, and many people were killed and facilities destroyed in Seoul. Park Chung-Hee, who gained power through the coup of 1961, was president until 1979 and served as a military officer during the Korean War. He was deeply concerned about overcrowding in Seoul and was directed to create a strong dictatorship. The initial installation of the green belt in Seoul in 1971 was based on his order (Kim, 1983; Son, 2005a, 2005b) and was considered not only a planning project but also a military one, with the purpose of deterring any Korean citizens from living close to the border. As shown in Figure 10.3, greenbelts were designated in the areas of 5 to 10 kilometers wide,

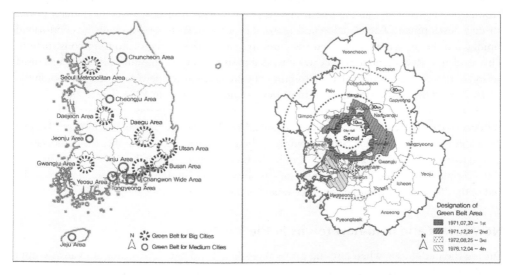

Figure 10.3   Green belt designations in the 1970s

Source: Adapted from Gyeonggi-Do (2015, pp. 5–6)

along the administrative boundaries of Seoul (most of them hills and mountains). By the 1970s, the SMA had extended to southwestern cities including Incheon, Anyang, Suwon, and Ansan. The southwestern part of the green belt reached a width of 30 kilometers.

By 1980, new developments in the SMA had been concentrated within Seoul's city boundary. Seoul was one huge metropolis, originating from a small city area connecting a 500-year-old city, smaller new developments, and shantytowns. These small urban areas could not accommodate the rapidly growing population. Throughout the 1960s and 1970s, the city developed suburban urban areas but allowed some undeveloped locations to remain. Land Readjustment Projects were used as the main new development means as joint projects with landowners, but it was difficult to satisfy the constant demands for land and housing.

To increase the land supply, the Korean government developed the underutilized Han river's floodplain and its surrounding area and constructed large apartment complexes of Banpo-Yeouido-Ichon-Jamsil-Apgujeong-Jamwon. The rapidly growing middle class concentrated on creating a new lifestyle in the modern living environment of apartments. These residential developments are still representative apartment complexes for high- and middle-income households in Seoul. Currently, they are attracting new attention with 30-story high-rise residential redevelopments.

Until the late 1980s, the new developments within Seoul were suburban; they currently are all connected and represent Seoul's metropolis. Although each new development was constructed individually, Seoul's urban planning has connected them all with transportation and various infrastructure services, and the market's demand has filled in the gap. The city of Seoul, which expanded rapidly beginning in 1960, reached the edges of green belt boundaries by the end of the 1980s, leading to a new phase of suburban developments in the SMA.

Housing and land development in the 1980s

The Land Readjustment Project, a major means of development in the 1970s, had certain limitations when it came to addressing the complex problems of coordinating individual rights. Housing supply for the middle class must be addressed by a military regime that aims for political stability (Son, 2003). The Korean government thus enacted the Residential Development Promotion Law in 1980, stating that it and its public corporation had eminent domain power for new development. The planning and approval process for such developments was performed rapidly, and the government supplied the lands to private developers for housing construction. This residential development project was intended to supply a large amount of housing in a short period of time. In addition, the Land Readjustment Project, a major new development tool in the 1970s and 1960s, imposed prohibitive measures around large cities in 1983 (Lee, 2008).

The underdeveloped areas that remained in Seoul until the end of the 1980s were newly developed as Residential Land Development projects, including Mokdong, Gaepo, Sangye, and Junggye. Despite this additional supply, there has been a rapid rise in housing prices since 1986; after the democratization movement in 1987, citizens' claim of rights increased. President Roh Tae Woo, elected in 1987 as the last president of the military regime, took measures to address the rise in housing prices at the end of the 1980s and announced the development of the first-generation new towns in April 1989.

New towns and mini-new towns in the 1990s

During the mid-1980s, when housing prices skyrocketed and few developable land areas within Seoul's boundaries were left, the Korean government's decision was to build new towns beyond the green belt. The new town construction plan was designed over approximately five months and released in full force in April 1989. Many city planners opposed the new development plan

Table 10.1 The first-generation new town developments in SMA

	Total	Bundang	Ilsan	Pyeongchon	Sanbon	Jungdong
Location	–	Seongnam-shi	Goyang-shi	Anyang-shi	Gunpo-shi	Bucheon-shi
Area (ha)	5,014	1,964	1,574	510	420	546
Population (thousand)	1,168	390	276	168	168	166
Gross density (inhabitants/ha)	233	199	175	329	399	304
Residential net density (inhabitants/ha)	678	614	525	870	893	883
Houses (thousands) (apartment %)	292 (96.2%)	97.6 (96.9%)	69.0 (91.4%)	42.0 (98.6%)	42.0 (98.6%)	41.4 (97.8%)
FAR (%)	–	184	169	204	205	226
Development period	–	1989.08–1992.12	1990.03–1992.12	1989.08–1992.12	1989.08–1992.12	1990.02–1992.12
Construction completion	–	1996.12.31	1995.12.31	1995.12.31	1995.1.31	1996.1.31
Land developer	–	Korea Land Corp.	Korea Land Corp.	Korea Land Corp.	Korea Housing Corp.	Bucheon-shi & the two Corp.

Source: Ministry of Land, Infrastructure, and Transport (MLIT, 2015, p. 196)

because of its development procedure and the differences from the previous planning consensus, but the government initiated the construction to reduce the dissatisfaction of the middle class.

Through the residential development project, the government built 300,000 houses for about 1.2 million residents in five new towns over eight years (see Table 10.1 and Figure 10.4). There were two large-scale towns in the first-generation of new towns, Bundang and Il-san, beyond the sacred green belt. They were also three times larger than the other new towns, and the planned population more than twice that of the others, reaching 390,000 and 276,000, respectively. New towns were originally intended to develop housing development in accordance with the purpose of construction, and government officials called it a Korean-style new town for promoting a housing-oriented new city.

Five new towns were built to address the demand for apartment complexes, all of which exceed 95 percent in terms of apartments, with Pyeongchon and Sanbon close to 99 percent. All five new towns have high floor area ratios from 1.69–2.26. These cities became the new utopia of the middle class in the 1990s; by the decade's end, the word *Bundang under Chun-dang* (heaven) became popular. Public facilities inaccessible in the old city were supplied; construction of stores and restaurants provided a location where the middle class could enjoy a new lifestyle.

Although the construction of new towns did not proceed until the late 1990s, the Korean government did not stop new developments. Small new developments using the Land Development Project in Gyeonggi province continued and were called "mini-new towns." Go and Choi (2013) calculated that mini-new towns in the SMA housed around 4.4 million people by 2009. Their development was concentrated around two large new towns, especially around Bundang due to the great accessibility to public transportation system and the regional transport networks of roads, railroads, and amenity facilities.

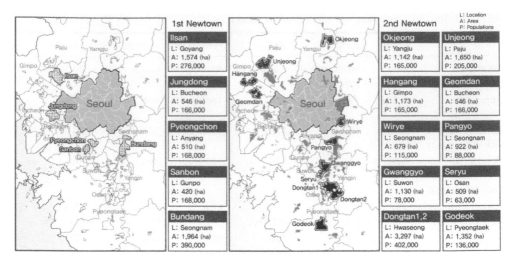

Figure 10.4 The first- and second-generation new town developments in the SMA

Source: Adapted from the Ministry of Land, Infrastructure and Transport (MLIT, 2015)

As a result of the recent development of the first-generation new towns and the mini-new towns, it has become quite common to see high-rise apartment complexes in the suburbs, surrounded by a small unit of unplanned developments. In the mid-1990s, these mini-new-towns were criticized as government-driven sprawl. As a result, debates began about the construction of a large-scale new town. However, for another new town construction using government policies, political power was needed, motivated by a rapid and worrisome rise in house prices.

In the late 1990s, housing prices in the SMA began to rise after the Asian Finance Crisis in 1997, and the Korean government again listened to new town proposals. From the early 2000s to 2009, the government laid out plans to build 11 first-generation towns in the SMA, representing the second-generation of new towns (see Table 10.2 and Figure 10.4).

The second-generation of new towns was completed after a comparatively long period of planning and construction, compared to the first new towns. The area of the first-generation new towns had increased over time, and the population density was relatively low. Among housing types, the proportion of apartments has dropped to less than 80 percent; more diverse housing styles have been provided, and new cities have been developed with higher amenity standards than previously.

Both the first- and second-generation of new towns focused on housing supply due to the rapid rise in housing prices. There were discussions about self-sufficiency at the time of the second new town's development, and each city has made various efforts to solve housing problems, although most of them are not satisfactory at present. Compared to the first-generation of new towns, the second-generation of new towns is likely to be improved in terms of self-sufficiency (Lee et al., 2015; Chang and Lee, 2006). Unlike the first-generation new town developments, which were within 20 and 25 kilometers of the CBD, the second-generation new towns were developed within 20 and 50 kilometers of the CBD. This development pattern is likely to increase commuting distance and time for workers, because most second-generation new towns still depend on the central city of Seoul.

However, Myung-Bak Lee, who was elected Korea's president in early 2008, promised during his campaign new apartments at 50 percent of the price of existing apartments and proposed affordable housing projects in some green belt areas that were not sensitive to environmental protection. This half-price apartment policy was fulfilled during his tenure under the name of

Table 10.2 The second-generation of new town developments

	Pangyo	Dong-tan1	Dong-tan2	Hangang	Unjeong	Gwang-gyo	Okjeong	Wirye	Godeok	Geom-dan	Segyo3
Location	Seongnam-shi	Hwaseong-shi	Hwaseong-shi	Gimpo-shi	Paju-shi	Suwon-shi	Yangju-shi	Seongnam-shi	Pyeongtaek-shi	Incheon-shi	Osan-shi
Area (ha)	922	904	2,397	1,173	1,650	1,130	1,142	679	1,352	1,812	509
Housing built (thousand)	29	41	111	59	78	31	59	46	54	92	23
Populations (thousand)	88	124	278	165	205	78	165	115	136	230	63
Gross density (pop./ha)	95	137	116	140	124	69	144	169	100	127	125
Period	2003–2010	2001–2008	2008–2015	2002–2012	2003–2014	2005–2011	2007–2013	2008–2015	2008–2013	2009–2016	2009–2016

Source: Ministry of Land, Infrastructure, and Transport (MLIT, 2015, pp. 197–198)

Bogeumjari (Nest) Housing (MLTMA, 2012). This housing supply was not significantly different from the second-generation new towns, focused on eminent domain and apartment-centered supply. The difference was mainly location: Bogeumjari was developed in previous greenbelts, but most of new towns were developed beyond or within the green belt. This housing development project was also planned to focus on public rental housing. Due to the subprime mortgage crisis at the end of 2008, housing prices in the SMA had been depressed and the public rental housing supply initiated by the government was not difficult to fill.

Since 1989, when the rapid increase of Gyeonggi province's population began, the government has been proceeding with large-scale development of new apartments to rapidly expand the housing supply. Until now, it has been difficult for the private sector to develop large-scale suburban land development because of the strong land use regulations. Private companies prefer constructing buildings by purchasing land supplied by the government and government investment institutions rather than land development (Ministry of Land, Transport and Maraime Affairs, 2011). As a result, it is difficult to identify low-density housing complexes in Gyeonggi province and its suburban and rural characteristics. Therefore, the most common new development pattern in Gyeonggi province is a high-rise apartment complex of 20 to 30 floors. The middle class also prefers the high-rise apartment complexes, because they provide good amenities and high quality of living with public and commercial facilities.

Suburban developments with the first- and second generation new towns have transformed the SMA to a more dispersed, polycentric urban form. With significantly higher housing prices in Seoul, new town developments had attracted middle-class households to suburban communities in Gyeonggi province. Table 10.3 shows the population migration from 2010 to 2016. The net in-migration of Seoul city from 2010 to 2016 was -852,012, which indicates that Gyeonggi province has accommodated most of the in-migration population from Seoul city.

Figure 10.5 indicates that net migration from Seoul to suburban communities in Gyeonggi province was substantial between 2010 and 2016. The destinations of migration are closely associated with new town developments. Figure 10.5 also shows that the SMA has been expanded to a radius of 50 kilometers from the CBD. However, the SMA's suburbanization pattern, initiated by the government housing supply policy, has caused longer commuting distance and time as well as a lack of transportation infrastructure in suburban areas. Most residents in the new towns have long commutes to Seoul due to the lower level of self-sufficiency. Although many new towns have improved self-sufficiency, with additional jobs and daily living facilities, most still depend on the central city of Seoul in terms of jobs, shopping, and leisure.

Table 10.3 Population migration in the SMA

Origin	Destination	2010	2011	2012	2013	2014	2015	2016	Total
Seoul	Gyeonggi	410,735	373,771	354,135	340,801	332,785	359,337	370,760	2,542,324
Seoul	Incheon	46,082	51,641	49,640	47,424	43,212	44,915	43,745	326,659
Gyeonggi	Seoul	285,963	272,407	254,175	246,464	249,701	239,557	234,357	1,782,624
Incheon	Seoul	35,670	33,008	32,216	33,017	34,380	33,570	32,486	234,347
Net In-migration in Seoul		-135,184	-119,997	-117,384	-108,744	-91,916	-131,125	-147,662	-852,012

Source: Authors' calculations from Korean Statistical Information Service (KOSIS), Population Census (2010–2016) (Korean Statistical Information Service, 2017a)

Note: The SMA has the central city of Seoul and surrounding cities and suburbs. Incheon is also a primary city that is located in the western part of Seoul city. Gyeonggi province has many cities and suburbs surrounding the central city of Seoul.

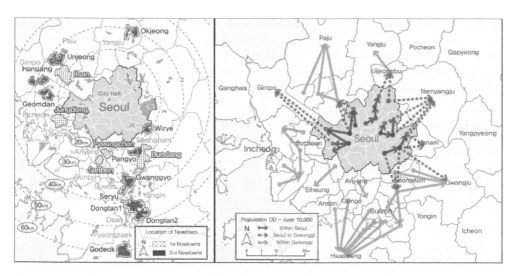

Figure 10.5 The first- and second-generation new town developments and population migration in the SMA

Table 10.4 Commuting distance, time, and volume in the SMA

Origin (residence)	Destination (job location)	Travel variables	1996	2002	2006	2010	Change (1996–2010)
Seoul	Seoul	Distance (km)	6.55	6.14	6.56	6.33	–3.36%
		Time (min.)	36.77	35.20	36.45	35.83	–2.56%
		Volume (1,000)	3109.33	3214.91	3254.85	3735.23	20.13%
Seoul	Gyeonggi/ Incheon	Distance (km)	19.60	20.87	21.36	21.13	7.81%
		Time (min.)	64.19	66.21	67.17	66.74	3.97%
		Volume (1,000)	310.32	400.96	446.84	549.70	77.14%
Gyeonggi/ Incheon	Seoul	Distance (km)	17.20	18.31	19.27	19.43	12.97%
		Time (min.)	61.67	63.73	65.48	65.71	6.55%
		Volume (1,000)	637.59	1009.81	1112.50	1173.14	84.00%
Gyeonggi/ Incheon	Gyeonggi/ Incheon	Distance (km)	7.61	6.56	7.31	8.55	12.35%
		Time (min.)	35.55	34.12	36.24	38.39	7.99%
		Volume (1,000)	2068.50	3202.49	3640.72	4413.85	113.38%
Total	Total	Distance (km)	8.68	8.64	9.33	9.70	11.75%
		Time (min.)	40.34	40.03	41.81	42.24	4.71%
		Volume (1,000)	6125.73	7828.17	8454.91	9871.91	61.15%

Source: Authors' calculations from Household Travel Survey (1996, 2002, 2006, 2010) in the SMA

Table 10.4 shows the changes of commuting distance, time, and volume among the central city of Seoul, Incheon city, and Gyeonggi province in the SMA from 1996–2010. Analysis results indicate that the average commuting distance and commuting time are 9.7 kilometers and 42.24 minutes in the SMA, respectively. In addition, the average commuting distance has increased from 8.68 kilometers in 1996 to 9.7 kilometers in 2010, and the average commuting time increased from 40.34 minutes in 1996 to 42.24 minutes in 2010. Table 10.4 also shows intra-metropolitan commuting patterns between Seoul and Gyeonggi/Incheon. Analysis results also indicate that commuting distance and time of intra-city commuting within Seoul have decreased over time, but commuting time and distance between Seoul and Gyeonggi/Incheon have increased from 1996–2010. In addition, commuting volume has increased between Seoul and Gyeonggi/Incheon, indicating the increasing trend of cross-commuting in the SMA.

Conclusion

We examined the growth pattern of the SMA over the past several decades in the context of government-initiated housing supply policies. The characteristics of suburbanization in the SMA are quite different from those of Western metropolitan regions. The dominant housing type in suburbanization has been high-rise apartment complexes in the SMA. In addition, the Korean government has played an important role in supplying suburban housing with new town developments.

The current spatial pattern of the SMA and the reasons behind such spatial patterns can be explained as follows. First, land use regulations in the SMA were so strong that Levittown-scale private developments have been virtually impossible. When the private developers carried out large new developments, it was difficult to get approval from the central and/or local governments. For example, the Housing Act, which regulates housing supply, obliged the government to comply with the supply criteria when a developer tries to build more than 20 houses. For large-scale suburban development by the private sector, considerable time and investment had to be employed, which is relatively rare.

Second, the government actively promoted the mass supply of apartments to cope with the rapid increase in housing demand. The Korean government has effectively implemented a massive housing supply system since the 1970s. Reclaiming lowlands from the Han River and its branches was a major means of supplying new land for apartment complexes before 1980. With the Housing Development Promotion Act of 1980, the state-owned company bought rural land with strong eminent domain power, procured much faster approvals for land developments, built infrastructures and facilities for urban life, and supplied urban lands to private construction and/or development companies to build, mostly, apartment complexes. This system has become the main land supply system in SMA and Korea, and has developed in other forms.

Third, the middle class has adapted to and now prefers the new lifestyle of apartments. People used to condemn apartment complexes in the 1970s, and until the early 1980s, apartment living was not the lifestyle of the middle class (Kahng, 1980, 1989). However, the middle class who wanted to get out of the old neighborhoods of Seoul and conventional detached houses with inconvenient facilities preferred apartments with relatively high residential and community facilities at a relatively low-cost per unit area. It was a more efficient choice for the middle class to live in high-rise apartments than in low-density single-family homes in previous neighborhoods.

Finally, during the 1990s and 2000s, the first- and second-generation new town developments continued to address the increasing demand of middle-class housing in the SMA, while new town developments were criticized as bedroom communities that depended on the central city of Seoul. Actually, intercity commuting distance and time in the SMA increased, which indicates the existence of spatial mismatches between Seoul city and surrounding suburban areas in terms of jobs and housing.

This study examined the nature of suburban growth, and the policies and plans to influence patterns of suburbanization in the SMA, focusing on the spatial pattern of suburban development with high-rise apartment complexes and the role of government-initiated housing policies in supplying decent housing for the middle class. During the rapid urbanization and industrialization of the period, the government-initiated massive housing supply was an efficient way to stabilize the housing market in a short period of time. However, government-initiated massive suburban development has also brought about some planning issues. First, suburban developments have been constructed far from the CBD because they need a large amount of undeveloped land. The first and second generations of new towns are located 20 to 40 kilometers from the city center. In addition, most first-generation new town developments are bedroom communities, which depend on the central city of Seoul, while the second-generation of new towns has tried to increase self-sufficiency. However, the levels of self-sufficiency are still low in the most of new towns, which leads to long commutes. Second, government-initiated suburban developments have uniform types of high-rise apartment complexes for the middle-class households. In other words, the new town developments could not address the increasing demand for affordable housing for low-income households in the SMA. Third, urban decline in the existing communities has emerged as an important planning issue under rapid suburbanization by government-initiated suburban development. Recently, the Korean government announced a shift in planning paradigm from urban development to urban regeneration. The urban regeneration policy, called the New Deal for Urban Regeneration, focuses on retrofitting existing urban areas with small-scale infill development and public participatory community development in residential areas.

Guide to further readings

Amsden, A.H. (1989) *Asia's Next Giant: South Korea and Late Industrialization*. New York: Oxford University Press.

Bae, C.C. and Jun, M.J. (2003) "Counterfactual Planning: What If There Had been No Greenbelts in Seoul?" *Journal of Planning Education and Research*, 22(4), pp. 374–383.

Hannah, L., Kim, K.H., and Mills, M. (1993) "Land Use Controls and Housing Prices in Korea." *Urban Studies*, 30(1), pp. 147–156.

Kim, C.H. and Kim, K.H. (2000) "The Political Economy of Korean Government Policies on Real Estate." *Urban Studies*, 37(7), pp. 1157–1169.

Korea Planning Association (2009) *Oral History of Urban Planning of a Half Century of Korea*. Seoul, Korea: Bo Sung Gak (in Korean).

Lee, G.H. (1999) *Public Concept of Land Ownership and New Towns*. Seoul, Korea: Samsung Economic Research Institute (in Korean).

References

Chang, J.S. and Lee, C.M. (2006) "A Study on the Change of Self-Containment Level of the Five New Towns in the Seoul Metropolitan Area." *Journal of Korean Planners Association*, 41(2), pp. 41–56 (in Korean).

Go, D.H. and Choi, C.G. (2013) "Commuting Distance and Mode Choice of Residents in the New Developments and the Existing Urban Areas in Gyeonggi-Do." *Journal of Korean Planners Association*, 48(2), pp. 83–106 (in Korean).

Gyeonggi-Do (2015) "Introduction to Greenbelt Policy." Department of Urban and Housing Development, Gyeonggi-Do, Korea.

Kahng, B.K. (1980) "'ROSARIO' Metropolis for Seoul 2001: A Strategic Frame for Joint Development along Subway System." In: Korea Planners Association (ed.) *The Year 2000: Urban Growth and Perspectives for Seoul*. Seoul: Korean Planners Association, pp. 492–511.

Kahng, B.K. (1989) "The Problems from Urban Planning Perspectives: Validity and Problems of the New Town Developments for Bundang and Ilsan." *Korean Society of Civil Engineers Magazine*, 37(4), pp. 74–79 (in Korean).

Kim, E.W. (1983) *A History: National Physical and Spatial Development in Korea*. Seoul, Korea: Daehak Doseo (in Korean).

Korean Statistical Information Service (2017a) "Population Census, 1966–2016." Available at: http://kosis.kr/index/index.jsp.

Korean Statistical Information Service (2017b) "International Statistical Yearbook, 1966–2015." Available at: http://kosis.kr.

Korean Statistical Information Service (2017c) "Trade Statistics by Standard International Trade Classification (SITC), 1966–2015." Available at: http://kosis.kr.

Lee, M.H. (2008) "Urban Development Related Laws and Special Law's System." In: Korea Planners Association (ed.) *Urban Development*. Seoul: Korean Planners Association, pp. 247–271 (in Korean).

Lee, S.G., Joo, M.J., and Ha, J.H. (2015) "An Analysis of Changes in Commuting Characteristics and Urban Spatial Structure of the First Generation New Town in the Seoul Metropolitan Area (1996–2010): Focused on Self-Containment and Centrality." *Journal of Korea Planning Association*, 50(5), pp. 5–22.

Ministry of Land, Infrastructure and Transport (MLIT) (2015) *A Handbook of Housing Administration Work*. Seoul: Ministry of Land, Infrastructure and Transport.

Ministry of Land, Transport and Maritime Affairs (MLTMA) (2011) *Development Restriction Area 1971–2011*. Seoul: Ministry of Land, Transport and Maritime Affairs (in Korean).

Ministry of Land, Transport and Maritime Affairs (MLTMA) (2012) *A Handbook of Bogeumjari Housing*. Ministry of Land, Transport and Maritime Affairs (in Korean).

Seoul Research Data Service (SRDS) (2017) "Seoul 2013 through Maps." Available at: http://data.si.re.kr/map-seoul-2013.

Son, J.M. (2003) *The Story of Seoul Urban Planning*. Paju, Korea: Hanul (in Korean).

Son, J.M. (2005a) *The Story of City in Korea for 60 Years 1*. Paju, Korea: Hanul (in Korean).

Son, J.M. (2005b) *The Story of City in Korea for 60 Years 2*. Paju, Korea: Hanul (in Korean).

Part III

Diversity, exclusion, and poverty in the suburbs

11

Queer suburbs
(Sub)urban spaces and sexualities in the Global North

Andrew Gorman-Murray and Catherine J. Nash

Introduction

The majority of the population in the Global North lives not only in urban areas – to put it more accurately, they live in the suburbs surrounding these cities. In Australia, for instance, four out of five people live in the suburbs comprising the country's sprawling metropolitan regions. Since these suburbs house the population's majority, they reflect its social and cultural diversity. However, fairly narrow imaginaries of suburban life prevail in the Global North, perhaps particularly in settler-societies with expansive, low-density suburbs such as Australia, Canada, and the United States. The particular social norms that underpin "expected" sexual identities and relationships are prominent, with the ideal of the heterosexual nuclear family often synonymous with both imaginaries of suburban life (Blunt and Dowling, 2006; Gorman-Murray, 2007) and their translation into urban planning and practice (Johnson, 2000; Howard, 2013). This chapter instead trains a lens on "other" sexualities and their relationships with suburban life, morphologies, and imaginaries. We focus on lesbian, gay, bisexual, transgender (LGBT), and queer experiences of the suburbs in the Anglophone Global North. We consider, on the one hand, the development of neighborhoods specifically associated with LGBT and queer identities and communities, and on the other hand, the experiences of LGBT and queer people in the context of "mainstream" suburbia. The latter remains underdeveloped in extant research agendas, but given growing political and social acceptance of LGBT and queer identities and relationships, demands greater attention.

Before moving on, we want to contextualize our interest in and approach to this topic. As geographers, we understand the production of knowledge to be situated geographically as well as socially. It is thus important to provide the geographical and scholarly context of our interest in suburbia, since this will aid readers' understanding of our approach. Since the early 2000s both authors have individually undertaken research on the historical and contemporary relationships between urban spaces and sexualities in our respective national fields, Australia and Canada (Gorman-Murray, 2006, 2007, 2013b; Gorman-Murray and Waitt, 2009; Nash, 2005, 2006, 2013b, 2014). Since 2012, we have collaborated on examining ongoing socio-spatial changes in gendered and sexual landscapes in each country's key global city, Sydney, and Toronto, in an effort to understand these changes within a transnational comparative context. We have examined the formation, transformation, and lived experience of urban neighborhoods linked to LGBT and queer communities, including the effects of planning, politics, and social mobility on inner-city suburbs (Gorman-Murray and Nash, 2014, 2016, 2017; Nash and Gorman-Murray, 2014, 2015a, 2015b). We extend our interests and approach in this chapter: thinking about urban life in the Anglophone Global North, we consider LGBT and queer experiences in relation to both inner-city "queer-friendly" suburbs and wider "mainstream" suburban landscapes.

It is important to define our key terms. Regarding suburbia, Johnson's (2006) adaption of Harris and Larkham's (1999) cross-national common elements of "a suburb" is useful: a largely residential location peripheral to an urban core, with a distinctive place identity and way of life; its landscape is often characterized by low-density, freestanding dwellings, and its social environment by heterosexual, nuclear family households and conventional gender roles (female homemakers, male breadwinners). Distinctions, sometimes more imaginative than factual, are also made between inner-city and middle-to-outer-ring suburbs, with inner-city suburbs perceived as more socially diverse and liberal than the supposedly homogenous and conservative suburbs further from the urban core (Johnson, 2000, 2006). LGBT – lesbian, gay, bisexual, transgender – is a now common acronym in the Global North that refers to a group of sexual and gender identities that contest heterosexual and cisgendered (i.e., conventionally masculine and feminine) norms. Queer is a more recent term (at least in its contemporary political and social usage) that questions sexual and gender binaries altogether (i.e., heterosexual/homosexual, man/woman) and urges appreciation of fluid and multiple sexual and gender subject positions. Some have taken up "queer" to challenge "fixed" LGBT politics and communities as well as mainstream heteronormativity, masculinity, and femininity. This distinction is important as it arguably has some bearing on the recent "decline" of inner-city "gayborhoods," discussed below (Nash, 2013b; Gorman-Murray and Nash, 2014).

The chapter proceeds as follows. We begin by discussing the development of gayborhoods, or inner-city suburbs associated with LGBT identities and communities (though often dominated by gay identities and communities). We then consider how wider suburbia – "mainstream" middle-to-outer-ring suburbia – has been understood and experienced by gay men in comparison to gayborhoods. However, gender intersects in different ways with sexual orientation in the construction of suburbia (Doan, 2015), so the third section brings attention to lesbian spaces across the inner and outer suburbs. Finally, we return to the inner-city to discuss changes in LGBT and queer urban spaces. Transformations and dissipations of some gayborhoods have been paralleled by the coalescence of new "queer neighborhoods" in cities such as Sydney and Toronto (Gorman-Murray and Nash, 2017; Nash and Gorman-Murray, 2015b). In our discussions, we draw on published research from diverse disciplines regarding urban spaces and sexualities in the Anglophone Global North.

Gayborhoods

Research on the development of inner-city gayborhoods in the Anglophone Global North is amongst the earliest and most enduring scholarship on urban spaces and sexualities. In disciplines such as urban sociology and urban geography, research on gayborhoods in U.S. cities emerged in the late 1970s and early 1980s (Castells, 1983; Levine, 1979). "Gayborhood" is a relatively new moniker for these inner-city suburbs; earlier terms, still sometimes used, include "gay ghetto," "gay village" and "gay enclave." All are used to describe a neighborhood that houses a gay residential, commercial, and service concentration – a spatial clustering of gay households along with leisure venues, businesses, and community organizations serving gay clientele. In one of the earliest discussions – but one that retains currency – Levine (1979, p. 364) suggested that "an urban neighborhood can be" seen as a gay ghetto, village, or enclave "if it contains gay institutions in number, a conspicuous and locally dominant gay subculture that is socially isolated from the larger community, and a residential population that is substantially gay." Soon after Levine's statement, in a ground-breaking case study, Castells (1983) demonstrated that gay men were intimately involved in the renovation of commercial spaces and housing stock in San Francisco's Castro area, effectively territorializing this neighborhood through property development. The Castro has been described as "the most famous example of a gay urban neighborhood" (Valentine, 2002, p. 146).

The work of Levine (1979) and Castells (1983) makes clear that early gayborhood development had specific gender, class, and race parameters, largely impelled by the interests of middle-class, white gay men (Knopp, 1995). While potentially exclusionary of lesbians, bisexuals, and trans people – a point to which we return later – the congruence of gayborhoods with middle-class white gay men in particular nevertheless allowed pioneering researchers to conceptualize reasons underpinning their development. Castells (1983, p. 161) argued that gayborhoods emerged in response to social, political, and economic imperatives, "to build up a new community at a financial and social cost that only 'moral refugees' are ready to pay." A gayborhood like the Castro provided a material (economic) and territorial (spatial) base for political organization and sub-cultural self-definition. Lauria and Knopp (1985) and Knopp (1990a) advanced these theoretical insights into gay men's territorialization of inner-city suburbs in U.S. cities. They identified place-making processes as a response to wider social oppression, transforming urban spaces "in such a way as to reflect gay cultural values and serve the special needs of individual gays vis-à-vis society at large" (Lauria and Knopp, 1985, p. 159). They neatly summarized the thesis thus: "Gays, in essence, have seized the opportunity to combat oppression by creating neighborhoods over which they have maximum control and which meet long-neglected needs" (Lauria and Knopp, 1985, p. 161). Gayborhood development thus provided a geographical base for rights' claims and electoral representation.

Knopp (1990a, 1990b) also examined the intertwining of gay territorialization of inner-city suburbs with gentrification and urban regeneration. These economic processes produced spatial marginalization and exclusion *within* LGBT populations along axes of gender, class, race, and age. It was mainly white, working-age, middle-class gay men who were able to muster the financial resources to buy and renovate residential and commercial properties, with surplus income and time to spend in the local businesses and venues created (Knopp, 1995, 1998). Castells (1983) also suggested that men, but not women, were inherently territorial in order to offer a reason for gay men's visible territorialization vis-à-vis lesbians. Adler and Brenner (1992), Rothenberg (1995), and Bouthillette (1997), among others, have shown this was, and is, not the case, and that lesbian urban communities certainly existed, although they took less palpable and commercial forms than gay men's. We expand upon this critique later. Subtle and overt exclusions also operated around race and age in the public venues and spaces of gayborhoods (Caluya, 2008; Gorman-Murray, 2013a). As a result of these social and economic processes, gayborhoods became spaces where certain groups of gay men fit, but fractions of the LGBT and queer communities have been marginalized – lesbians, bisexuals, trans people, as well as non-white, non-wealthy, and older gay men.

While early insights focused on well-known U.S. examples, including San Francisco's Castro and New York's Greenwich Village, since the 1990s scholarship has expanded within and beyond the U.S., with studies investigating gayborhood development in Los Angeles, U.S. (Forest, 1995), Chicago, U.S. (Winkle, 2015), Vancouver, Canada (Bouthillette, 1997), Toronto, Canada (Nash, 2005), Sydney, Australia (Murphy and Watson, 1997), Manchester, U.K. (Quilley, 1997), and London, U.K. (Collins, 2004), inter alia. Collins's (2004) evolutionary model of gay village development is important. Based on observations of Soho, London, he devised a four-stage model of gayborhood evolution, which underscored entwined economic, social, and political forces propagating gay urban spaces. His model describes the dynamism of gayborhoods, which typically begin with a "beachhead" gay bar in a marginal inner-city location, bringing in gay clientele; this attracts additional gay venues, consolidating a commercial cluster that draws a gay residential population along with a diversity of secondary businesses and services catering to this population. However, in the fourth stage, gayborhood success encourages both "mainstream" colonization and property price increases that may force out gay residents and institutions, leading to decline. The model has been applied elsewhere, with particular focus on understanding the "final stage." For instance, Ruting (2008) and Lewis (2015), for Sydney and Washington, DC, respectively, interrogate diverse trajectories of "decline," including assimilation, dissipation, dispersal, fragmentation, relocation, and replacement. Collins and Drinkwater (2017) revisited the model and integrated these complications. We revisit the geographies of gayborhood decline later; first, we discuss gay lives in "mainstream" suburbia.

Gay suburbia

If gayborhoods are perceived as "gay territories" enabling security and freedom, however diffuse and permeable, then we must also consider from what and where these territories have been supposedly sequestered. While the urban/rural binary is one well-established spatial trope – where the countryside is seen as oppressive and the city as liberating (Weston, 1995; Bech, 1997) – another, perhaps more crucial, juxtaposition is posited between the "queer" inner-city and "heteronormative" suburbia. With reference to Sydney, for instance, both Hodge (1995) and McInnes (2001) have highlighted how inner-city gay territories and identities are constructed in relation to middle-to-outer-ring suburbia, often by the city's gay media. In this relational geography, the suburbs become the "non-gay" other of gayborhoods like the Oxford Street "pink triangle," which encompasses parts of the inner-city suburbs of Darlinghurst, Surry Hills, and Paddington (cf. Kirby and Hay, 1997 on Adelaide). This spatial imaginary has a problematic dual function. On the one hand, it normalizes the inner-city as "safe" gay territory, which is often not the case, as this visible gay concentration can attract anti-gay violence (Tomsen, 2009). On the other hand, it elides the lives and experiences of gay men (and other LGBT and queer people) living in the suburbs.

Unfortunately, this has been mirrored in research: with fewer visible gay venues, services, and residents, less scholarly effort has been invested in understanding suburban gay lives (Doan, 2015). Nevertheless, work by sociologists and geographers in the 1970s, 1980s and 1990s was attentive to the role of sexuality and intimacy in suburbia, underscoring the heterosexual ideals suffusing "suburban dreams." Game and Pringle (1979) and Johnson (1993), for instance, demonstrate how suburban development in Australia was impelled by interlocking desires for opposite-sex marriage, nuclear family formation, and homeownership, rendering the suburbs both material and imaginative sites of heterosexual marriage, home life, and reproduction. Perhaps it is no wonder that the suburbs have been cast as the gayborhood's "other." But as Hodge (1995) equally demonstrates for Australian suburbia, many gay men have "suburban dreams" and make their homes there. Since the 1980s, a small but slowly growing number of studies have begun to investigate suburban gay lives in the Anglophone Global North (Dines, 2009; Tongson, 2011).

Perhaps the earliest study was Lynch's (1987, p. 13) ethnographic research on "the lifestyles of white, middle-class suburban homosexuals" in a Californian city. Through his four-year field study, together with a comparison to extant accounts of gayborhoods, he found that these suburban gay men were less "out" than inner-city denizens, and adhered to "suburban dreams" of building a career, owning a home, and obtaining a life partner. He surmised that the middle-class suburban environment inhibited homosexual identity formation and that suburban gay men were individualistic and assimilationist, concluding that "the lure of large cities and gay ghettos has faded" (Lynch, 1987, p. 13). His findings prefigured later conceptual developments: homo-normativity, or assimilation to coupled, domestic, consumerist, middle-class lifestyles (Duggan, 2003); and the decline of inner-city gayborhoods as cultural, political, and economic touchstones for gay communities (Ghaziani, 2014). Kirkey and Forsyth (2001) further investigated differences in suburban and inner-city gay identity work in the U.S., and Brekhus (2003) highlights the diversity of gay suburban lives. Brekhus's (2003) ethnographic study of a New Jersey suburb revealed different modes of gay identity work in suburbia. "Commuters" travel to gay leisure venues to be "out," but adopt "generic" attributes in suburban environments; "integrators" live openly gay lives in the suburbs, but see this as one part of their holistic identities. In comparison, Kirkey and Forsyth's (2001) interviews with gay men living on the "metropolitan edge," in Massachusetts' Connecticut River Valley, found a "gay life" that was nonurban and home-centered but also positively integrated with the wider community.

Studies of gay suburbia in Australia have been significant, too. Wotherspoon's (1991) history of Sydney's gay subculture describes twentieth-century suburban gay life as clandestine, but also notes how rising car ownership post-World War II enabled easier travel both between suburbs and to the emerging Oxford Street gayborhood. In a ground-breaking work on gay men in Sydney's western suburbs, Hodge (1995, pp. 41, 43) argued that "broadening . . . research into suburban spaces . . . can diversify our understanding of the relationship between sexuality and space and may . . . reveal strategies for overcoming homophobia that may not have been clear in research on gay territories."

While the suburban imaginary has "played a role in constructing gay spaces (and therefore gay identities) by becoming [inner-city] gay Sydney's 'other,' there is also a diversity of sexual practices and sexual identities in the suburbs" (Hodge, 1995, p. 41). He argued, "Gay men in western Sydney have a different relationship to the 'wider heterosexual world' than that experienced by the inner-city gay community" through the affordances of the suburban home (Hodge, 1995, p. 46). The home was, and is, a site in which gay relationships and identities can be affirmed, and through which relations with friends and neighbors meant "bridges could be built between straight and gay" (Hodge, 1995, p. 46). Gorman-Murray (2007, 2012) furthered this work, highlighting the paradoxical role of the suburban home for gay men. Using interview data, he found that the home is a place where gay identities, relationships, and rights' claims can be publicly asserted – to friends, family, and communities, for instance – thus queering the "suburban dream" and reinforcing Hodge's (1995) suggestion of "building bridges" through the home. Simultaneously, he contended that homes are not utterly private, but public sites where gay lives are surveilled by neighbors, salespeople, and tradespeople. Through this tension, we might understand the suburban home as a site where gay identities and rights are continuously negotiated.

Lesbian urban spaces

Research into lesbian urban spaces, like work on gay suburbia, has offered a corrective to the early focus on gayborhood development. Moreover, this work has countered the early centering of (some) gay men's experiences in social and geographical research, and brought a more nuanced gendered lens to our understanding of the relationship between sexuality and space. Beginning in the late 1980s, researchers began to examine processes of urban territorialization by lesbians.

This was partly a response to problematic gendered assertions Castells (1983, p. 140) made in his earlier study of gay territorialization in the Castro, in which he argued, "Lesbians, unlike gay men, tend not to concentrate in a given territory, but establish social and interpersonal networks." Work by Winchester and White (1988), Adler and Brenner (1992), and Valentine (1995) on Paris, and an unnamed American and British city, respectively, contested these assumptions. Lesbians sometimes concentrate in certain neighborhoods, but these communities are often less visible than gay male enclaves because lesbians, like heterosexual women, have less access to capital than gay or heterosexual men, and "because a fear of male violence deters their willingness to have an obvious presence in the landscape" (Valentine, 2002, p. 148). Castells (1983, p. 140) also assumed that lesbians were "more concerned with the revolution of values than with the control of institutional power," and this has some support in later work. Valentine (2002, p. 148) contends that due to "the influence of feminism lesbian 'communities' have tended to be more radical, politicized, and less materially oriented than gay men, which has stymied the development of businesses and bars run for, and by, women."

Since the 1990s, researchers have analyzed a range of lesbian neighborhoods in the Anglophone Global North, particularly in U.S. and Canadian cities. A number of more-or-less visible lesbian enclaves have been discussed, with their development and morphology differing from gayborhoods. As Valentine (2000, p. 3) argues, "Lesbians do create spatially concentrated communities but . . . these neighborhoods are often composed of clusters of lesbian households and sometimes countercultural institutions such as alternative bookstores or cooperative stores rather than commercial bars and institutions." Rothenberg's (1995) study of Park Slope, in New York City's borough of Brooklyn, was an insightful riposte to Castell's (1983) suggestion that lesbians do not territorialize while highlighting the distinct nature of lesbian neighborhoods and their formation. The intersection of the women's movement with early waves of gentrification provided a socio-spatial context for lesbians to engage in "sweat equity" housing development in Park Slope, attracting a cluster of lesbian households – similar to early gay gentrification in the Castro (Castells, 1983). However, the reasons for their continued concentration differed: "word-of-mouth, not statistical information, is what lures women to a 'lesbian neighborhood'" (Rothenberg, 1995, p. 169). The "power of lesbian social networking" (p. 177) was crucial for the consolidation of a lesbian neighborhood, captured in the title of Rothenberg's piece: "And she told two friends."

Other work, particularly from Canada, has progressed our understanding of relationships between lesbian urban spaces and gayborhoods. Bouthillette's (1997, p. 214) comparison of "the lesbian neighborhood versus the gay male enclave" in Vancouver, Canada – respectively, the Drive and the West End – highlighted divergent characteristics of lesbian and gay inner-city concentrations. The Drive was understood as symbolizing "grounded multiculturalism and family living in inner-city Vancouver" (p. 215) vis-à-vis body-conscious commercialism evident in the West End (cf. Lo and Healy, 2000). Similarly, in her research on Montreal's lesbian community, Podmore (2001, 2006) contends that lesbian place-making differs from gay men's, occurring in "spaces of difference" – marginal inner-city areas marked by demographic diversity and mixed land uses, such as the Plateau neighborhood. Countering visibility as central to territorialization, like Rothenberg (1995), she documents vibrant social networks operating in "spaces of difference" that create quasi-underground lesbian communities. In their work on lesbian geographies in Toronto (Canada) and Sydney (Australia), Nash and Gorman-Murray (2015a) extend these ideas by discussing the highly mobile nature of lesbian urban spaces, which seem to disperse and coalesce in different inner-city suburbs with more frequency than gay urban spaces. In Sydney, for instance, lesbian community spaces have moved across the inner-city since the 1980s, from Darlinghurst to Leichhardt to Newtown and adjacent suburbs. As Podmore (2001, p. 335) contends, understanding lesbian geographies requires recognition of how "gender mediates geographies of sexualities in the urban landscape."

Extant research suggests that lesbians are more likely than gay men to reside and be visibly present in the middle-to-outer-ring suburbs. Forsyth (1997) provided a critical early pointer to this in her study of the lesbian concentration in Massachusetts' Connecticut River Valley area, in and around the small city of Northampton. She examined how lesbian households and services were distributed in the Valley, finding visible concentrations in fringe suburban and peri-urban areas, as well as a lesbian "service core" in Northampton. Forsyth (1997, p. 57) suggested this "raises questions about the general diversity of the semirural fringe" and "the relative locations of lesbians and gay men." Using 2006 census data and geographic information systems, Gorman-Murray and Brennan-Horley (2011) mapped the relative location and concentration of same-sex couple family households in Australia using location quotients (i.e., residential concentrations as a proportion of the national average), disaggregated for female and male couples (a proxy for lesbian and gay couples). They found that female couples were more diffused than male couples and more likely to reside in above-average concentrations in suburban, outer-metropolitan, and peri-urban locations. From this we can at least suggest that lesbian couples are more likely to be suburbanites than gay couples. Likewise, Doan's (2015, p. 119) study of Tallahassee (U.S.) found "a clear preference among gay men for residence in or near downtown Tallahassee compared to the stronger preference among lesbians for housing near the urban fringe." There are multinational inferences, then, that lesbians are more likely than gay men to take up home in the suburbs in the Anglophone Global North.

Queer neighborhoods

In this final section, we return to inner-city gayborhoods and consider their future. A number of urban sociologists, geographers, and historians across various countries – the U.S. Canada, Australia, and Britain – have begun to note that long-established gayborhoods appear to be in "decline" (Collins, 2004; Ruting, 2008; Reynolds, 2009; Brown, 2014; Ghaziani, 2014; Collins and Drinkwater, 2017). As long ago as 1987, Lynch suggested that the attraction of "gay ghettos" was fading, but the twenty-first century has seen this become a prominent concern: as Brown (2014, p. 457) argued in his review of geographical work on gayborhoods, they have been perceived as "receding in size, scope, and function." Concerns have been raised in the media and in research about the future of some long-standing gayborhoods, including San Francisco's Castro, New York's Greenwich Village, London's Soho, Toronto's Church-Wellesley Village and Sydney's Oxford Street, and the implications for community and health service provisions (Rosser et al., 2008).

Gayborhood decline has been attributed to a number of diverse factors. Changing economic landscapes comprise one critical reason. For instance, Ruting's (2008) analysis of Sydney's Oxford Street, Doan and Higgins's (2011) of Atlanta's Midtown, and Collins and Drinkwater's (2017) of gay villages in England all identified adverse effects from ongoing waves of gentrification. On the back of initial "sweat equity" gentrification undertaken by LGBT people, super-gentrification by (often) "mainstream" investors and wealthy buyers has impelled ever-rising house prices and residential and commercial rents. Continuous cost increases for residential and commercial properties have become prohibitive for some, forcing LGBT residents (especially renters) and businesses out of gayborhoods. Simultaneously, super-gentrification inhibits new LGBT constituents – especially those with fewer financial means – from moving into gayborhoods. The result has been a deconcentration of LGBT households, leisure venues, and services in cities in the U.S., Canada, Britain, and Australia. At the same time, other social changes have reduced the imperative for LGBT and queer people to relocate to gayborhoods. Some research has suggested that more liberal social attitudes, together with new legal and political gains, have made it easier for LGBT and queer people to continue living openly in "mainstream" suburban and regional spaces (Gorman-Murray and Waitt, 2009; Nash, 2013a, 2013b; Collins and Drinkwater, 2017), stemming migration to gayborhoods. Other research has argued that the rise of online and mobile digital technologies – dating and community websites and apps – has made it easier for LGBT and queer people to find partners and form friendships and communities online, beyond physical meeting spaces afforded by gayborhoods (Batiste, 2014; Roth, 2016; Collins and Drinkwater, 2017). For example, Roth (2016, p. 441), referencing Tongson (2011), contends, "Grindr's [a gay dating and hook-up app] model of community has . . . more in common with the 'remote intimacies' of the suburbs . . . than with the classic forms of urban gay life." Furthermore, some researchers proffer generational changes in identity politics, and that identities such as "gay" and "lesbian" are yielding to "post-gay" and "queer" sensibilities, with these subjects eschewing the fixity of gay villages (Brown, 2004; Reynolds, 2009; Nash, 2013a, 2013b).

Yet, the desire for socio-spatial concentration cannot be dismissed: entwined with processes of decline, recent research in the Anglophone Global North has identified new "alternative" neighborhoods with LGBT and queer concentrations, distinct from "traditional" gayborhoods, materializing in other inner-city suburbs. Researchers from Australia, Canada, Britain, and the U.S. have documented the development of some of these "queer neighborhoods," including Newtown, Sydney (Gorman-Murray and Waitt, 2009; Gorman-Murray and Nash, 2014), Parkdale, Toronto (Nash, 2013a; Nash and Gorman-Murray, 2015b), Shoreditch and Vauxhall, London (Andersson, 2009, 2011) and Berkeley, Oakland/San Francisco (Compton and Baumle, 2012). Our work has focused on understanding the coalescence of queer neighborhoods such as Newtown in Sydney and Parkdale in Toronto, including their distinction from and ongoing connections to traditional gayborhoods in these cities (Gorman-Murray and Nash, 2014, 2016, 2017; Nash and Gorman-Murray, 2014, 2015a, 2015b). While imperfect, we termed these "queer neighborhoods" not just to distinguish them from "gayborhoods," but because we, along with researchers cited previously, found that these alternative urban spaces are perceived and often experienced as enabling a more expansive range of gendered and sexual subjects than gayborhoods, and are thus differently woven into the urban fabric.

While acknowledging ongoing tensions, arguably queer neighborhoods encompass a broader LGBT constituency than gayborhoods – including differences of gender, ethnicity, class, and age (Andersson, 2009; Compton and Baumle, 2012; Nash, 2013a; Nash and Gorman-Murray, 2015a) – as well as potential for more cohesive articulations between LGBT and mainstream communities (Gorman-Murray and Waitt, 2009; Gorman-Murray and Nash, 2016). As noted, Newtown, in Sydney's inner western suburbs, is an example. Newtown and surrounding suburbs have been posited in Australia's LGBT media as the "alternative" center (to Oxford Street) of Sydney's LGBT community, housing a range of leisure venues (Gorman-Murray, 2006). Since the 1990s, a number of community services have moved from Oxford Street to Newtown, following a growing LGBT residential concentration (Gorman-Murray and Nash, 2014). In residential terms, Newtown houses the most concentrated lesbian population in Australia, and a visible trans and genderqueer community, alongside a gay community seen as more politicized and less image-conscious than Oxford Street (Nash and Gorman-Murray, 2015a, 2015b). Meanwhile, liberal politics, local settings, community organizations, and government initiatives foster mainstream recognition for LGBT people (Gorman-Murray and Waitt, 2009; Gorman-Murray and Nash, 2017). We stress that we do not offer the example of Newtown as an "ideal type" free of tension and exclusion – for instance, it is now confronting the challenges of super-gentrification and homophobic violence around commercial leisure spaces – but rather as an instance of a "queer neighborhood" that is emerging as an alternative to gayborhood decline.

Conclusion

On face value, a focus on urban spaces and sexualities might seem narrow for some readers. However, in this chapter we hope we demonstrated that LGBT and queer subjectivities and communities are tightly woven into – and help produce – a diverse range of urban spaces in the Anglophone Global North. Certainly, we have traversed a wide terrain, from well-known gayborhoods in inner-city suburbs, to gay lives in middle-to-outer-ring suburbia, to the development of lesbian spaces across the urban landscape, to the decline of gayborhoods and coalescence of new queer neighborhoods. In all cases, urban spaces associated with LGBT and queer people are never about the experiences of sexuality isolated from other social axes. Intersections with gender, class, race, and age also shape LGBT urban spaces, including internal parameters of inclusion and exclusion. Tensions and conciliations with mainstream communities are crucial too: LGBT urban spaces have changed over time, from territorial reactions to social and political exclusion, to place-based communities responding to shifting social attitudes, legal rights, and potentials for inclusion. Given these complex and ever-changing relationships between urban spaces and sexualities, this remains fertile ground for ongoing research.

Guide to further reading

Doan, P. (ed.) (2015) *Planning and LGBT Communities*. New York: Routledge.
Ghaziani, A. (2014) *There Goes the Gayborhood?* Princeton, NJ: Princeton University Press.
Stein, M. (2012) *Rethinking the Gay and Lesbian Movement*. New York: Routledge.

References

Adler, S. and Brenner, J. (1992) "Gender and Space: Lesbians and Gay Men in the City." *International Journal of Urban and Regional Research*, 16(1), pp. 24–34.
Andersson, J. (2009) "East End Localism and Urban Decay: Shoreditch's Re-Emerging Gay Scene." *The London Journal*, 34(1), pp. 55–71.
Andersson, J. (2011) "Vauxhall's Post-Industrial Pleasure Gardens: 'Death Wish' and Hedonism in 21st-Century London." *Urban Studies*, 48(1), pp. 85–100.
Batiste, D.P. (2014) "0 Feet Away: The Queer Cartography of French Gay Men's Geo-Social Media Use." *Anthropological Journal of European Cultures*, 22, pp. 111–132.
Bech, H. (1997) *When Men Meet: Homosexuality and Modernity*. Chicago, IL: University of Chicago Press.
Blunt, A. and Dowling, R. (2006) *Home*. London: Routledge.
Bouthillette, A. (1997) "Queer and Gendered Housing: A Tale of Two Neighbourhoods in Vancouver." In: Ingram, B., Bouthillette, A. and Retter, Y. (eds.) *Queers in Space: Communities, Public Places, Sites of Resistance*. Seattle: Bay Press, pp. 213–232.
Brekhus, W. (2003) *Peacocks, Chameleons, Centaurs: Gay Suburbia and the Grammar of Gay Identity*. Chicago, IL: University of Chicago Press.
Brown, G. (2004) "Cosmopolitan Camouflage: (Post-)Gay Space in Spitalfields, East London." In: Binnie, J., Holloway, J., Millington, S. and Young, C. (eds.) *Cosmopolitan Urbanism*. Oxon: Routledge, pp. 130–145.
Brown, M. (2014) "Gender and Sexuality II: There Goes the Gayborhood?" *Progress in Human Geography*, 38(3), pp. 457–465.
Caluya, G. (2008) "'The Rice Steamer': Race, Desire and Affect in Sydney's Gay Scene." *Australian Geographer*, 39(3), pp. 283–292.
Castells, M. (1983) *The City and the Grassroots: A Cross-Cultural Theory of Urban Social Movements*. Berkeley, CA: University of California Press.
Collins, A. (2004) "Sexual Dissidence, Enterprise, and Assimilation: Bedfellows in Urban Regeneration." *Urban Studies*, 41(9), pp. 1789–1806.

Collins, A. and Drinkwater, S. (2017) "Fifty Shades of Gay: Social and Technological Change, Urban Deconcentration, and Niche Enterprise." *Urban Studies*, 54(3), pp. 765–785.

Compton, D.R. and Baumle, A.K. (2012) "Beyond the Castro: The Role of Demographics in the Selection of Gay and Lesbian Enclaves." *Journal of Homosexuality*, 59(10), pp. 1327–1355.

Dines, M. (2009) *Gay Suburban Narratives in American and British Culture: Homecoming Queens*. Basingstoke: Palgrave Macmillan.

Doan, P.L. (2015) "Understanding LGBTQ-Friendly Neigborhoods in the American South." In: Doan, P.L. (ed.) *Planning and LGBTQ Communities: The Need for Inclusive Queer Spaces*. London: Routledge, pp. 111–123.

Doan, P.L. and Higgins, H. (2011) "The Demise of Queer Space? Resurgent Gentrification and the Assimilation of LGBT Neigborhoods." *Journal of Planning Education and Research*, 31(1), pp. 6–25.

Duggan, L. (2003) *The Twilight of Equality? Neoliberalism, Cultural Politics, and the Attack on Democracy*. Boston, MA: Beacon Press.

Forest, B. (1995) "West Hollywood as Symbol: The Significance of Place in the Construction of a Gay Identity." *Environment and Planning D: Society and Space*, 13(2), pp. 133–157.

Forsyth, A. (1997) "'Out' in the Valley." *International Journal of Urban and Regional Research*, 21(1), pp. 38–62.

Game, A. and Pringle, R. (1979) "Sexuality and the Suburban Dream." *Australian and New Zealand Journal of Sociology*, 15(2), pp. 4–15.

Ghaziani, A. (2014) *There Goes the Gayborhood?* Princeton, NJ: Princeton University Press.

Gorman-Murray, A. (2006) "Imagining King Street in the Gay/Lesbian Media." *M/C Journal: A Journal of Media and Culture*, 9(3). Available at: http://journal.media-culture.org.au/0607/04-gorman-murray.php.

Gorman-Murray, A. (2007) "Contesting Domestic Ideals: Queering the Australian Home." *Australian Geographer*, 38(2), pp. 195–213.

Gorman-Murray, A. (2012) "Queer Politics at Home: Gay Men's Management of the Public/Private Boundary." *New Zealand Geographer*, 68(2), pp. 111–120.

Gorman-Murray, A. (2013a) "Liminal Subjects, Marginal Spaces and Material Legacies: Older Gay Men, Home and Belonging." In: Taylor, Y. and Addison, M. (eds.) *Queer Presences and Absences: Time, Future and History*. Basingstoke: Palgrave Macmillan, pp. 93–117.

Gorman-Murray, A. (2013b) "Straight-Gay Friendships: Relational Masculinities and Equalities Landscapes in Sydney, Australia." *Geoforum*, 49, pp. 214–223.

Gorman-Murray, A. and Brennan-Horley, C. (2011) "The Geography of Same-Sex Families in Australia: Implications for Regulatory Regimes." In: Gerber, P. and Sifris, A. (eds.) *Current Trends in the Regulation of Same-Sex Relationships*. Sydney: Federation Press, pp. 43–64.

Gorman-Murray, A. and Nash, C.J. (2014) "Mobile Places, Relational Spaces: Conceptualizing Change in Sydney's LGBTQ Neighborhoods." *Environment and Planning D: Society and Space*, 32(4), pp. 622–641.

Gorman-Murray, A. and Nash, C.J. (2016) "LGBT Communities, Identities and the Politics of Mobility: Moving from Visibility to Recognition in Contemporary Urban Landscapes." In: Brown, G. and Browne, K. (eds.) *The Routledge Research Companion to Geographies of Sex and Sexualities*. London: Routledge, pp. 247–253.

Gorman-Murray, A. and Nash, C.J. (2017) "Transformations in LGBT Consumer Landscapes and Leisure Spaces in the Neoliberal City." *Urban Studies*, 54(3), pp. 786–805.

Gorman-Murray, A. and Waitt, G. (2009) "Queer-Friendly Neighborhoods: Interrogating Social Cohesion across Sexual Difference in Two Australian Neighborhoods." *Environment and Planning A*, 41(12), pp. 2855–2873.

Harris, R. and Larkham, P.J. (1999) "Suburban Foundation, Form and Function." In: Harris, R. and Larkham, P.J. (eds.) *Changing Suburbs: Foundation, Form and Function*. New York: Routledge, pp. 1–31.

Hodge, S. (1995) "'No Fags Out There': Gay Men, Identity and Suburbia." *Journal of Interdisciplinary Gender Studies*, 1(1), pp. 41–48.

Howard, C. (2013) "Building a 'Family-Friendly' Metropolis: Sexuality, the State, and Postwar Housing Policy." *Journal of Urban History*, 39(5), pp. 933–955.

Johnson, L.C. (1993) "Text-Ured Brick: Speculations on the Cultural Production of Domestic Space." *Australian Geographical Studies*, 31(2), pp. 201–213.

Johnson, L.C. (2000) *Placebound: Australian Feminist Geographies*. Melbourne: Oxford University Press.

Johnson, L.C. (2006) "Style Wars: Revolution in the Suburbs?" *Australian Geographer*, 37(2), pp. 259–277.

Kirby, S. and Hay, I. (1997) "(Hetero)Sexing Space: Gay Men and 'Straight' Space in Adelaide." *The Professional Geographer*, 49(3), pp. 295–305.

Kirkey, K. and Forsyth, A. (2001) "Men in the Valley: Gay Male Life on the Suburban-Rural Fringe." *Journal of Rural Studies*, 17(4), pp. 421–441.

Knopp, L. (1990a) "Some Theoretical Implications of Gay Involvement in an Urban Land Market." *Political Geography Quarterly*, 9(4), pp. 337–352.

Knopp, L. (1990b) "Exploiting the Rent-Gap: The Theoretical Significance of Using Illegal Appraisal Schemes to Encourage Gentrification in New Orleans." *Urban Geography*, 11(1), pp. 48–64.

Knopp, L. (1995) "Sexuality and Urban Space: A Framework for Analysis." In: Bell, D. and Valentine, G. (eds.) *Mapping Desire: Geographies of Sexualities*. London: Routledge, pp. 149–164.

Knopp, L. (1998) "Sexuality and Urban Space: Gay Male Identity Politics in the United States, the United Kingdom, and Australia." In: Fincher, R. and Jacobs, J.M. (eds.) *Cities of Difference*. New York: Guildford Press, pp. 149–176.

Lauria, M. and Knopp, L. (1985) "Toward an Analysis of the Role of Gay Communities in the Urban Renaissance." *Urban Geography*, 6(2), pp. 152–169.

Levine, M.P. (1979) "Gay Ghetto." *Journal of Homosexuality*, 4(4), pp. 363–377.

Lewis, N.M. (2015) "Fractures and Fissures in 'Post-Mo' Washington, DC: The Limits of Gayborhood Transition and Diffusion." In: Doan, P.L. (ed.) *Planning and LGBTQ Communities: The Need for Inclusive Queer Spaces*. London: Routledge, pp. 56–75.

Lo, J. and Healy, T. (2000) "Flagrantly Flaunting It? Contesting Perceptions of Locational Identity among Urban Vancouver Lesbians." In: Valentine, G. (ed.) *From Nowhere to Everywhere: Lesbian Geographies*. Binghamton: Harrington Park Press, pp. 29–44.

Lynch, F.R. (1987) "Non-Ghetto Gays: A Sociological Study of Suburban Homosexuals." *Journal of Homosexuality*, 13(4), pp. 13–42.

McInnes, D. (2001) "Inside the Outside: Politics and Gay and Lesbian Spaces." In: Johnston, C. and van Reyk, P. (eds.) *Queer City: Gay and Lesbian Politics in Sydney*. Annandale: Pluto Press, pp. 164–178.

Murphy, P. and Watson, S. (1997) *Surface City: Sydney at the Millennium*. Annandale: Pluto Press.

Nash, C.J. (2005) "Contesting Identity: The Struggle for Gay Identity in Toronto in the Late 1970s." *Gender, Place and Culture*, 12(1), pp. 113–135.

Nash, C.J. (2006) "Toronto's Gay Village (1969 to 1982): Plotting the Politics of Gay Identity." *The Canadian Geographer*, 50(1), pp. 1–16.

Nash, C.J. (2013a) "Queering Neighborhoods: Politics and Practice in Toronto." *ACME: An International Journal for Critical Geographies*, 12(2), pp. 193–213.

Nash, C.J. (2013b) "The Age of the 'Post-Mo'? Toronto's Gay Village and a New Generation." *Geoforum*, 49, pp. 243–254.

Nash, C.J. (2014) "Consuming Sexual Liberation: Gay Business, Politics and Toronto's Barracks Bathhouse Raids." *Journal of Canadian Studies*, 48(1), pp. 82–105.

Nash, C.J. and Gorman-Murray, A. (2014) "LGBT Neighborhoods and 'New Mobilities': Towards Understanding Transformations in Sexual and Gendered Urban Landscapes." *International Journal of Urban and Regional Research*, 38(3), pp. 756–772.

Nash, C.J. and Gorman-Murray, A. (2015a) "Lesbian Spaces in Transition: Insights from Toronto and Sydney." In: Doan, P.L. (ed.) *Planning and LGBTQ Communities: The Need for Inclusive Queer Spaces*. London: Routledge, pp. 181–198.

Nash, C.J. and Gorman-Murray, A. (2015b) "Recovering the Gay Village: A Comparative Historical Geography of Urban Change and Planning in Toronto and Sydney." *Historical Geographies*, 43, pp. 84–105.

Podmore, J.A. (2001) "Lesbians in the Crowd: Gender, Sexuality and Visibility along Montreal's Boul. St-Laurent." *Gender, Place and Culture*, 8(4), pp. 333–355.

Podmore, J.A. (2006) "Gone 'Underground'? Lesbian Visibility and Consolidation of Queer Space in Montreal." *Social & Cultural Geography*, 7(4), pp. 595–625.

Quilley, S. (1997) "Constructing Manchester's 'New Urban Village': Gay Space in the Entrepreneurial City." In: Ingram, B., Bouthillette, A. and Retter, Y. (eds.) *Queers in Space: Communities, Public Places, Sites of Resistance*. Seattle: Bay Press, pp. 275–292.

Reynolds, R. (2009) "Endangered Territory, Endangered Identity: Oxford Street and the Dissipation of Gay Life." *Journal of Australian Studies*, 33(1), pp. 79–92.

Rosser, B.R.S., West, W., and Weinmeyer, R. (2008) "Are Gay Communities Dying or Just in Transition? Results from an International Consultation Examining Possible Structural Change in Gay Communities." *AIDS Care*, 20(5), pp. 588–595.

Roth, Y. (2016) "Zero Feet Away: The Digital Geography of Gay Social Media." *Journal of Homosexuality*, 63(3), pp. 437–442.

Rothenberg, T. (1995) "'And She Told Two Friends': Lesbians Creating Urban Social Space." In: Bell, D. and Valentine, G. (eds.) *Mapping Desire: Geographies of Sexualities*. London: Routledge, pp. 165–181.

Ruting, B. (2008) "Economic Transformations of Gay Urban Spaces: Revisiting Collins' Evolutionary Gay District Model." *Australian Geographer*, 39(3), pp. 259–269.

Tomsen, S. (2009) *Violence, Prejudice and Sexuality*. London: Routledge.

Tongson, K. (2011) *Relocations: Queer Suburban Imaginaries*. New York: New York University Press.

Valentine, G. (1995) "Out and about: Geographies of Lesbian Landscapes." *International Journal of Urban and Regional Research*, 19(1), pp. 96–111.

Valentine, G. (2000) "Introduction: From Everywhere to Nowhere: Lesbian Geographies." In: Valentine, G. (ed.) *From Nowhere to Everywhere: Lesbian Geographies*. Binghamton: Harrington Park Press, pp. 1–9.

Valentine, G. (2002) "Queer Bodies and the Production of Space." In: Richardson, D. and Seidman, S. (eds.) *Handbook of Lesbian and Gay Studies*. London: Sage, pp. 145–160.

Weston, K. (1995) "Get Thee to a Big City: Sexual Imaginary and the Great Gay Migration." *GLQ: A Journal of Lesbian and Gay Studies*, 2(3), pp. 253–277.

Winchester, H.P.M. and White, P.E. (1988) "The Location of Marginalised Groups in the Inner City." *Environment and Planning D: Society and Space*, 6(1), pp. 37–54.

Winkle, C. (2015) "Gay Commercial Districts in Chicago and the Role of Planning." In: Doan, P.L. (ed.) *Planning and LGBTQ Communities: The Need for Inclusive Queer Spaces*. London: Routledge, pp. 21–38.

Wotherspoon, G. (1991) *City of the Plain: History of a Gay Sub-Culture*. Sydney: Hale and Iremonger.

12

Inequality and poverty in the suburbs

The case of metropolitan Cairo

Erina Iwasaki

Introduction

Cairo is one of the world's largest megacities and has long attracted scholars, writers, and travelers for its historical dimension. Recent attention has also drawn attention to Cairo as an exploding metropolis with sprawling suburban "unplanned" or "slum" areas, labeled *ashwa'iya* (Kipper and Fischer, 2009; Sims, 2003; Sims, 2010). An "unplanned" area is a residential area that has developed outside of regulations; such areas are estimated to contain around 62 percent of Cairo's metropolitan population and occupy 53 percent of the built residential areas in Greater Cairo (Sims and Séjourné, 2000, p. 4). They include residential areas that were originally agricultural land on which buildings were subsequently constructed, and government-owned land on which buildings were constructed by *wada' yad* ("put hands" on the land). As land transactions were typically informal affairs, the term "unplanned" refers to the lack of building permits, and thus denotes the problem of tenure insecurity that exists in the "unplanned areas."

The image of an unplanned area is commonly associated with the same characteristics as slums, such as a lack of infrastructure and public services, and the poverty that exists in such areas (Iwasaki, 2015). This image of poverty and its associated social ills has been reinforced, particularly since the 1990s, by the political protests that took place in the Cairo suburbs (Bayat and Denis, 2000; Dorman, 2009). Although several studies have been conducted recently on the "unplanned areas," they are limited to geographical or political studies concerned with urbanization or social unrest. Due to a lack of data about these places, we do not really know how poor these areas are, and we lack a profile of poor residents. National data, on which most income and poverty estimations rely, do not allow for analysis below the levels of regions or governorates.

The main objectives of this chapter are to estimate the poverty in Cairo, focusing on transient poverty, in order to provide insights into the state of suburban Greater Cairo. This chapter attempts to evaluate and clarify the state of poverty in the three unplanned areas in Greater Cairo as a case study using panel data. The panel data used in this chapter were collected via an exhaustive household survey conducted as part of joint research project between Hitotsubashi University (Tokyo) and CAPMAS. The survey (2005 Urban Household Survey) was conducted in 2005, and its follow-up survey was conducted in 2010 and 2011 to construct the panel data (Central Agency for Public Mobilization and Statistics, 2006, 2013a, 2013b).

In the following sections of this chapter, the data and survey areas are presented, and then estimations of poverty are put forth in the two study areas. Next, the determinants of the two types of poverty, transient and chronic poverty, are explained. The chapter concludes with a discussion of poverty in the metropolitan area of Cairo.

Overview of the survey and study areas

Study areas

The two study areas are the Bigam sub-district, belonging to the Shubra Khaima district in the Qalyubiya governorate, and the Zinin sub-district, belonging to the Bulaq Dakrur district in the Giza governorate (see Figure 12.1). The two areas are located in the suburb of Cairo where residential development has taken place especially since the late 1970s, as is demonstrated in the percentage of the buildings under construction.

Both Bigam and Zinin were originally villages surrounded by an agricultural field that existed for a long time. After the late 1970s, these agricultural fields changed rapidly to residential areas (Kato et al., 2004, 2005). Old residential areas that in the past had clear boundaries between the residential areas and agricultural land are surrounded by the recently built apartment buildings, and we hardly can distinguish where the original villages were.

Figure 12.2 displays the year of these buildings' construction in these two study areas, and they are informative of this urbanization. The year of construction, defined as when the ground was originally built, is gathered and mapped on the digital map elaborated by the fieldwork during the exhaustive survey in 2005. Major actors of this urbanization are the residents themselves. Most of the households moved from the densely populated center of Cairo or came from rural areas. They bought the agricultural land from the farmers directly or through local real estate agents, and later built houses with a first floor, then added a second floor, and so on. The flats are either given to their children after marriage, or they are rented to young couples seeking an apartment in which to start their new lives.

Sampling

In the 2005 Urban Household Survey, the two sub-districts were chosen using a systematic sampling method. First, we randomly selected one district from each of the governorates of Cairo, Qalyubiya, and Giza, which constitute Greater Cairo, and then we randomly selected one sub-district from each district. Because one sub-district in Greater Cairo has a large population, and there is no precise information available about the number and location of households in the "unplanned area" to enable us to conduct sampling, we chose one study area for each sub-district based on the criteria of location near the administrative border, and we conducted a preparatory survey to establish an exhaustive list of households.

All the households listed in the chosen location were surveyed using the questionnaire. As a result, households covered by the exhaustive survey totaled 1,420 in Bigam, and 1,974 in Zinin. The fieldwork was conducted from February to July 2005. The follow-up survey was carried out between December 2010 and January 2011. The panel households were selected from the households surveyed in the 2005 Urban Household Survey. The fieldwork continued until the number of panel households reached 600 in each area, and these were surveyed again using a questionnaire similar to that from the 2005 Urban Household Survey. The panel data were subsequently constructed by matching the household identification numbers of the 2010/2011 Panel Household Survey with those of the 2005 Urban Household Survey.

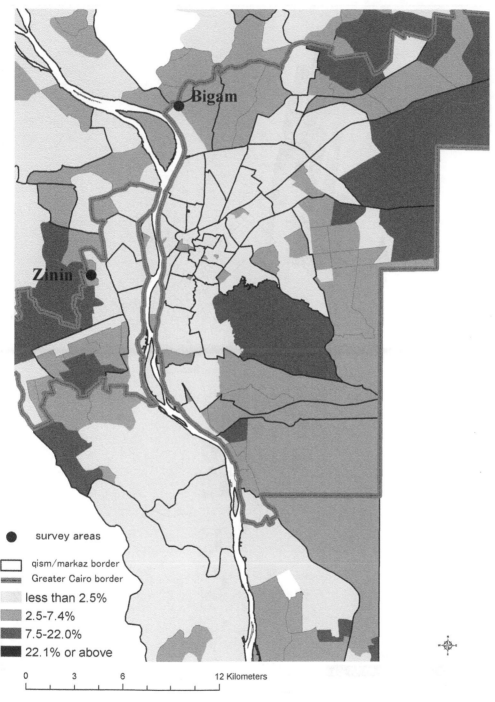

survey areas

qism/markaz border
Greater Cairo border

less than 2.5%
2.5-7.4%
7.5-22.0%
22.1% or above

0   3   6   12 Kilometers

Figure 12.1   Percentage of the buildings under construction in Greater Cairo
Source: 1996 Building Census

Figure 12.2   Year of construction of the surveyed areas in a) Bigam and b) Zinin

Source: 2005 Urban Household Survey

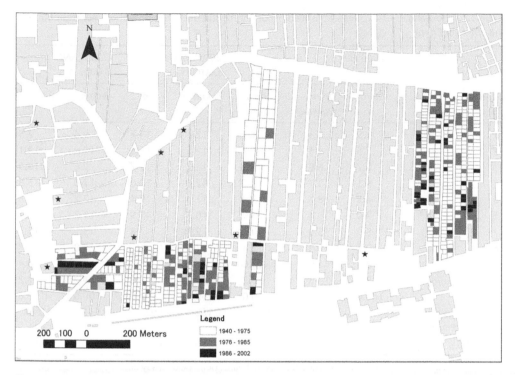

Figure 12.2 (Continued)

Poverty estimation at district level

Table 12.1 shows the proportion of households compared with the household consumption distribution for the whole of Egypt; this information was obtained from the Household Income, Expenditure, and Consumption Survey (HIECS), where it was officially used for consumption and poverty estimation in Egypt in 2005. Although Greater Cairo has a higher income level compared with the rest of country, there was a considerably large income disparity within it; the households in the urban parts of the Qalyubiya and Giza governorates have lower consumption levels compared with those of the Cairo governorate. The households in these two urban areas, with consumption levels similar to the lowest three deciles in the whole of Egypt, made up 14.8 and 13.2 percent, respectively, of the households surveyed in these two areas in the HIECS in 2005.

Table 12.1 also shows the distribution of the households living in the two study areas, classified along the household consumption deciles of the whole of Egypt in the HIECS in 2005. Because the definition of household consumption in our survey is the same as the HIECS, it is possible to compare the consumption level of the two study areas with the HIECS. Household consumption in our survey is the sum of annual consumption, including imputed rent.

The comparison clearly reveals that the study areas had remarkably low consumption levels compared with the average values for Greater Cairo. In fact, the proportions of households in the study areas comparable with the lowest three deciles of the whole of Egypt were 38.8 and 46.8 percent, respectively, in Bigam and Zinin. Therefore, although the households in Greater Cairo enjoy a high standard of living on average when compared with the rest of Egypt, there is a large disparity between the households inside the city, and these two study areas have considerably

Table 12.1 Households of two areas classified according to the deciles of household consumption for the whole of Egypt

Decile	2005 HIECS			2005 MASTER SAMPLES		2005 PANEL SAMPLES	
	Cairo	Urban Qalyubiya	Urban Giza	Bigam	Zinin	Bigam	Zinin
1	1.9	3.3	4.1	12.1	15.8	9.6	7.5
2	2.5	4.6	3.7	12.5	15.3	12.8	13.1
3	3.7	6.9	5.4	14.2	15.7	9.6	13.5
4	5.1	8.1	5.3	12.6	12.0	12.6	13.5
5	6.2	8.1	7.0	11.4	10.4	12.2	10.8
6	7.0	10.4	8.9	10.8	7.9	11.7	11.1
7	8.9	13.0	10.1	9.8	7.8	13.1	10.3
8	12.9	15.3	12.4	7.2	6.2	7.8	8.7
9	18.5	17.0	18.0	6.3	6.3	6.4	8.4
10	33.2	13.3	25.2	3.1	2.6	4.2	3.2
Total	100.0	100.0	100.0	100.0	100.0	100.0	100.0
Average consumption (Egyptian Pounds)				2,183	2,039	2,147	2,096
(Number)				5,985	8,355	2,616	2,498

Source: Iwasaki and El-Laithy (2013, p. 175), 2005 & 2010/2011 Panel Household Survey data.

lower consumption levels compared not only with the average of Greater Cairo but also with the whole of Egypt (Iwasaki and El-Laithy, 2013).

The panel households have relatively more homogeneous consumption levels than the overall households surveyed in 2005, and they were distributed less in the lowest deciles than the overall sample households. However, their consumption levels were similar to the overall sample households, and they lagged behind the overall households in urban Qalyubiya and Giza.

Estimation of poverty in the study areas

Estimation of consumption-based poverty using different lines

Table 12.2 summarizes the consumption-based poverty indices in the study areas using the method of poverty line measurement. The headcount index of the study areas using the household-specific poverty line was estimated to be 18.4 percent for Bigam, and 29.0 percent for Zinin in 2005. According to the World Bank (2002, p. 19), the regional headcount index in 2005 based on the HIECS, and using the household-specific lower poverty line, was 5.7 percent for the urban governorates in which the Cairo governorate is located, 9 percent for the lower urban governorates, 18.6 percent for the upper urban governorates, and 19.6 percent for all of Egypt. Thus, it is obvious that poverty in these two study areas is much more pronounced than the national data can tell us (Iwasaki and El-Laithy, 2013).

The poverty rates for the panel households were slightly higher than those for the overall sample households. The rates were 22.4 and 24.0 percent in Bigam and Zinin in 2005, respectively. This difference probably reflects age and household composition, because the households in the panel survey have older household heads and a larger proportion of adults than do the master

Table 12.2 Estimation of poverty indices in metropolitan Cairo

Poverty line		MASTER SAMPLES		PANEL SAMPLES			
		2005		2005		2010/11	
		Bigam	Zinin	Bigam	Zinin	Bigam	Zinin
Per capita lower poverty line (calculated from HIECS)							
	Poverty rate (P0)	20.6	24.0	25.8	21.5		
Household-specific poverty rate							
Food poverty line		1.9	4.0	2.8	2.9	11.0	7.9
Lower poverty line	Poverty rate (P0)	18.4	29.9	22.4	24.0	48.6	39.3
	Poverty gap (P1)	3.2	5.6	4.2	4.3	11.2	8.6
	Poverty severity (P2)	0.8	1.7	1.1	1.2	3.8	2.7
Upper poverty line		57.8	69.4	58.7	62.0	81.7	76.8
Number of samples (persons)		5,985	8,355	2,616	2,498	2,579	2,367

Source: 2005 and 2010/2011 Panel Household Survey data, 2005 Urban Household Survey data, 2005 Household Income, Expenditure, and Consumption Survey (HIECS).

Note: For the regional and "household-specific" poverty line in 2005, due to lack of data, we used the cost of calories evaluated by the *(World Bank, 2002)* based on the 2005 HIECS data. The food, lower, and upper poverty lines are calculated based on this methodology, using the cost of a minimum food bundle in each region, taking into account household size, gender, and age composition. The regional lower poverty line, on the other hand, is the per capita poverty line computed using the consumption expenditure data in the HIECS. For 2005, it was estimated to be 1,435.5 Egyptian Pounds (LE) per capita for Bigam, and 1,418.2 LE per capita for Zinin. It is the poverty line computed at the district (qism) level; the Shubra Khaima I district for Bigam, and the Bulaq Dakrur district for Zinin.

samples in the 2005 Urban Household Survey. These demographic factors raise the household-specific poverty line. In comparison with 2005, the poverty rate rose surprisingly: two times higher, reaching 48.6 percent in Bigam, and 39.3 percent in Zinin in 2010/2011.

Estimation using the upper poverty line raises the headcount index to such an extent that the majority of households in the study areas fall beneath this poverty line. In 2005, more than half of the households in these survey areas were already under the upper poverty line. In 2010/2011, this rate rose to 81.7 percent for Bigam, and 76.8 percent for Zinin, implying that many households that were non-poor in 2005 currently fell under the poverty line.

The poverty gap index (P1), which measures the depth of poverty, and the poverty severity index (P2), which measures the degree of inequality in distribution below the poverty line, also indicate the degradation of poverty. In 2005, similar characteristics were found in other regions; many households in the two study areas were concentrated around the poverty line. This means that many households were living at very limited consumption levels, and that small changes in consumption that were caused, for example, by a rise in the cost of living or the addition of new household members, would have easily affected the poverty incidence. In 2010/2011, both the depth and the severity of poverty increased.

Transition of consumption-based poverty between 2005 and 2010/11

One of the advantages of using panel data is that this enables study of the income poverty's dynamics and to estimate poverty mobility. The key concepts for understanding the dynamics of income poverty are "chronic" and "transient" poverty. Chronic (transient) poverty refers to a long-term (short-term) inability to meet minimum requirements.

Table 12.3 Distribution of five-year panel households by poverty status in 2010 in metropolitan Cairo

| | BIGAM | | ZININ | |
	Number	%	Number	%
Chronic poor	87	14.5	63	10.5
Transient poor	181	30.2	137	22.8
Previously poor	46	7.7	67	11.1
Never poor	286	47.7	335	55.6
Total	600	100.0	602	100.0

Source: 2005 and 2010/2011 Panel Household Survey data.

Here, we classify as "chronic poor" those who were income poor in the 2005 Urban Household Survey and remained so according to the 2010/2011 Panel Household Survey. Those classified as "transient poor" were non-poor in 2005, but they fell into poverty in 2010/2011. "Previously poor," on the other hand, are the households that were income poor in 2005, but they escaped from poverty in 2010/2011. "Never poor" are the categories of households whose consumption level was above the lower poverty line in both 2005 and 2010/2011.

The results using the lower poverty line show that more than a majority of the poor households were newly poor (or transient poor) and only fell below the poverty line in 2010/2011. From 2005 to 2010/11, about one-fifth (20.4 percent) of the panel households were considered chronic poor, whereas 10.4 percent in Bigam and 22.8 percent in Zinin were transient poor (see Table 12.3). Around 7.7 percent and 11.1 percent were previously poor, whereas 47.7 percent and 55.6 percent were considered never poor in the two areas, respectively. Interestingly, more than six out of ten households, on average, that were classified as poor in 2010/2011 were previously non-poor in 2005. That is more than double those who escaped from poverty in the same period and also exceeds by far those who stayed in chronic poverty.

The most important point to note here is that even though the number of households below the lower poverty line increased between 2005 and 2010/2011, not all the households that were poor in 2010/2011 remained poor. There were movements into and out of poverty. Although one out of ten households remained poor in 2010/2011 (chronic poor), 7.7 percent from Bigam and 11.1 percent from Zinin had their welfare improve to the extent of being categorized as above the lower poverty line in 2010/2011. Conversely, 30.2 and 22.8 percent of the total households in the two areas had their welfare decline from being above the poverty line in 2005 to below the lower poverty line in 2010/2011. In these two areas, entries into poverty exceeded exits largely out of poverty between 2005 and 2010/2011.

Characteristics of the poor

Profile of the poor

To examine the patterns of escaping poverty or falling back into poverty, we first look at the poor's profile in the 2010/2011 data, using basic indicators of household characteristics.

Table 12.4 summarizes a profile of residents in metropolitan Cairo. A profile of the chronic poor, using the lower poverty line, is similar to that found in urban and rural Egypt (Datt et al., 2001; Lokshin et al., 2010; World Bank, 2007; United Nations World Food Programme, 2013). Among the chronic poor, only 6.9 percent of households in Bigam and 15.9 percent in Zinin had

Table 12.4 Household characteristics of panel households, by poverty status, in metropolitan Cairo

	BIGAM					ZININ				
	NONPOOR		POOR			NONPOOR		POOR		
	Never poor	Previously poor	Transient poor	Chronic poor	Total	Never poor	Previously poor	Transient poor	Chronic poor	Total
Household heads										
Gender of head										
Female	17.1	13.0	5.5	6.9	11.8	26.6	22.4	19.7	15.9	23.4
Male	82.9	87.0	94.5	93.1	88.2	73.4	77.6	80.3	84.1	76.6
Age of head										
Less than 35	6.3	6.5	8.3	5.8	6.8	6.9	4.5	5.8	3.2	6.0
Aged 35–45	30.1	23.9	38.7	19.5	30.7	19.4	17.9	31.4	11.1	21.1
Aged 45–54	30.8	26.1	28.2	50.6	32.5	25.7	37.3	29.9	55.6	31.1
Above 55	32.9	43.5	24.9	24.1	30.0	48.1	40.3	32.9	30.2	41.9
Education of head										
Illiterate	23.8	30.4	16.0	29.9	22.8	28.7	29.9	29.2	41.3	30.2
Read and write	9.8	17.4	18.8	19.5	14.5	19.1	20.9	13.1	23.8	18.4
Primary/preparatory	14.7	17.4	16.6	23.0	16.7	14.6	16.4	21.9	14.3	16.5
Secondary	34.6	23.9	38.1	24.1	33.3	25.7	29.9	29.9	19.1	26.4
University and above	17.1	10.9	10.5	3.5	12.7	11.9	3.0	5.8	1.6	8.5
Work status of head										
Waged worker	64.5	68.3	72.7	74.7	68.7	55.6	61.4	62.2	78.6	60.4
Self-employed	16.9	22.0	18.0	14.1	17.3	15.6	7.0	13.5	10.7	13.6
Unemployed	0.4	0.0	2.5	0.0	1.0	0.7	3.5	3.4	0.0	1.6
Out of labor force	18.2	9.8	6.8	11.3	13.1	28.2	28.1	21.0	10.7	24.5
Social security of head										
With pension	82.5	76.1	70.2	74.7	77.2	87.8	88.1	78.8	69.8	83.9
Without pension	17.5	23.9	29.8	25.3	22.8	12.2	11.9	21.2	30.2	16.1

(Continued)

Table 12.4 (Continued)

| | BIGAM | | | | | ZININ | | | | |
| | NONPOOR | | POOR | | | NONPOOR | | POOR | | |
Households	Never poor	Previously poor	Transient poor	Chronic poor	Total	Never poor	Previously poor	Transient poor	Chronic poor	Total
Household size										
1–3 persons	38.5	34.8	7.7	6.9	24.3	51.6	41.8	16.8	4.8	37.7
4 persons	26.9	23.9	28.2	14.9	25.3	23.6	22.4	24.8	23.8	23.8
5 persons	28.3	30.4	35.9	52.9	34.3	14.9	22.4	36.5	28.6	22.1
6 persons	4.6	8.7	22.1	17.2	12.0	5.4	6.0	16.8	30.2	10.6
7 persons or more	1.8	2.2	6.1	8.1	4.0	4.5	7.5	5.1	12.7	5.8
Labor										
Dependency rate	55.2	41.7	81.2	48.2		49.5	35.5	70.5	38.7	
Labor participation rate	44.3	50.0	42.8	41.5		40.7	46.3	42.9	49.0	
Education										
Proportion of illiterates	19.1	21.3	13.0	20.7		24.5	20.8	19.9	25.3	
Proportion with university-level education	15.9	8.0	8.6	5.2		11.1	10.8	8.6	5.2	
Number of government employees	0.3	0.3	0.2	0.2		0.3	0.2	0.2	0.3	
Number of samples (households)	286	46	181	87		335	67	137	63	

Source: 2005 and 2010/2011 Panel Household Survey data.

Note: "Employment contract" indicates whether the household head has an employment contract or not. "Social security" is judged by whether the household head currently receives, or will receive in the future, a pension. The dependency rate is the proportion of members aged younger than 15 years or older than 64 years to the total number of household members. The labor participation rate is the proportion of workers to the total number of household members aged at least 15 years. The proportion of illiterates and the proportion with university-level education are measured in relation to the total number of household members aged at least 15 years. The number of samples includes all household members, except for sector and industry of employment, which is for household members who have heads working as wage workers. Consumption poverty is based on the lower poverty line.

female heads, whereas among never poor households, 17.1 percent in Bigam and 26.6 percent in Zinin had female heads. Although all groups of households had predominantly male heads, our results imply that many of the households with female heads, who were mostly divorced or widowed, were consistently non-poor. This implies that the proportion of households with female heads seems to increase with income.

Heads of the chronic and transient poor's households were relatively more concentrated in the working-age group than heads of the previously and never poor households. The chronic poor have household heads clustered in the 45–54 age class, and the transient poor in the 35–45 age class.

Interestingly, the household heads' educational level differs between the chronic and transient poor. The proportion of chronic poor households decreases as the educational level of the head increases. The reverse is not true among the transient poor, and their educational level does not differ significantly from those who are non-poor, except for the university-level. In Bigam, 17.1 percent of the never poor and in Zinin, 11.9 percent of the never poor were headed by those with university-level education, whereas the overall proportion of the transient poor headed by those with university-level education was 10.5 percent in Bigam and 5.8 percent in Zinin.

Moreover, the proportion of the chronic poor to total panel households whose heads were engaged in waged work is slightly higher than the proportion of those whose heads were self-employed (at 10.8 percent). Such differences between poor and non-poor, in terms of employment, are small, as is pointed out in Datt and Jolliffe (1999), and Datt et al. (2001). Most of these chronic poor are clearly concentrated in the private sector. Those in Bigam are engaged in manufacturing and construction, whereas for those in Zinin, the chronic and transient poor are engaged in the private sector without social security coverage, as shown in the proportion of "without pension," but do not differ much from the non-poor in economic activity.

On average, the chronic poor households were composed of six members and the transient poor households comprised five members. The never poor households, on the other hand, were usually composed of no more than three members. The chronic poor had a lower dependency ratio than the non-poor, which averaged around 48.2 in Bigam and 38.7 in Zinin. The transient poor, on the other hand, had a remarkably higher dependency ratio, at 81.2 and 70.5 in Bigam and Zinin, respectively.

Concerning asset ownership, the chronic and transient poor do not differ from the non-poor with regard to owning a washing machine, refrigerator, telephone/cellphone, and television set. These are common assets for both the poor and the non-poor. Concerning cellphones and computers, fewer than 10 percent of these households owned cellphones and fewer than 20 percent owned computers. This was the case for both the poor and the non-poor in 2005, but in 2010/2011, about 90 percent of households, including the poor ones, owned one or more cellphones, and 40 percent owned computers in both areas. In terms of access to basic amenities such as electricity, safe water, and a sanitary toilet facility, nearly all the households, including the poor ones, have had access to these facilities since 2005.

Concerning the condition of housing and number of rooms, the percentages were similar for the poor and the non-poor. Most of the households in Bigam and Zinin live in a three- or four-room apartment: one room serves as both a lounge and dining area, and two other rooms are bedrooms.

On the other hand, homeownership clearly differs between the poor and the non-poor (see Table 12.5). The number of chronic poor households that rent an apartment is nearly double that of the never poor in both areas. This holds true also for the transient poor and, to a lesser extent, the previously poor. This implies that households that own an apartment are less likely to fall into poverty.

Table 12.5 Ownership of residence by poverty status in metropolitan Cairo

	BIGAM				ZININ			
	NONPOOR		POOR		NONPOOR		POOR	
	Never poor	Previously poor	Transient poor	Chronic poor	Never poor	Previously poor	Transient poor	Chronic poor
Owned	21.0	13.0	9.9	8.1	15.5	10.5	2.9	3.2
Rented	38.8	58.7	57.5	67.8	29.0	37.3	59.9	61.9
Inherited	18.9	15.2	11.1	8.1	38.8	46.3	24.8	25.4
Other	21.3	13.0	21.6	16.1	16.7	6.0	12.4	9.5
Total	100.0	100.0	100.0	100.0	100.0	100.0	100.0	100.0
Number	286	46	181	87	335	67	137	63

Source: 2005 and 2010/2011 Panel Household Survey data.

Note: "Inherited" indicates that the residents live in an apartment owned by a parent. "Other" cases include situations such as a residence provided by a company/institution or a "special gift" by a relative or a person without a parental relationship.

Model and variables used

To examine the poverty dynamic in detail, a binominal logit regression was conducted. There are several models to estimate the effects using the panel data, such as the fixed effects model and the random effects model. However, because the time points were only two (2005 and 2010/2011), and because of the small sample size, this study adopted the ordinary logit model. A logit model analyzes the probability of being poor (= 1) or not (= 0). Here, it is used to identify the probability of 1) falling into the categories of consumption poor (transient poor, i.e., a household that was non-poor in 2005 but fell under the lower poverty line in 2010/2011) in comparison with the never poor households that were non-poor in 2005 and 2010/2011; and 2) exiting the categories of consumption poor (previously poor, i.e., a household that was poor in 2005 but raised its consumption level above the lower poverty line in 2010/2011) in comparison with the chronic poor households that were poor in both 2005 and 2010/2011. The probability of being poor is calculated by the formula: $\log(P_{ij}/P_{i1}) = \beta_{0j} + \beta_{1j}x_i$.

This formula uses the odds' log, $\log[P(y = 1)/(1 - P(y = 1))]$, called the logistic transformation. The model is expressed as $\text{logit}[P(x)] = \alpha + \beta_1 x_1 + \beta_2 x_2 + \ldots + \beta_p x_p$.

The probability of being in a particular poverty category is expressed in terms of the odds ratio. The ratio is computed and expressed as $[100(e\beta - 1)]$. A coefficient greater than 1 indicates that the variable in question increases the probability of falling into poverty from non-poor (transient poor) and escaping from poor status (previously poor). Coefficients of less than 1 indicate the opposite result, keeping all other characteristics constant.

Variables used in the analysis are those included in the poverty profile. These variables are the household-based determinants of poverty as hypothesized in the literature on Egypt and other developing countries (El-Laithy et al., 2010). For the regression of the poor, the variables for household characteristics are gender (male = 1, female = 0), age and its squared term, educational level (illiterate, read and write, primary, preparatory, secondary, above secondary, university), household size and its squared term, labor participation rate, dependency rate, and absence of employment contracts. Work status of household heads (waged worker, self-employed, unemployed, out of the labor force), sector of waged employment (government, public, private, other), industry of waged employment (manufacturing, construction, commerce, administration, other) were not included in the model, because the correlations of poverty status with these factors were extremely small.

In addition, variables were included to capture the effects of the change in household structure between 2005 and 2010/2011. These included: 1) Household size (the change in household size obtained by household size in 2010 minus household size in 2005); 2) Dependency rate (the change in dependency rate obtained by dependency rate in 2010 minus dependency rate in 2005); and 3) Labor participation rate obtained by labor participation rate in 2010 minus labor participation rate in 2005.

Results

The results are summarized in Table 12.6, which displays the findings for two study areas. The regression was conducted separately for each study area, but the outcomes did not vary significantly. The logistic results provide support for the description of the poverty profile. In summary, the demographic factor, the household composition, and the assets are the strongest factors that affect whether the households fall into or escape from poverty.

The results show that level of education is associated with transient poverty. When the household head is illiterate, it is difficult for the household to exit poverty (chronic poverty). In other words, the household head who can read and write can more easily exit poverty. The same applies to the transient poor ("poverty in,"). Thus, the results confirm the role of education as the main factor determining poverty status, similar to the findings for urban and rural Egypt (Datt and Jolliffe, 1999; World Bank, 2002). A household is likely to fall into consumption poverty when the household head is either illiterate, can read and write but has no schooling, or has only a primary-level education, as compared with household heads with preparatory level education.

It is notable that a larger household size has the strongest effect on the probability of falling into poverty (transient poverty). Looking at the change between 2005 and 2010/2011, the reduction of household size has the strongest impact on escaping from poverty, and the increase in household size has the strongest impact on falling into poverty, as shown in the odds value and Z-value of household size (2010–2005). Thus, the demographic factor is the most important factor in explaining why households are poor in consumption-based poverty. This strong and positive effect of household size on consumption-based poverty probably reflects the household consumption distribution. Because many households in the study areas are concentrated around the poverty line, an additional increase of a household member is very likely to risk the household falling into consumption-based poverty.

Participation in the labor force and the dependency rate are significant factors for transient poverty. When the households have a smaller labor force, they tend to fall into poverty, as is reflected in the significantly negative Z-value of labor participation rate (2010–2005). For those who escaped from poverty ("poverty exit,"), the coefficients of dependency rate and dependency rate (2010–2005) are both positive. This may reflect the fact that the households with children aged above 15 years are likely to remain in poverty.

The quality of employment measured by its informality affects poverty in a minor way, because absence of an employment contract is insignificant. This may be because the workers in the two survey areas are homogeneous in terms of wage level, regardless of types of work.

Assets measured by residence ownership have a remarkably strong effect on poverty. When the household does not own the residence, the probability of falling into poverty (transient poverty) is significantly higher, and the opposite is true for escaping poverty (poverty exit). This implies that the households that rent their apartment are obliged to reduce consumption in order to pay the rent.

Table 12.6 Regression results of the binomial logit models for poverty in metropolitan Cairo

		POVERTY OUT		POVERTY IN	
		Odds Ratio	z-value	Odds Ratio	z-value
Status in 2005	*Gender of household head*				
	Male	0.710	−0.660	1.177	0.540
	Age of household head				
	Age	1.014	0.110	0.977	−0.410
	Age squared	1.000	−0.050	1.000	0.400
	Education of household head				
	Illiterate	0.434	−1.790*	1.563	1.780*
	Read and write	1.011	0.020	1.315	1.130
	Primary	0.367	−1.300	1.162	0.420
	Secondary	0.883	−0.190	1.343	0.960
	Above secondary	0.183	−1.320	0.723	−0.870
	University	3.685	1.360	0.462	−2.680***
	Household size				
	Household size	0.975	−0.030	1.878	2.030**
	Household size squared	0.927	−1.050	1.004	0.120
	Labor participation and dependency				
	Labor participation rate	1.006	0.610	0.991	−1.950**
	Dependency rate	1.007	1.660*	0.998	−1.170
	Informality of employment of household members				
	No contract	1.033	0.090	1.189	0.940
	Asset				
	Residence ownership	3.084	2.550***	0.348	−5.090***
Change between 2005 and 2010					
	Household size (2010–2005)	0.334	−5.170***	2.107	7.130***
	Dependency rate (2010–2005)	1.010	2.390**	0.999	−0.510
	Labor participation rate (2010–2005)	1.269	1.270	0.696	−3.190***
	Constant	2.165	0.220	0.103	−1.700*
	Pseudo R-squared	0.198		0.146	
	LR chi-square (18)	70.410		167.430	
	Log likelihood	−142.781		−489.496	
	Prob > chi-square	0.000		0.000	
	Number of samples (households)	261		877	

Source: 2005 and 2010/2011 Panel Household Survey data.

Notes: The symbols ***, **, and * indicate significance at the 1 percent, 5 percent, and 10 percent levels, respectively. Gender is male (= 1) and female (= 0); 3) The base variable for the educational level is preparatory level. Labor participation rate is the proportion of workers to the total number of household members at least 15 years old. The dependency rate is the proportion of household members younger than 15 years or old than 64 years to the total number of household members aged between 15 and 64 years. "No contract" refers to whether a household member has an employment contract (= 0) or not (= 1). Residence ownership is owner (= 1) and rented or other cases who do not own their residence (= 0). Household size (2010–2005), dependency rate (2010–2005), and labor participation rate (2010–2005) are the variables displaying the change between 2005 and 2010/2011 that are measured by the number in 2010/2011 minus that for 2005. Poverty in is poor in 2010/2011 (= 1) and non-poor in 2010/2011 (= 0) relative to the total of non-poor in 2005; 10) Poverty out is non-poor in 2010/2011 (= 1) and poor in 2010/2011(= 0) relative to the total of poor in 2005.

Conclusion

This chapter examined the state of poverty in two study areas in the suburban Greater Cairo as a case study, using panel data. Our findings suggest that consumption-based poverty in these unplanned areas was at a high level in 2005, and they increased to a surprising extent in 2010/2011. In this period, the majority of households in the study areas came to be clustered between the lower and upper poverty lines. Thus, in these unplanned areas, poverty has been much more pronounced than the average data indicate and deteriorated in the latter half of the 2000s, the period just before the Arab Spring in 2011.

Although the suburban areas are not homogenous, the two survey areas share common features in terms of the nature of poverty; it is transient poverty that is dominant in these areas. In suburban Greater Cairo, where the majority of households live in apartments, ownership of their residence is a very important factor for maintaining their consumption level. For those who do not have such asset, they are vulnerable to even a small change; change due to demographic factors such as an addition of household member, or due to external shocks such as price increase.

Guide to further reading

Sims, D. (2012) *Understanding Cairo: The Logic of a City Out of Control.* Cairo: The American University in Cairo Press.

References

Bayat, A. and Denis, E. (2000) "Who Is Afraid of Ashwaiyyat? Urban Change and Politics in Egypt." *Environment and Urbanization*, 12(2), pp. 185–199.

Central Agency for Public Mobilization and Statistics (CAPMAS) (2006) *Income, Expenditure, and Consumption Survey 2004/2005.* 6 Vols. Cairo: CAPMAS (in Arabic).

Central Agency for Public Mobilization and Statistics (CAPMAS) (2013a) *Ahamm mu'šširat baht al-dakhl wa al-infāq wa al-istihlāk 2012/2013.* November.

Central Agency for Public Mobilization and Statistics (CAPMAS) (2013b) *Trend in Income and Expenditure.* November (in Arabic).

Datt, G. and Jolliffe, D. (1999) "Determinants of Poverty in Egypt: 1997." FCND Discussion Paper, No. 75, International Food Policy Research Institute. Available at: www.ifpri.org.

Datt, G., Jolliffe, D., and Sharma, M. (2001) "A Profile of Poverty in Egypt." *African Development Review*, 13(2), pp. 202–237.

Dorman, W. J. (2009) "Informal Cairo: Between Islamist Insurgency and the Neglectful State?" *Security Dialogue*, 40(4–5), pp. 419–441.

El-Laithy, H., Lokshin, M., and Banerji, A. (2010) "Poverty and Economic Growth in Egypt, 1995–2000." World Bank Policy Research Working Paper, No. 3068, World Bank, Washington, DC.

Iwasaki, E. (2015) "Income Distribution in Rural Egypt: A Three-Village Case." *Journal of African Studies and Development*, 7(1), pp. 15–30.

Iwasaki, E. and El-Laithy, H. (2013) "Estimation of Poverty in Greater Cairo: Case Study of Three 'Unplanned' Areas." *African Development Review*, 25(2), pp. 173–188.

Kato, H., Iwasaki, E., and El-Shazly, A. (2004) "Internal Migration Patterns to Greater Cairo: Linking Three Kinds of Data: Census, Household Survey, and GIS." *Mediterranean World*, 17, pp. 173–212.

Kato, H., Iwasaki, E., El-Shazly, A., and Goto, Y. (2005) "Migration, Regional Diversity, and Residential Development on the Edge of Greater Cairo: Linking Three Kinds of Data: Census, Household-Survey and Geographical Data-with GIS." In: Okabe, A. (ed.) *GIS-Based Studies in the Humanities and Social Sciences.* London: Routledge.

Kipper, R. and Fischer, M. (eds.) (2009) *Cairo's Informal Areas: Between Urban Challenges and Hidden Potentials.* Portugal: Norprint SA.

Lokshin, M., El-laithy, H., and Benerji, A. (2010) "Poverty and Economic Growth in Egypt, 1995–2000." *Journal of African Studies and Development*, 2(5), 150–165.

Sims, D. (2003) "The Case of Cairo, Egypt." *Understanding Slums: Case Studies for the Global Report 2003*, UN-Habitat & UCL/DPU: 20. Available at: www.ucl.ac.uk/dpuprojects.

Sims, D. (2010) *Understanding Cairo: The Logic of a City Out of Control*. Cairo: The American University in Cairo Press.

Sims, D. and Séjourné, M. (2000) *Residential Informality in Greater Cairo: Typologies, Representative Areas, Quantification, Valuation and Causal Factors*. Cairo: ECES, ILD.

United Nations World Food Programme (WFP) (2013) *The Status of Poverty and Food Security in Egypt: Analysis and Policy Recommendations*. Preliminary Summary Report. Available at: http://documents.wfp.org/stellent/groups/public/documents/ena/wfp257467.pdf.

World Bank (2002) *Arab Republic of Egypt: Poverty Reduction in Egypt, Diagnosis and Strategy*. Joint Report with the Ministry of Development. Egypt: World Bank and Ministry of Development.

13

Social exclusion and multiethnic suburbs in Sweden

Magnus Dahlstedt and David Ekholm

It will be divided at all times, wherever you go, wherever you are, it is divided. So, I don't want to sound negative, but it feels like there will always be divisions. I don't know reason why these divisions exist, there is something mysterious about them.

Siana

Introduction

In recent years, tensions and conflicts in the suburban Swedish landscape have attracted increased attention. Media reports on youth uprisings, burning cars, and stones thrown at police and rescue vehicles have put the focus on urban peripheries, particularly on suburban youth as the subject of social disorder and disintegration (Stigendal, 2016). Consequently, the urban peripheries and suburban youth are formed as a subject of social change and social policy interventions.

These reports conflict with previous images of Sweden and Swedish welfare policy as successful in terms of its universal outreach and arrangements aiming for equality, diversity, and social inclusion (Schierup and Ålund, 2011). In Sweden, as well as internationally, the problems of marginalization have given rise to intense debates about the challenges or possible decline of multiculturalism and the need for strategies promoting integration and social solidarity (Dahlstedt and Neergaard, 2016). These conflicts are played out in relation to fundamental transformations of Swedish welfare state policy, where welfarist governing from the social point of view has gradually been more and more influenced by advanced liberal governing (Schierup et al., 2006; Larsson et al., 2012). Since the 1990s, Swedish social policy has been characterized by a shift from equality to freedom of choice, from redistribution to activation, from collective rights to individual responsibilities (Dahlstedt, 2009).

These shifts have had a range of consequences such as increasing social and economic divisions and intensifying polarization in Swedish cities (Schierup et al., 2014). Once more, public attention has been drawn to suburban areas previously known as part of the Million-program, a large-scale housing project initiated in late 1960s as part of broader universal and state-centered welfare policies, providing rental apartments for the broad population (Molina, 1997). From the start, these urban areas of rental departments have been portrayed in terms of deviance and as sites of risk and social problems, tensions and conflicts – in the 1970s with a focus on class, in the 1980s and 1990s with a focus on ethno–cultural difference and otherness (Ristilammi, 1993). In the new millennium, these suburban areas and their residents were primarily characterized in terms of exclusion – as excluded and outside the rest of society, "areas of exclusion" (*utanförskapsområden*) and consequently, in political terms, "the problem of the outside" (*utanförskapsproblemet*) (Davidsson, 2010; Dahlstedt, 2015). In the following chapter, we examine the "mysterious" drawing of boundaries and divisions by using the above expression by Siana. These boundaries are approached as expressions of social exclusion, particularly relevant for forming the lives of youth in multi-ethnic suburbs in Sweden.

In this chapter, we problematize the current discourses of social exclusion and segregation in suburban Sweden by providing an overview of contemporary Swedish research, further illustrated by ongoing research of the social exclusion of youth in the suburban landscape. First, we outline the conceptual debate on social exclusion in suburban Sweden and, in relation to this, we outline a theoretical approach to social exclusion based on social rights and substantial citizenship. Second, we use these concepts to approach four dynamics of social exclusion and their effects in the urban peripheries of Sweden today: spatial exclusion, poverty, education, and political participation. Third, on the basis of this overview of Swedish research, we elaborate on interventions aimed at social inclusion and social change emerging in recent Swedish social policy. Fourth, we discuss current discourses on suburban social exclusion, and provide alternative frames of interpretation of the dynamics and effects of urban polarization.

Perspectives on social exclusion

The concept of exclusion has been part of Swedish political discourse ever since the 1990s, but it was normalized in the beginning of the new millennium, particularly in the 2006 election, where the center-right Alliance for Sweden succeeded in defining the main political challenge in Sweden as a choice between work or exclusion, activation or passive welfare benefits (Davidsson, 2010). Along this discourse, the distressed suburban areas have been portrayed as a problem, "areas of exclusion," characterized by a particular mentality of welfare dependency, alienation, distrust, and political passivity. Although a high level of unemployment is seen as an urgent problem in these suburban "areas of exclusion," once it has taken shape the specific morality that these socioeconomic conditions generate is seen as having a dynamic of its own (Dahlstedt, 2015). Thus, the "culture of exclusion" is separated from the wider political, social, and material context. In sum, the suburban residents are themselves made responsible for their own exclusion (cf. Schierup et al., 2014). Moreover, such conceptualization of the suburban areas as containers of problems of exclusion makes possible certain social interventions specifically targeting the areas and their residents (Ekholm, 2016).

This discourse on social exclusion, on boundaries between the inside and the outside, constitutes a means of governing, whereby social change is geographically located at the areas of the outside. Interventions are promoted, focusing not least on social pedagogic means providing and facilitating social inclusion, guiding individuals and groups from the outside into the inside (cf. Dahlstedt and Ekholm, 2017). In this discourse, exclusion is approached as *static conditions* in need of intervention, rather than in terms of *social processes* and *dynamics*. In order to observe and understand social exclusion, we focus on citizenship in terms of formal as well as substantial rights (civil, political, and most notably social rights), forming opportunities for youth as well as other residents to participate in the societal community as equals (cf. Schierup et al., 2006). Central to forming both opportunities and limitations for participation and inclusion are symbolic representations of reality, of individuals and groups as well as geographic areas, not least the multi-ethnic suburbs of Sweden (Sernhede, 2011). By approaching the situation in suburban "areas of exclusion" in relation to citizenship and social participation, four main dynamics can be identified in contemporary Swedish research, forming the conditions referred to in terms of "social exclusion," boundaries between the inside and the outside. In the following, we will outline some of the main lines in research, with a focus on spatial exclusion, poverty, education, and political participation.

The dynamics of social exclusion

In the last few decades, economic inequalities have increased in Sweden, creating geographic divisions in the urban landscapes. On the outskirts of larger and mid-sized cities in Sweden, there is a steady relative increase of socially vulnerable residents. These patterns of segregation have to a high degree become synonymous with ethnic segregation, as the concentration of migrants and residents with foreign backgrounds in these suburban areas are comparably high (NBHW, 2010).

Spatial exclusion: the formation of the suburb

The Swedish urban and suburban geography is today characterized by economic, social, and ethno-cultural segregation (Stigendal, 2016). The relative decrease in public housing and rental apartments in cities in general, and the concomitant concentration of subsidized public housing in urban outskirts, have provided a notable precondition for spatial segregation. Families and residents with foreign backgrounds and residents with poor financial situations have limited opportunities to obtain other forms of housing; consequently, the concentration of these residents and families in public housing of the Million-program is steadily increasing (Andersson, 2008). This situation creates general divisions between the native Swedish population and people of foreign backgrounds. This, in turn, gives rise to difficulties promoting integration and socialization and, furthermore, difficulties for people of foreign backgrounds in gaining access to arenas where they may meet and socialize with native Swedes (Mukhtar-Landgren, 2012).

A driving force behind the ethno-cultural spatial segregation is "Swedish avoidance," rather than "Swedish flight." Low levels of in-migration of native Swedes, and higher levels of in-migration of migrants and people of foreign background into distressed suburban areas, explain long-term ethno-cultural segregation (Bråmå, 2006). In relation, it is becoming a relatively rare experience among native Swedes to live in areas characterized by "Swedish avoidance," which furthermore provides a breeding ground for mythologies about otherness and deviant life in distressed suburbia (Andersson, 2008).

There is strong covariance between the ethno-cultural and the socioeconomic composition of geographical areas in Swedish cities. Whereas areas dominated by homogenous native Swedish residents have generally good financial resources, multi-ethnic areas are generally financially weak. Most notably, the residents in multi-ethnic suburbs suffer from a diversity of social problems, such as unemployment, poor health, and a lack of participation in formal political institutions (NBHW, 2010).

Moreover, various representations of the "suburb" are produced in media and political discourse, shaping dramatic images of the suburb as a place of otherness and deviancy (Ristilammi, 1993; Dahlstedt, 2005) characterized by chaos, tensions and conflicts, risks and problems (Ekholm, 2016). Residents themselves seldom articulated the dominant representations of the suburbs, as they lack the privilege to articulate and present life in the suburbs in their own words. Here, not least mass media has the privilege of articulating the "suburbs" as a site of otherness, differentiated from the rest of Swedish society, creating boundaries between the normal and Swedish inside, and the deviant and the excluded outside (Dahlstedt, 2005). Such discourse enables the location of conflicts and problems to the outside, the urban peripheries of the multi-ethnic suburb (Tedros, 2008). As illustrated in an interview with Dimen, a young male living in a multi-ethnic suburb in Sweden, these discourses may have a range of consequences for those portrayed as living in the excluded outside, not least in terms of feelings of alienation and a sense of unfair treatment by dominant Swedish society.

> So, when I watch TV, the media describe the Area as a ghetto . . . it's really horrible . . .
> they take pictures and film the shabbiest place. There are shabby places all over the country.

> If they come and really look at the Area, in the schools, fields, everywhere, they will not see gangs destroying. That's not how it is so. There are gangs everywhere. I cannot deny that gangs does not exist in the Area, because they do. But I think the Area is a wonderful place, like many other places in the city and in Sweden.

In relation to media reports and dominant political discourse, suburban youth are learning to conceive of society as consisting of an "inside" and an "outside," inhabited by "them" and "us," where they may also come to view themselves as other, deviant and the subject of exclusion (Sernhede, 2011). Such mass media and political discourses of the suburb may – as in the preceding excerpt – not at all be legitimate by the suburban youth. At the same time, youth need to relate to and develop strategies in order to deal with them, in terms of identification or dis-identification, acceptance or resistance (Andersson, 2003; León Rosales, 2010). In the process of spatial segregation, the discursive and the material intertwine, producing boundaries between inside and outside society, manifested in the form of cultural expressions, identifications, and experiences as well as socioeconomic materiality.

Poverty and exclusion

Economic vulnerability is linked to a range of welfare problems such as unemployment and school failure, which in turn result in health problems and premature death. The risk of suffering from a plurality of welfare problems and vulnerabilities rose after the financial crisis in 2008, particularly among residents in distressed suburbs (NBHW, 2010). Moreover, increasing differences in income and economic resources lead to reduced social mobility, particularly for those with the highest incomes and the strongest resources – they do not generally risk downward social mobility and their position remains quite stable over generations. Growing up in homes lacking economic resources increases the risk of being poor as an adult and lower opportunities to become a high-income earner substantially. Social positions are transferred from parents to children in families, and exposure to social problems and risk of social exclusion, as well as opportunities of social inclusion, are dependent upon the parents' social position as well as the residential area (Bäckman and Nilsson, 2011; NBHW, 2010). When inequalities are persistent, poverty and vulnerability are reproduced and often spatially located to the suburban geography, segregation, and imagined boundaries between the included and the excluded are consolidated and difficult to transcend (Fritzell, 2011).

Furthermore, spatial segregation leads to unequal opportunities in the labor market. For instance, youth with foreign backgrounds are particularly vulnerable to unemployment and more often perform jobs with low education requirements (Schierup et al., 2015). Here, the risk is not only unemployment but also temporary and part-time work, which may not provide sufficient income or qualify one for social insurance (Mörtvik, 2014). Along with stricter policies on state-administered social insurance qualifications (and levels of insurance), the need for municipal means-tested social support has increased, further localizing public responsibility for and individualizing economic security. Long-term means-tested social support creates a perceived lack of personal integrity and autonomy, as well as exclusion from prevalent consumption, social relations, and networks. Moreover, relations to public authorities tend to become stigmatized. In all, this creates a "double powerlessness," accumulating and reinforcing social and welfare problems and limiting the ability to manage them (Angelin, 2009). These socioeconomic conditions are crucial among youth for self-confidence and hopes for the future. Youth in socio-economically distressed suburbs tend to be less optimistic about the future than youth in other areas. Consequently, negative attitudes about the future reinforce future social exclusion (Alm, 2014). Mara, a young woman living in one multi-ethnic suburb in Sweden, reflects upon the future and her life trajectory, considering her status as a working-class woman of foreign background.

> I'm a woman from Thailand and I don't have a Swedish surname. Few people can pronounce my last name and nobody knows how to spell it [laughter]. I'm also working class. Sure, some people are fighting for equality, that immigrants should have the same opportunities and some are struggling for the working class, but I am at the bottom in all three categories. . . . I'm a little afraid of the future. If there are a thousand job applicants and I am one of them, how far will I get? It's a little dark, but I hope it will get better. If you live with that dream, there is still a chance.

Here, Mara illustrates how contemporary life conditions form her situation not only today but also in the future, as her plans are perceived as restricted by the fact she is a young, working-class woman "from Thailand." At the same time, for Mara, as well as for other youth living in suburban areas, there are hopes and dreams for the future. Education is one of the main interventions made in Sweden's welfarist social policy to promote social mobility, compensating for social inequalities and for realizing individual hopes and dreams for the future.

Education and exclusion

There are strong connections between youth's socioeconomic backgrounds and their educational progress, where youth from vulnerable families are generally more likely to receive weak or incomplete grades and drop out of school, and less likely to continue studying at university (Social report, 2010). As these factors are geographically distributed, the urban patterns of socioeconomic segregation constitute a basis for educational inequalities, further reinforced by a range of neoliberal policy measures, not least of which are the free school choice reforms and the establishment of private schools introduced in the early 1990s (Dahlstedt and Trumberg, 2017). With this reform pupils are allowed to choose their school in a system where private schools are publically funded through a voucher system. Consequently, schools compete for pupils and their education voucher, forming a "winner-schools in the inner-city and loser-schools in the suburbs" dynamic (Bunar and Kallstenius, 2008, p. 11). Here, integration is possible for pupils with high social and cultural capital. At the same time, such individual mobility and integration through active school choices consolidates the composition of pupils in suburban schools (Ambrose, 2017). Suburban pupils and their parents acknowledge the poor learning conditions in their residential schools, which are characterized by an absence of students who speak Swedish as a first language; by changing schools there is hope for better opportunities and access to "the Swedish culture" (Bunar, 2009). In the process of choosing schools, the dynamics of choice are underpinned and guided by different schools' reputations. In a focus group with suburban primary school pupils, the following dialog played out, illustrating the importance of reputation.

Pupil 6: Well, I think it's important that the school has a good status. Because if it has a bad status, then maybe they think you cannot . . . [. . .] But it is not *really* that important, because it's the school, in fact. So, it's how *you* think about it that matters, more than what status it has. But it is still important.

Pupil 10: If you for example attend the Southern school, then you know, well, you're not good.

Pupil 7: Yeah, it pops up, like automatically. One really shouldn't think like that, but you do.

It is obvious the way a school is perceived by society defines its ability to be competitive.

In line with policy changes focusing on pupils and parents' active choices and responsibilities, there is an increasing emphasis on the need to be active in choosing a school and provide a supportive home environment for learning. Thus, unequal home conditions form yet another basis for reinforced inequalities in terms of school performance (Dahlstedt, 2009). Current conditions of urban segregation in Swedish cities, together with the freedom of choice reform, contribute to unequal educational opportunities (Sernhede, 2011). This diminishes schools' traditional welfarist role, where education is seen as a "public good," and schools constitute meeting places and arenas compensating for unequal opportunities and overcoming social divisions (Trumberg, 2011). The role of education, in turn, has a range of implications for the participation of society in general.

Exclusion from political participation

Residents in distressed suburbs are considerably underrepresented in municipal and regional councils as well as in the national parliament. There are certain informal rules and conventions that make it difficult for people with foreign backgrounds to participate on equal terms in political parties. To obtain a political position, it seems crucial to be familiar with "Swedish meeting culture" and informal social norms, and have established social networks. The lack of representation means there are unequal opportunities for the residents to *speak for* themselves in political debate as well as *speak about* life in the distressed suburbia (Dahlstedt, 2005, 2015).

When it comes to youth in multicultural suburbs, they are portrayed in political discourse as either passive, due to their lack of participation in political and civil society, and inability to make active choices, or as misguided through trouble-making activity (Ekholm, 2017a, 2017b; Kings, 2011).

When democratic institutions, as a result of segregation, seem unavailable to a certain group of people and democracy seems reserved for those with resources, it may create reserved attitudes towards democratic institutions among those experiencing social exclusion (Dahlstedt, 2005).

Besides formal political practices and parliamentary participation, civil society operates as an arena for civil and political participation in multi-ethnic suburbs, involving not least youth (Kings, 2011). Consequently, reserved attitudes towards formal politics, the lack of opportunities for democratic participation, and an inability to make their voices heard among suburban youth, create a space for the development of new social movements. Social disorder, in terms of throwing stones against police and rescue services, has been interpreted as an illustration of reserved attitudes towards a reserved Swedish society (Stigendal, 2016). In response to recent uprisings among youth, suburban organizations such as The Megaphone and The Panthers have had an important role in speaking for and claiming the rights of suburban youth (Schierup et al., 2014; Ålund and Lèon Rosales, 2017). Among these organizations, Suburbs Against Violence has organized a number of demonstrations proposing alternative ways of representing life in suburbia:

> Here in our suburbs, people live in cages, with poor life chances. And things are getting worse and worse, from a social point of view. Is it strange that people here are turning inwards and even killing each other? The violence is just a symptom and we are here to protest against the causes and construct our own resistance to these causes, regardless of what politicians and officials have to say.
>
> *(Kitimbwa Sabuni)*

Leisure time activities such as culture, music, and sports provide forms of political and civil participation where youth have the opportunity to speak for themselves. For instance, hip-hop has been emphasized as a crucial means of cultural expression, giving voice to experiences of suburban life among the youth (Sernhede, 2011). Sports practices may also provide opportunities where youth may experience community and develop their dreams of progress and success (León Rosales, 2010).

On the promotion of social inclusion

Notably, these are some main features of social exclusion in suburban Sweden as presented in contemporary research and experienced and dealt with by some of the youth exposed to social exclusion. So, how do governing agencies respond to these forms of social exclusion in multiethnic suburbs and how can social inclusion and social change be promoted in Sweden today? Based on current research on youth in suburban Sweden, we will briefly illustrate three isolated yet significant interventions to counter the social exclusion of youth in these areas. For the interventions illustrated, and for a range of other interventions in Swedish social policy today, there is an emphasis on reaching out to residents of the "areas of exclusion," its youth, their families, and their communities, based on a desire to include them in Swedish society.

In the last decade, sports have been promoted as a means of social change, promoting social inclusion and as an alternative to delinquent behaviors. Accordingly, sport-based interventions are a common feature in Sweden today, performed as a collaboration between public and private actors. Sports, in particular, are tied to specific spatial locations. It is performed at a specific ground or venue, and it has a limited and residential outreach. Sport-based interventions imbue a certain pedagogic rationality, with a certain focus on empowerment and activation, learning to take responsibility and make the "right" decisions in life, promoting discipline and encouraging assimilation into Swedish norms and behaviors. Centerpieces of sport-based interventions are local community role models, embodying the will, conduct, and actions promoted, enacting activation, discipline, and transformation into Swedish-ness in their very appearance (cf. Ekholm, 2017a, 2017b; Dahlstedt and Ekholm, 2017).

Other interventions work on youth by targeting the families living in "areas of exclusion." One particular intervention studied is Home-get-togethers, initiated to establish meeting places where migrant mothers can meet representatives from the police, emergency services, and the municipality in the relaxed and familiar environment of the mothers' own homes. In this setting, the intervention aims to develop trusting relationships between the mothers and the local actors, relationships that are personal and effective, rather than juridical, thus blurring the boundaries between private and public, professional relationships and friendship. By such interventions, the families are made reachable targets for various learning activities, where there is a possibility to promote change among the mothers and their children (cf. Dahlstedt and Lozic, 2017).

In "areas of exclusion," schools are another site where a wide range of interventions are initiated, commonly based on cooperation between local actors such as the police, social workers, and civil society organizations. Among these school-based interventions, there is a strong focus on education for security. Through collaboration between various actors, a range of activities throughout Sweden aims to change the students' norms – and those of the teachers and eventually the parents – as a means of preventing crime and contributing to security in school, the local community and society at large. Through a wide range of pedagogic techniques, with a focus on reflection, activation, and responsibility, students are invited to challenge their past and develop new norms (cf. Dahlstedt and Hertzberg, 2011; Dahlstedt and Foultier, 2017).

Although these three examples are separate interventions and do not constitute the full range played out in Sweden's current suburban context, they are illustrative of a broader tendency in Swedish social policy, which could be approached and understood in terms of the following recurring features: first, the interventions target the local domains where youth are reachable

– their leisure activities, their families and private homes, and their local schools. Here, the domain of the "suburb" is constructed by means of the interventions promoted. Accordingly, the "suburb" is not a predetermined geographic area where interventions are played out, but first and foremost a discursive formation made possible by these interventions, and – in turn – making certain interventions possible. Second, by certain interventions in these domains, and in collaboration with a range of actors, it's possible to aim pedagogic interventions at shaping the subjectivities and behaviors of the youth. Third, these interventions, in turn, make wider social change possible. The skills and competencies attained are supposed to be transferred into other spheres of suburban life; thus, youth are themselves made agents of social change. Accordingly, youth themselves attain competencies for inclusion, which are distributed by means of social relations, in turn, creating social inclusion in the wider societal community.

Conclusively, while multi-ethnic suburbs are exposed to a range of processes of social exclusion, these and a multitude of social policy interventions contribute to separating them from the rest of society. In line with such rationality, the cause of exclusion is located at the delimited "area of exclusion"; simultaneously, it is formed as the area for potential solutions to this problem. Here, social problems as well as potential solutions are located at the particular urban periphery and moreover attributed to its youth and residents.

Conclusion

The exposition in this chapter raises some concerns about the dominant ways contemporary Swedish welfare policy conceptualizes and responds to problems of social exclusion more widely.

In line with the transformations of social policy in recent decades, more and more responsibility for managing risk, social problems, and social exclusion has been placed on individuals, their families, and their communities. This is illustrated in current discourses on multi-ethnic suburbs in Sweden as sites of social problems as well as social change. In these discourses, the focus is put on exclusion in terms of static *conditions* located to the excluded "outside," while the *processes* causing social inequality (manifested in terms of the above-described four dynamics of social exclusion) are left out of the limelight and not addressed in locally based social policy interventions. Locating the problems of inequality and segregation to the outside, to the suburban "areas of exclusion," may solve a political problem – that of the "problem of the outside" (*utanförskapsproblemet*) – however, without seriously addressing the social processes and problems caused by the dynamics of exclusion (expounded on previously).

Certainly, Sweden has an obvious problem with social exclusion that is displayed and acted out particularly in the suburban landscapes. However, this situation needs not to be addressed by locating problems to the residents or geographies excluded and separated from society. Rather, this situation needs to be addressed as the consequences of structural processes creating inequality and segregation and, thus, as concerns for society as a whole. Here, there is a need to approach social change from the social point of view. If there are political wills to promote integration and overcome boundaries between the included and the excluded, processes of segregation and exclusion forming these boundaries need to be addressed, not only the subjectivities and geographies of the excluded being redressed.

Guide to further reading

Andersson, R. (2013) "Reproducing and Reshaping Ethnic Residential Segregation in Stockholm." *Geografiska Annaler, Series B, Human Geography*, 95(2), pp. 163–187.
Baeten, G. and Listerborn, C. (2015) "Renewing Urban Renewal in Landskrona, Sweden." *Geografiska Annaler, Series B, Human Geography*, 9(3), pp. 249–261.

Behtoui, A. (2013) "Social Capital and Stratification of Young People." *Social Inclusion*, 1(1), pp. 46–58.

Bråmå, Å. (2006) "White Flight?" *Urban Studies*, 43(7), pp. 1127–1146.

Bunar, N. (2010) "The Geographies of Education and Relationships in a Multicultural City." *Acta Sociologica*, 53(2), pp. 141–159.

Johansson, M., Salonen, T., and Righard, E. (2015) "Urban Marginalisation in Scandinavian Cities." In: Righard, E., Johansson, M. and Salonen, T. (eds.) *Social Transformations in Scandinavian Cities*. Lund: Nordic Academic Press, pp. 281–288.

Schierup, C.-U. and Ålund, A. (2011) "The End of Swedish Exceptionalism?" *Race & Class*, 53(1), pp. 45–64.

Sernhede, O. (2011) "School, Youth Culture and Territorial Stigmatization in Swedish Metropolitan Districts." *Young*, 19(2), pp. 159–180.

Tunström, M. and Bradley, K. (2014) "Opposing the Postpolitical Swedish Urban Discourse." In: Metzger, J., Allmendinger, P. and Oosterlynck, S. (eds.) *Planning against the Political*. New York: Routledge, pp. 69–84.

Urban, S. (2009) "Is the Neighbourhood Effect an Economic or an Immigrant Issue?" *Urban Studies*, 46(3), pp. 583–603.

References

Alm, S. (2014) "Ungas framtidstro." In: Olofsson, J. (ed.) *Den långa vägen till arbetsmarknaden*. Lund: Studentlitteratur, pp. 297–310.

Ålund, A. and Lèon Rosales, R. (2017) "Renaissance from the Margins." In: Ålund, A., et al. (eds.) *Reimagineering the Nation*. Frankfurt am Main: Peter Lang, pp. 351–374.

Ambrose, A. (2017) *Att navigera på en skolmarknad*. Stockholm: Stockholm University.

Andersson, Å. (2003) *Inte samma lika*. Eslöv: Symposion.

Andersson, R. (2008) "Skapandet av svenskglesa områden." In: Turner, L.M. (ed.) *Den delade staden*. Umeå: Boréa, pp. 119–160.

Angelin, A. (2009) *Den dubbla vanmaktens logik*. Lund: Lund University.

Bäckman, O. and Nilsson, A. (2011) "Social exkludering i ett livsförloppsperspektiv." In: Alm, S., Bäckman, O., Gavanas, A. and Nilsson, A. (eds.) *Utanförskap*. Stockholm: Dialogos förlag, pp. 142–163.

Bråmå, Å. (2006) "White Flight?" *Urban Studies*, 43(7), pp. 1127–1146.

Bunar, N. (2009) *När marknaden kom till förorten*. Lund: Studentlitteratur.

Bunar, N. and Kallstenius, J. (2008) *Valfrihet, integration och segregation i Stockholms grundskolor*. Stockholm: Stockholm City.

Dahlstedt, M. (2005) *Reserverad demokrati*. Umeå: Boréa.

Dahlstedt, M. (2009) *Aktiveringens politik*. Malmö: Liber.

Dahlstedt, M. (2015) "Discourses of Employment and Inclusion in Sweden." In: Righard, E., et al. (eds.) *Transformation of Scandinavian Cities*. Lund: Nordic Academic Press, pp. 61–82.

Dahlstedt, M. and Ekholm, D. (2017) "Inclusion as assimilation." *International Society for Third-Sector Research Conference Working Papers Series*, 10.

Dahlstedt, M. and Foultier, C. (2017) "Förändringens agenter." *Arkiv*, forthcoming.

Dahlstedt, M. and Hertzberg, F. (2011) "Den entreprenörskapande skolan." *Pedagogisk forskning i Sverige*, 16(3), pp. 179–198.

Dahlstedt, M. and Lozic, V. (2017) "Problematizing Parents." In: Ålund, A., et al. (eds.) *Reimagineering the Nation*. Frankfurt am Main: Peter Lang, pp. 209–233.

Dahlstedt, M. and Neergaard, A. (2016) "Crisis of Solidarity?" *Critical Sociology*. Advance online publication. DOI: 10.1177/0896920516675204.

Dahlstedt, M. and Trumberg, A. (2017) "Towards a New Education Regime." In: Ålund, A., et al. (eds.) *Reimagineering the Nation*. Frankfurt am Main: Peter Lang, pp. 189–208.

Davidsson, T. (2010) "Utanförskapelsen." *Socialvetenskaplig tidskrift*, 17(2), pp. 149–169.

Ekholm, D. (2016) *Sport as a Means of Responding to Social Problems*. Linköping: Linköping University.

Ekholm, D. (2017a) "Mobilising the Sport-Based Community." *Nordic Social Work Research*, 7(2), pp. 155–167.

Ekholm, D. (2017b) "Sport-Based Risk Management." *European Journal for Sport and Society*, 14(1), pp. 60–78.

Fritzell, J. (2011) "Fattig och rik i Sverige." In: Alm, S., Bäckman, O., Gavanas, A. and Nilsson, A. (eds.) *Utanförskap*. Stockholm: Dialogos förlag, pp. 27–52.

Kings, L. (2011) *Till det lokalas försvar*. Lund: Arkiv.

Larsson, B., Letell, M., and Thörn, H. (2012) "Transformations of the Swedish Welfare State." In: Larsson, B., et al. (eds.) *Transformations of the Swedish Welfare State*. New York: Palgrave Macmillan, pp. 3–22.

León Rosales, R. (2010) *Vid framtidens hitersta gräns*. Botkyrka: Mångkulturellt centrum.

Molina, I. (1997) *Stadens rasifiering*. Uppsala: Uppsala University.

Mörtvik, R. (2014) "Unga utanför." In: Olofsson, J. (ed.) *Den långa vägen till arbetsmarknaden*. Lund: Studentlitteratur, pp. 79–90.

Mukhtar-Landgren, D. (2012) *Planering för framsteg och gemenskap*. Lund: Lund University.

The National Board of Health and Welfare [NBHW] (2010) *Social Report 2010*. Stockholm: The National Board of Health and Welfare.

Ristilammi, P.-M. (1993) *Rosengård och den svarta poesin*. Stehag: Symposion.

Schierup, C.-U. and Ålund, A. (2011) "The End of Swedish Exceptionalism?" *Race & Class*, 53(1), pp. 45–64.

Schierup, C.-U., Ålund, A., and Kings, L. (2014) "Reading the Stockholm Riots." *Race & Class*, 55(3), pp. 1–21.

Schierup, C.-U., Hansen, P., and Castles, S. (2006) *Migration, Citizenship and the European Welfare State*. Oxford: Oxford University Press.

Schierup, C.-U., Krifors, K., and Slavnic, Z. (2015) "Social Exclusion." In: Dahlstedt, M. and Neergaard, A. (eds.) *International Migration and Ethnic Relations*. London: Routledge, pp. 200–226.

Sernhede, O. (2011) "School, Youth Culture and Territorial Stigmatization in Swedish Metropolitan Districts." *Young*, 19(2), pp. 159–180.

Stigendal, M. (2016) *Samhällsgränser*. Stockholm: Liber.

Tedros, A. (2008) *Utanför storstaden*. Gothenburg: Gothenburg University.

Trumberg, A. (2011) *Den delade skolan*. Örebro: Örebro University.

14

Uneven development and the making of Rio de Janeiro

Anjuli N. Fahlberg

Brazil's troubled initiation into the global arena

The tensions of inclusion and exclusion in Rio de Janeiro cannot be understood outside of the city's historic role as Brazil's capital until 1960 and its continued significance as a major urban economic and political hub. Beginning in the 1800s when Brazil embarked on a long process of modernization and state-building, Rio de Janeiro served as a critical site for negotiating citizenship by determining which members of society would be entitled to full political, civil, and social rights, and which populations would be excluded from the privileges of formal belonging (Carvalho, 2013). Although voting rights expanded and contracted throughout the nineteenth and twentieth centuries, high rates of poverty and illiteracy prevented women and poor men (including former slaves) from direct political participation in Brazil's emerging democracy. Political rights were further curtailed after the 1964 military coup, which ended open elections and rights to free speech and assembly for the following 20 years.

In 1985, Brazil's modern democratic project gained full steam as the dictatorship collapsed and elections were finally opened to all Brazilian citizens. Brazil's 1988 constitution – widely touted as one of the most progressive constitutions in the world – further expanded the legal, civil, and social rights of poor citizens. While these changes have paved the way for universal healthcare, workers' rights, and increased investments in education, inequality across Brazil remains high. The endurance of traditional regional political machines and the lack of economic investment in citizens' social rights have largely prevented Brazil's urban poor from accessing political institutions or from claiming the political and civil rights guaranteed by the constitution (Fischer, 2008; McCann, 2008). As a result, Brazil has been labeled a "disjunctive democracy," in which many citizens – particularly poor black populations – are unable to access their rights due to lack of social services and effective law enforcement (Caldeira and Holston, 1999).

Despite these internal political and social struggles, in the last 20 years Brazil has experienced a dramatic ascension into the global economic arena as a result of its investment in commodities production. Under the leadership of Luiz "Lula" Ignácio da Silva of the leftist Worker's Party (or PT, Partido dos Trabalhadores), Brazil was able to pay back much of its national debt, stabilize the domestic economy, and become one of the largest exporters of raw materials in the world. At the same time, Lula invested in several welfare programs, including the controversial "Bolsa Familia," which provided a monthly stipend to poor families and brought millions of Brazilians out of extreme poverty. Brazil's status as an emerging superpower was solidified when it was selected to host the 2014 World Cup and the 2016 Summer Olympics, both important signifiers of global status.

Sadly, Brazil has struggled to hold onto the gains of the last two decades. Corruption – an endemic feature of Brazil's political landscape – contributed to the downfall of Petrobras, Brazil's state-controlled oil company, when several of its top executives were imprisoned for money laundering and graft in the Lava Jato (Carwash) scandal beginning in 2014. Then, in August 2016, Lula's successor, President Dilma Rousseff, was impeached after misappropriating funds to hide budget deficits during a time when dozens of other political leaders were also facing charges for bribery and graft. The political takeover of the more conservative PMDB (the Partido do Movimento Democratico do Brasil, or the Brazilian Democratic Movement Party) with the rise of former Vice President Michel Temer, coupled with the economic recession that followed, led to a wave of austerity measures. This political transition erased many of the social gains made by the poor and working classes during the reign of the Worker's Party. This fiscal conservatism, along with growing distrust in Brazil's political elite, provoked widespread protest among the working classes across Brazil's urban centers that increased tension between social classes and conflicts with police (Vicino and Fahlberg, 2017). Rio de Janeiro, as the host of several World Cup soccer matches and the Summer Olympics and the site of many protests and class struggles, has been at the center of these conflicts.

The making of inequality in Rio de Janeiro

In both past and present, Rio de Janeiro played a central role in Brazilian economics, politics, and culture. It began as one of the main shipping ports for exports from Brazil to Portugal, became the capital in 1763 – over 50 years before Portugal declared Brazil's independence in 1822 – and has been a key military and logistics center (Lessa, 2000). One consequence of Rio's central role in nation-making was that urban concerns were often of secondary consideration to local politicians. This issue was exacerbated by the lack of a directly elected mayor until 1960, which left the Federal Senate in charge of municipal laws and resource allocation. As Osorio et al. (2015) argue, Rio de Janeiro's historic role as a national city led to haphazard urban planning and public policies, and contributed to economic, social, and infrastructural decline after it lost its status as the nation's capital to Brasilia. Relative to Brazil's other major urban centers, Rio de Janeiro's population has experienced fewer improvements in employment, healthcare, and education since then.

Despite these setbacks, Rio de Janeiro continues to grow at a dramatic pace (Vicino, 2017). In 2015, the city had an estimated 6.5 million residents within its limits, and a total of 12.2 million inhabitants across the metropolitan area, making it the second largest metropolitan region in Brazil (after São Paulo) (G1 do Rio, 2015). Like other urban centers in Brazil, Rio de Janeiro benefited from the economic stability that set in in the 1990s and has experienced relatively steady growth in extractive and industrial production and exports, particularly oil, and more recently in the service, commerce, and industrial transformation sectors (FGV Projetos, 2011). In 2009, the rate of illiteracy across Rio de Janeiro reached a low of 3.97 percent. The city also witnessed a significant decline in the infant mortality rate and improvements in housing quality, internet and telephone access, and transportation.

While Rio de Janeiro has witnessed an overall improvement in its human development indicators, it continues to struggle with two interconnected issues: spatial segregation and public insecurity. Over one-fifth (22 percent) of Rio's population resides in substandard housing areas popularly termed favelas, which grew at a rate of 27.5 percent between 2000 and 2011 (compared to a growth rate of 3.4 percent for the rest of the city) (Hurrell, 2011). Though there is dramatic variability between favelas (Cavallieri and Vial, 2012), these areas tend to have higher rates of poverty, illiteracy, infant mortality, and homicides than more formal residential areas. They also have higher concentrations of blacks and *pardos*, meaning "brown" or "mixed-race."

The first favelas emerged in the late 1800s and early 1900s, populated primarily by freed slaves and poor migrants from Brazil's rural northeastern states. Most of these informal settlements were constructed on the hills bordering plantations or commercial areas where dwellers were employed. During the early 1900s, city officials mostly ignored these informal settlements and offered few public policies to improve their inhabitants' quality of life or enforce their rights. Over the course of the twentieth century, the relationship between the city and its favela population changed in two significant ways. For one, as favelas grew as a result of rapid urbanization without accompanying social supports for poor city dwellers, policymakers came to see favelas as a visual and moral blight on the city. Forced evictions became common as the city attempted to rid itself of poor residents – particularly those deemed politically subversive during the dictatorship – by destroying their homes and displacing homeless families to more distant areas (Valladares, 2005).

At the same time, favela residents experienced growing access to legal rights, first under the nationalist and leftist regimes of Getúlio Vargas and Juscelino Kubitschek between the 1930s and 1950s, and then under the new democratic regimes following the fall of the dictatorship in the 1980s and 1990s. As Brodwyn Fischer (2008) has argued, however, the gains made by Rio's urban poor in legal and political rights resulted in little material progress due to a lack of investment in the social services – education, healthcare, basic infrastructure, etc. – necessary to access and advocate for their rights. While favela residents' mobilization efforts between the 1950s and 1970s were somewhat successful in earning legal claim to their houses, their neighborhoods remained largely neglected by city officials and therefore struggled to attain the same development standards as the rest of the city.

Beginning in the 1970s, a second issue came to characterize favela living: the establishment of the drug trade. As a major South American port, Rio de Janeiro became a key site along the global drug trade route, which proliferated along with increased trade between Brazil, the United States, and Europe. At the same time, recently released political dissidents and petty criminals imprisoned during the dictatorship united to establish the first drug faction – the Comando Vermelho, or the CV – within Rio's favelas (Amorim, 1993). The lack of police presence in favelas, and the proximity of favelas to many wealthy neighborhoods and tourist sites, made these neighborhoods ideal spaces to store, package, and sell drugs (de Souza, 2005).

Under pressure from the United States to support the war on drugs, Rio de Janeiro's military police launched an aggressive, decades-long campaign to combat the drug trade. With the exception of a few isolated community policing efforts, Rio's police adopted extremely brutal tactics that provoked heavy armed combat with drug traffickers but did little to decrease their control over favelas. While the recent "Pacifying Policing Units" – which employed a combination of military tactics and community policing principles – were able to temporarily expel the drug trade from some occupied favelas during a few years preceding the Olympics, the long battle between the state and drug traffickers has resulted in homicide rates in favelas that parallel those of many warzones (Fahlberg and Vicino, 2016; Zaluar, 2007). Although public insecurity is an issue across the city, the ongoing armed struggles between the police and the drug trade is a feature unique to favela living that shows little hope for improvement in the near future. Let us now explore the contemporary markings of inequality and exclusion between favelas and the rest of the city.

Spatialities of inequality in Rio de Janeiro

Despite their proximity to the city's main downtown areas, the residents of favelas have been systematically excluded from the benefits and privileges of urban living, becoming de facto "outsiders" within the city. The favelas' exclusion from the city can be largely seen as an interplay between the real and the imagined, the production of material inequalities based on socially con-structed differences. Although the polarization of Rio de Janeiro exists along several axes, six are addressed here: economic, political, social, geographic, security, and symbolic.

Economic exclusion

Two of the clearest divides between favela residents and other city residents are income dispari-ties and their struggles to access the formal job market. A comparison between some of Rio's largest favelas and bordering neighborhoods found that non-favela neighborhoods had average household incomes between five and 20 times higher than favelas (Rocha et al., 2011). This is owed to several factors, including the concentration of low-skilled workers within favelas, as well as discrimination against favela residents within the formal job market. Furthermore, Ribeiro and Telles (2011) argue that the exclusion of Rio's poor from real estate and consumer markets – a feature increasingly visible as the city sought to insert itself into the global economy – further exacerbating economic polarization between favelas and the city. The decrease in the availability of industrial jobs in the 1980s gave way to a growing informal work sector and contributed to the favelas' dramatic expansion, as the urban poor lost their ability to afford houses in working-class neighborhoods and relied upon informal employment opportunities available within favelas.

Political exclusion

There are several ways favela residents struggle to access the city's political institutions in order to advocate for their individual and collective needs. For one, favela residents are severely under-represented in municipal government. Of the 51 city counselors elected in the 2016 municipal elections, for instance, only one – Marielle Franco – was a favela resident. Additionally, the *milicia* – criminal vigilante groups composed of retired police officers and firefighters who impose high taxes on many favelas and other poor neighborhoods in exchange for protection from drug traffickers – exert a great deal of control over local elections in order to ensure that their operations continue unhindered (Gombata, 2014). The needs of favelas are often neglected within this context. Even when policymakers set aside funding for social or economic projects in favelas, the local political institutions in charge of overseeing the implementation of these projects are often ineffective as a result of systemic corruption, the coercive control of drug traffickers, and local clientelistic practices (Arias, 2006; Gay, 2006). As a consequence, resources are often allocated disproportionately within favelas, and many programs are either abandoned before completion or deteriorate quickly due to the misappropriation of funds.

Social exclusion

Although Rio de Janeiro, like Brazil in general, has seen steady improvement in citizens' access to education, healthcare, and cultural and leisure activities, favela residents remain less connected to public institutions and continue to suffer deficiencies in many of these areas relative to other city residents. According to a 2008 study, only 2.5 percent of favela residents across the city had attended some college, compared to 24 percent in the rest of Rio de Janeiro (Agencia Brasil, 2010). Racial dynamics must also be mentioned. Although Brazil has not suffered from the same types of institutionalized racism as the United States, there remain nonetheless strong racial disparities across the country that, in Rio de Janeiro, have taken on a distinctly spatial composi-tion: favela residents are more likely to be black or *pardo* than other city dwellers. Not only does this mean that the city's black and *pardo* populations are more likely to struggle with the many issues that accompany favela living, but they are also much more likely than their white urban counterparts to be mistreated or killed by the police (Amnesty International, 2015), discriminated against by employers, and perceived as criminals.

Geographic exclusion

Favela residents experience elevated vulnerabilities related to the geography of these environments. For one, Rio de Janeiro is prone to flooding, which often results in landslides along favela hillsides where vast deforestation during the twentieth century and poorly engineered housing structures facilitated erosion (Jacobi, 2016). Favelas on flat land, such as the City of God, have suffered devastating floods as a result of dysfunctional drainage and sewage canals filled with trash. Another significant form of inequality is the challenge of access to the city. Many households were built far up in the hills where there are no paved roads. While cars and buses are able to access the lower hillside areas of major favelas thanks to more recent urbanization projects, most residents living in hilly favelas must carry groceries, appliances, children, and all other belongings up dozens (or hundreds) of steps. As a result, residents of the upper areas of hillside favelas struggle to access the transportation necessary to get to work, school, or leisure activities outside of the favela, despite their neighborhood's proximity to downtown areas. Those with physical disabilities are often unable to leave their homes. In 2010, the municipal government constructed cable cars in the Complexo do Alemão, one of the largest favela complexes in the city, in order to facilitate residents' mobility. Residents, however, complained that state investment in the project – an estimated R$210 million (or USD$70 million) – would have been better spent on updating the sewage system or improving security (Richardson, 2017). Shortly after the Olympics, the government stopped paying operating costs for the cable cars, and now these sit empty as a symbol of a temporary, but ultimately failed state intervention.

Security

Rio de Janeiro is widely perceived by its residents as a dangerous and violent city. High rates of homicide, theft, muggings, and physical assaults are exacerbated by sensationalized media stories that reinforce fear and residents' sense of vulnerability (Penglase, 2007). Machado da Silva (2004) has argued that Rio de Janeiro is governed by a "violent social order," wherein power is determined by those with access to weapons and violent networks. However, like other social problems, the risk of violence is not evenly distributed across the city. Young black men are much more likely to be victims of homicide than any other group (Zaluar, 2007). Favela residents are also most likely to suffer ongoing disruptions to everyday life imposed by drug traffickers and other criminal groups seeking to maintain territorial control (Arias and Rodrigues, 2006; Penglase, 2009; da Silva, 2008). Disorder and vulnerability in favelas has been amplified by aggressive policing practices that prioritize frequent, irregular invasions into favelas that often result in shootouts with drug traffickers and high casualty rates (Misse, 2011).

Symbolic exclusion

Although the preceding features demonstrate that there are numerous material differences between the lived experiences of favela residents and those of other city dwellers, these differences are in many ways a product of socially constructed divisions that continue to influence policy-making and societal beliefs about "favelados" – the pejorative term for favela residents. As Penglase (2014) highlights, the urbanization and formalization of many favelas has helped to decrease some of the historic divisions between many favelas and other city neighborhoods. In fact, a recent study by Cavallieri and Vial (2012) found that fewer than 10 percent of all housing in neighborhoods popularly considered favelas was "subnormal" (i.e., informal and precarious). What remain as the two most poignant features among neighborhoods labeled "favelas" are a history of informal housing and outsiders' ongoing perception of favelas as places of moral depravity and criminality.

Many scholars have suggested that favelas should be understood as symbolic demarcations of difference and inequality, spaces of "otherness" through which belonging and inclusion in the city are defined and affirmed (Burgos, 2005; Silva et al., 2009; Valladares, 2005). In her seminal book *The Invention of the Favela*, Licia Valladares (2005) describes the century-long process, beginning in the early 1900s, through which policymakers, borrowing from and adding to popular beliefs about the inferiority of Rio's urban poor, worked to institutionalize difference between favelas and the rest of the city. Social scientists, Valladares (2005) argues, have been complicit in the social construction of spatialized difference through theories that attributed urban poverty, informality, and other forms of inequality to cultural and moral differences, a perspective that still persists in academia, public policies, and social attitudes, both in Rio de Janeiro and around the world.

Uneven government interventions in favelas

While the disparities between favelas and other city neighborhoods may give credence to the popular assumption that the government has been largely absent from favelas, both the local and national governments have been extremely present in Rio's favelas for decades. At times, this presence has been more harmful than beneficial, such as in the case of police interventions in favelas. However, investments in favela urbanization and the urban poor's social rights have contributed to increased opportunities for social and economic mobility among favela residents.

The police

Police presence in favelas has been a subject of great contestation. Until the 1980s, the police primarily entered favelas in order to evict tenants, tear down shacks, and arrest petty criminals. Once the drug trade began to proliferate, the police launched an aggressive security campaign that emphasized sudden, short-term invasions often resulting in shootouts with drug traffickers and the brutal assault of local residents, but this initiative had little effect on decreasing the drug trade's power over the local territory. As a result, the state's presence in favelas has often been viewed as mostly pernicious, promoting a penalization of poverty rather than supporting the urban poor (Wacquant, 2008). There have been some notable exceptions to this approach. The *Grupo de Policiamento em Áreas Especiais* (GPAE), or the Policing Group in Special Areas, was launched in 2000 in two Rio favelas in an effort to decrease aggressive policing practices, reduce access to and use of guns among local residents, and prevent youth from joining the drug trade. Though the program had some notable successes, it was disorganized and susceptible to political influence and deteriorated quickly. In 2007, the federal government committed to supporting changes in policing practices by investing in police re-training and promoting community policing efforts in poor neighborhoods. Rio de Janeiro was the largest recipient of funding from the project, known as PRONASCI, the *Programa Nacional de Segurança Púbica com Cidadania*, or the *National Program for Public Security with Citizenship*, although the program was discontinued as a result of political conflicts (Ruediger, 2013).

Perhaps the most famous policing program in Rio was the UPP – the *Unidade de Policia Pacificadora*, or *Pacifying Policing Units*. Designed and run by Rio de Janeiro's governor and Secretary of Security, the intervention intended to expel drug traffickers from Rio's favelas by deploying specially trained military police that combined aggressive military tactics and weapons with "policing of proximity" practices to gain and maintain control over select favelas. Though the intervention weakened after the 2016 Summer Olympics, it temporarily subdued drug traffickers, lowered homicide rates in favelas by as much as 75 percent, increased social services in targeted neighborhoods, and promoted better relations between the state and the favela (Abdala, 2011; Cano et al., 2012; Menezes, 2017; Oosterbaan and van Wijk, 2015). The UPP offers an example of how state security measures can promote polarization (Fahlberg and Vicino, 2016) as well as inclusion among favela residents.

Urbanization efforts

The state has played an important role in the uneven development of its informal settlements. While a detailed description of all government projects in favelas fall outside the scope of this chapter, a few deserve mention. Launched in 1994 with funding from the Inter-American Development Bank, the *Favela-Bairro* (Favela-Neighborhood) program is the largest slum-upgrading project in Latin America to date. Focused primarily on medium-sized favelas, which house approximately 60 percent of the favela population, the city claimed the program benefited over half a million people by 2006 by improving infrastructure, transportation systems, housing, and public spaces, building new schools and community centers, and investing in job trainings, among other efforts (Fiori et al., 2000). In 2010, the municipal government implemented the "Programa Morar Carioca" – the "Program Living Carioca" ("Carioca" is the popular term for residents of Rio de Janeiro), which aimed at expanding urbanization and integration efforts within favelas.

The federal government has also helped support investments in Rio's favelas. In 2007, the federal government implemented the "PAC" (Programa de Aceleração do Crescimento, or the Program for Growth Acceleration), which prioritized investments in infrastructure, such as housing construction and improvements, and building water and sewage systems, new schools, and health clinics. "Minha Casa Minha Vida," or "My Home My Life," has funded public housing projects in several favelas since it began in 2009, and thousands of favela residents receive financial support from the welfare program "Bolsa Familia." Additionally, race and income-based quotas in public universities, coupled with federal investments in scholarships, have given favela residents greater access to higher education and created a small but growing group of university graduates from favelas, who have greater access to social mobility than has ever been historically possible (Valladares, 2017).

While these recent programs and funding streams have helped urbanize many favelas and provided needed assistance to poor residents, they are not without problems. Fiori and Brandão (2009) argue that many of the state's social interventions into favelas have been conducted in a piecemeal, disorganized, "schizophrenic" nature that neglected to structure favela-based projects around larger urban planning efforts. Furthermore, political corruption and bureaucratic red tape have led to extremely costly slum-upgrading projects often characterized by poor quality and cheap materials that deteriorated shortly after the project's conclusion (Perlman, 2010). Finally, in many neighborhoods, violent criminal networks play a role in managing the allocation of government resources, which results in the uneven provision of benefits.

The fringes within the city

Rio de Janeiro's favelas offer several important implications for how we understand urban development in the contemporary Global South. For one, Rio de Janeiro's favelas do not easily lend themselves to categorization (as urban, suburban, peri-urban, etc.). On the one hand, their continued exclusion from broader urban systems and privileges reflects the peripheralization of many squatter settlements across the Global South. Yet, decades of state urbanization efforts have helped raise the quality of living among many favela residents and given them some of the resources needed for integration and social mobility. Rather than attempt to classify them as a particular socio-spatial agglomeration, favelas are best defined by the processes that make and remake them. Like many other poor neighborhoods, favelas are dynamic spaces where both material and symbolic realities are continually recreated and reaffirmed within the modern urban landscape. The mechanisms by which they are excluded from the city and the opportunities through which favela residents integrate into the broader urban fabric continually readjust to local and global forces.

These dynamic processes of development and regression also complicate assumptions about polarization within global and emerging cities. Much of the recent literature in urban studies has tended to view cities – particularly those in the Global South – as increasingly divided as a result of neoliberal economic and political policies that decrease state-based social services and limit employment opportunities among low-skilled workers (Friedmann and Wolff, 1982; Sassen, 2011). States, it is suggested, are gradually surrendering their role in the development of poor areas (Wacquant, 2009). Yet, in Rio de Janeiro, state presence in favelas has increased dramatically since Brazil's rise to democracy and shift to neoliberal policies. Although Rio remains a highly unequal city, the urban poor have witnessed several improvements as a result of government intervention. Many of the challenges they face are also a result of the state's presence, particularly policing. In Rio de Janeiro, poverty and exclusion are not products of state abandonment but of uneven, unsustainable, and often dysfunctional government projects that both improve residents' quality of living while also creating new challenges.

Finally, the tensions and opportunities of uneven development have helped produce a wide array of identities and practices within favelas that at once reaffirm and challenge favelas' exclusion from the city. Residents have created strong local economic and social systems that foster informal employment and support systems necessary for adapting to security issues and the lack of dependable public services (da Silva 2008). While many of these systems provide a critical safety net not afforded by the government, informal political dynamics also serve to reinforce local inequalities and power dynamics (Arias, 2006; da Silva, 2008). Additionally, favela residents have been extremely engaged in mobilization efforts to demand increased legal and human rights. These campaigns were critical in obtaining many of the services described previously and in securing greater protections for residents' rights (Fischer, 2008; McCann, 2014). Favela residents have also been outspoken in broader movements for identity-based rights around race, gender, sexual orientation, and religion, with a particular focus on the ways that discrimination in these areas exacerbates the challenges of favela living. Through these efforts, activists help to connect favelas with social justice networks across and beyond the city.

Ultimately, uneven development has made favelas into dynamic spaces that must constantly adapt to the costs and opportunities of unpredictable government intervention. While the challenges of transition often serve to retrench historic forms of inequality, they also create new practices and possibilities. As Brazil grapples with economic recession and a chaotic national political arena, future research should focus on the examination of how federal and municipal policies affect the socio-spatial dynamics of favelas, their integration into the broader urban fabric, and the strategies of adaptation, resilience, and action constructed by favela residents to make the most of emerging opportunities.

Guide to further reading

Reid, M. (2014) *Brazil: The Troubled Rise of a Global Superpower.* New Haven, CT: Yale University Press.

References

Abdala, V. (2011) "Rio, 40: A Verdade Sobre a UPP: Um Histórico Dos Registros de Violência E de Confrontos Armados Nas Favelas 'Pacificadas'." *Rio*, 40. Available at: http://rioponto40.blogspot.com/2011/08/verdade-sobre-upp-um-historico-dos.html [Accessed 26 August 2016].

Agencia Brasil (2010) "Escolaridade Nas Favelas Do Rio Cresceu Pouco, Aponta FGV – Educação – iG." *Último Segundo*, August 31. Available at: http://ultimosegundo.ig.com.br/educacao/escolaridade-nas-favelas-do-rio-cresceu-pouco-aponta-fgv/n1237766482389.html [Accessed 4 April 2017].

Amnesty International (2015) *You Killed My Son: Homicides by Military Police in the City of Rio de Janeiro.* Rio de Janeiro, Brazil: Anistia Internacional Brasil.

Amorim, C. (1993) *Comando Vermelho, a História Secreta Do Crime Organizado.* Rio de Janeiro, RJ: Editora Record.

Arias, E.D. (2006) "Trouble En Route: Drug Trafficking and Clientelism in Rio de Janeiro Shantytowns." *Qualitative Sociology*, 29(4), pp. 427–445.

Arias, E.D. and Rodrigues, C.D. (2006) "The Myth of Personal Security: Criminal Gangs, Dispute Resolution, and Identity in Rio de Janeiro's Favelas." *Latin American Politics and Society*, 48(4), pp. 53–81.

Burgos, M.B. (2005) "Cidade, Territórios E Cidadania." *DADOS – Revista de Ciencias Sociais*, 48(1), pp. 189–222.

Caldeira, T.P.R. and Holston, J. (1999) "Democracy and Violence in Brazil." *Comparative Studies in Society and History*, 41(4), pp. 691–729.

Cano, I., Borges, D., and Ribeiro, E. (2012) *O Impacto Das Unidades de Polícia Pacificadora (UPPs) No Rio de Janeiro.* São Paulo: Fórum Brasileiro de Segurança Pública.

Carvalho, B. (2013) *Porous City: A Cultural History of Rio de Janeiro.* Oxford: Oxford University Press.

Cavallieri, F. and Vial, A. (2012) *Favelas Na Cidade Do Rio de Janeiro: O Quadro Populacional Com Base No Censo 2010.* Rio de Janeiro: Insituto Pereira Passos.

da Silva, L.A.M. (2004) "Sociabilidade Violenta: Por Uma Interpretação Da Criminalidade Contemporânea No Brasil Urbano." *Sociedade E Estado*, 19(1), pp. 53–84.

da Silva, L.A.M. (2008) *Vida sob cerco: violência e rotina nas favelas do Rio de Janeiro.* Rio de Janeiro: Nova Fronteira.

de Souza, M.L. (2005) "Urban Planning in an Age of Fear: The Case of Rio de Janeiro." *International Development Planning Review*, 27(1), pp. 1–19.

Fahlberg, A. and Vicino, T. (2016) "Breaking the City: Militarization and Segregation in Rio de Janeiro." *Habitat International*, 54, pp. 10–17.

FGV Projetos. (2011) *Indicadores de Desenvolvimento Econômico & Social Do Estado Do Rio de Janeiro 1997–2009.* Fundação Getulia Vargas. Available at: https://bibliotecadigital.fgv.br/dspace/handle/10438/7863 [Accessed 9 May 2017].

Fiori, J. and Brandão, Z. (2009) "Spatial Strategies and Urban Social Policy: Urbanism and Poverty Reduction in the Favelas of Rio de Janeiro." In: Hernández, F., Kellett, P. and Allen, L.K. (eds.) *Rethinking the Informal City: Critical Perspectives from Latin America.* New York: Berghahn Books, pp. 181–206.

Fiori, J., Riley, E., and Ramirez, R. (2000) *Urban Poverty Alleviation through Environmental Upgrading in Rio de Janeiro: Favela Bairro.* Development Planning Unit, University College London. Available at: http://share.nanjing-school.com/dpgeography/files/2014/05/Favela-Bairro-Report-1f64qi1.pdf [Accessed 12 May 2017].

Fischer, B.M. (2008) *A Poverty of Rights: Citizenship and Inequality in Twentieth-Century Rio de Janeiro.* Stanford, CA: Stanford University Press.

Friedmann, J. and Wolff, G. (1982) "World City Formation: An Agenda for Research and Action." *International Journal of Urban and Regional Research*, 6(3), pp. 309–344.

G1 do Rio (2015) "Região Metropolitana Do Rio Tem 12, 2 Milhões de Habitantes, Diz IBGE." August 28. Available at: http://g1.globo.com/rio-de-janeiro/noticia/2015/08/regiao-metropolitana-do-rio-tem-122-milhoes-de-habitantes-diz-ibge.html.

Gay, R. (2006) "The Even More Difficult Transition from Clientelism to Citizenship: Lessons from Brazil." In: Fernández-Kelly, P. and Shefner, J. (eds.) *Out of the Shadows: Political Action and the Informal Economy in Latin America*. State College, PA: Pennsylvania State University Press.

Gombata, M. (2014) "O poder da milícia nas eleições do Rio de Janeiro." *CartaCapital*, September 30. Available at: www.cartacapital.com.br/politica/o-poder-da-milicia-nas-eleicoes-do-rio-de-janeiro-1597.html [Accessed 13 May 2017].

Hardoy, J.E. and Satterthwaite, D. (2014) *Squatter Citizen: Life in the Urban Third World*. New York: Routledge.

Hurrell, F. (2011) "Rio Favela Population Largest in Brazil Daily." *The Rio Times*. December 23. Available at: http://riotimesonline.com/brazil-news/rio-politics/rios-favela-population-largest-in-brazil/ [Accessed 13 May 2017].

Jacobi, P.R. (2016) "City Tour: Rio de Janeiro." *disP: The Planning Review*, 52(2), pp. 6–13.

Lessa, C. (2000) *O Rio de Todos Os Brasis*. Rio de Janeiro: Record.

McCann, B. (2008) *The Throes of Democracy: Brazil since 1989*. Halifax, Nova Scotia, London, and New York, NY: Zed Books.

McCann, B. (2014) *Hard Times in the Marvelous City: From Dictatorship to Democracy in the Favelas of Rio de Janeiro*. Durham, NC: Duke University Press.

Menezes, P.V. (2017) "Os Rumores Da 'pacificação': A Chegada Da UPP E as Mudanças Nos Problemas Públicos No Santa Marta E Na Cidade de Deus." *Dilemas – Revista de Estudos de Conflito E Controle Social*, 7(4), pp. 665–684.

Misse, M. (2011) *Autos de Resistência: Uma Anlásie Do Homicídios Cometidos Por Policiais Na Cidade (2001–2011)*. Núcleo de Estudos da Cidadania, Conflito e Violência Urbana. Rio de Janeiro: Universidade Federal do Rio de Janeiro.

Oosterbaan, S. and van Wijk, J. (2015) "Pacifying and Integrating the Favelas of Rio de Janeiro An Evaluation of the Impact of the UPP Program on Favela Residents." *International Journal of Comparative and Applied Criminal Justice*, 39(3), pp. 179–198.

Osorio, M., Martins De Melo, L., Versiani, M.H., and Wernek, M.L. (2015) *Uma Agenda Para O Rio de Janeiro: Estratégias E Políticas Públicas Para O Desenvolvimento Socioeconômico*. Rio de Janeiro, RJ: Editora FGV.

Penglase, R.B. (2007) "Barbarians on the Beach: Media Narratives of Violence in Rio de Janeiro, Brazil." *Crime, Media, Culture*, 3(3), pp. 305–325.

Penglase, R.B. (2009) "States of Insecurity: Everyday Emergencies, Public Secrets, and Drug Trafficker Power in a Brazilian 'Favela'." *PoLAR: Political and Legal Anthropology Review*, 32(1), pp. 47–63.

Penglase, R.B. (2014) *Living with Insecurity in a Brazilian Favela: Urban Violence and Daily Life*. New Brunswick, NJ: Rutgers University Press.

Perlman, J. (2010) *Favela: Four Decades of Living on the Edge in Rio de Janeiro*. New York: Oxford University Press.

Ribeiro, L.C. and Telles, E. (2011) "Rio de Janeiro: Emerging Dualization in a Historically Unequal City." In: Marcuse, P. and Kempen, R.V. (eds.) *Globalizing Cities: A New Spatial Order?* New York: John Wiley and Sons, Inc., pp. 78–94.

Richardson, C. (2017) "Rio Hits the Brakes on Controversial Favela Cable Car." *DW.COM*, February 3. Available at: www.dw.com/en/rio-hits-the-brakes-on-controversial-favela-cable-car/a-37368291 [Accessed 15 May 2017].

Rocha, A. (2005) *Cidade cerzida: a costura da cidadania no morro Santa Marta*. Rio de Janeiro: Museu da República.

Rocha, L.P., Pessoa, M., and Machado, D.C. (2011) "Discriminação Espacial No Mercado de Trabalho: O Caso Das Favelas No Rio de Janeiro." *Centro de Estudos sobre Desigualdade e Desenvolvimento, Universidade Federal Fluminense*. Available at: www.uff.br/econ/download/tds/UFF_TD287.pdf [Accessed 4 April 2017].

Ruediger, M.A. (2013) "The Rise and Fall of Brazil's Public Security Program: Pronasci." *Police Practice and Research*, 14(4), pp. 280–294.

Sassen, S. (2011) *Cities in a World Economy*. Thousand Oaks, CA: Sage Publications.

Silva, J.S., Barbosa, J.L., Biteti, M.O., and Lannes, F. (eds.) (2009) *O Que é a Favela Afinal?* Rio de Janeiro, RJ: Observatório de Favelas.

Simon, D. (2008) "Urban Environments: Issues on the Peri-Urban Fringe." *Annual Review of Environment and Resources*, 33(1), pp. 167–185.

Valladares, L.D.P. (2005) *A invenção da favela: do mito de origem a favela*. Rio de Janeiro: FGV.

Valladares, L.D.P. (2017) "Educação E Mobilidade Social Nas Favelas Do Rio de Janeiro: O Caso Dos Universitários (graduandos E Graduados) Das Favelas." *Dilemas – Revista de Estudos de Conflito E Controle Social*, 2(5–6), pp. 153–172.

Ventura, Z. (1994) *Cidade Partida*. São Paulo: Companhia das Letras.

Vicino, T.J. (2017) "The City in Brazil." In: Short, J. (ed.) *A Research Agenda for Cities*. Northampton, MA and Cheltenham, UK: Edward Elgar Publishing, pp. 182–194.

Vicino, T.J. and Fahlberg, A. (2017) "The Politics of Contested Urban Space: The 2013 Protest Movement in Brazil." *Journal of Urban Affairs*, 39(7), pp. 1001–1016.

Wacquant, L. (2008) "The Militarization of Urban Marginality: Lessons from the Brazilian Metropolis." *International Political Sociology*, 2(1), pp. 56–74.

Wacquant, L. (2009) *Punishing the Poor: The Neoliberal Government of Social Insecurity*. Durham, NC: Duke University Press.

Zaluar, A. (2007) "Crimes and Violence Trends in Rio de Janeiro, Brazil." *UN Habitat*. Available at: http://cn.unhabitat.org/downloads/docs/GRHS.2007.CaseStudy.Crime.RiodeJaneiro.pdf [Accessed 11 December 2015].

15

Dividing the metropolis
The political history of suburban incorporation in the United States

Jon C. Teaford

Early history of municipal incorporation

Americans share a devotion to grassroots rule, but they do not share a uniform system of local government. Instead, each state has its own laws and traditions regarding the creation and powers of subordinate units, a reality that complicates the history of municipal incorporation in the United States. With its system of town government, New England stands apart from the rest of the nation. As early as 1816 the Supreme Judicial Court of Massachusetts ruled that the state's towns were municipal corporations (Williams, 1985, pp. 372–373, 421). Since the entire state was divided into towns, there were no unincorporated areas. Boston's metropolitan area has, then, been a collection of municipalities from the very beginning of European settlement in the seventeenth century. The Boston-area suburban units of Cambridge, Watertown, Sudbury, Concord, Dedham, Hingham, Medford, and Weymouth all date from the 1630s. New towns were created simply by a division of existing towns rather than being carved from unincorporated areas. The city of Boston itself expanded beyond its original boundaries by consolidating with existing adjacent towns, absorbing in the mid-nineteenth century the seventeenth-century creations of Charlestown, Dorchester, and Roxbury.

The remainder of the United States was divided into incorporated and unincorporated areas. Incorporated communities exercised the municipal powers specified by each state, whereas unincorporated areas were the domain of rural township and county authorities. State legislatures determined the procedures whereby an unincorporated community could achieve the status of municipal corporation. During the early nineteenth century, state legislatures generally granted municipal charters through the passage of special legislation. Each aspiring municipality needed to petition the state legislature for passage of a law granting the specific locality municipal status. Because of the large number of petitions for municipal incorporation, however, one state legislature after another abdicated direct responsibility for incorporating communities by adopting general legislation defining the procedure for incorporation. These general laws devolved the task of creating municipalities to county authorities and the local electorate. Petitions for incorporation would be submitted to a county court or governing board, which would then submit the question to the electorate in the area seeking incorporation. If the voters approved, incorporation was granted.

No matter whether through special legislation or a general statutory procedure, municipal status was easily achieved. Nineteenth-century American lawmakers regarded municipal incorporation as a basic right to self-government rather than a privilege bestowed on a few fortunate communities. If a community's voters wanted incorporation, lawmakers were willing to accede to their judgment. The result was a bumper crop of new municipalities, with the single state of Ohio claiming 419 municipal corporations as early as 1868.

As the population of America's metropolitan areas soared in the late nineteenth and early twentieth centuries, the urban fringe nurtured an especially large number of nascent municipalities. Chicago's Cook County witnessed the creation of ten municipalities during each decade from 1860 to 1890. In the 1890s, the number of new municipal corporations rose to 26, followed by 14 between 1900 and 1910. By the latter date 66 municipal corporations existed in Cook County. Pittsburgh's Allegheny County nearly matched this figure with 65 municipalities; 50 municipal corporations clustered in Bergen County, New Jersey, across the Hudson River from New York City; rapidly developing Los Angeles County could already claim 25 city governments.

The motives for, and circumstances surrounding, incorporation differed, but a common factor was the desire for improved services among suburban residents. In 1886, the Pittsburgh suburb of Wilkinsburg sought incorporation to provide such amenities as sewerage, fire protection, and street paving and lighting. Wary of the prospect of higher taxes, an incorporation opponent claimed that those seeking municipal status were "city fellers" who wanted to introduce new-fangled city improvements – such as sewers, paved streets, public water supply, and bath tubs – at the taxpayers' expense, "fellers who are too lazy to bring a tub up from the cellar on a Saturday night" (Teaford, 1979, p. 27). Four years later the Denver suburb of Elyria petitioned for incorporation because of poor drainage, the absence of electric street lights, an inadequate supply of drinking water, an overcrowded school, and the "widespread nuisance from the stench of dead animals" (Teaford, 1979, p. 27). One community after another cited poor or nonexistent services as a compelling reason for self-rule through incorporation.

Especially serious were the complaints about potential or real lawlessness and immorality. In Los Angeles County a number of communities incorporated to take advantage of California's local option liquor law and exercise the power to ban saloons. In 1886, Santa Monica citizens seemed motivated by the general disorder plaguing the seaside suburban community. According to one observer, incorporation would enable the town "to check the lawlessness of the Los Angeles hoodlums, male and female, who resort thither by the hundreds every Sunday. These toughs committed some of the most atrocious outrages against law and decency" (Bigger and Kitchen, 1952, p. 82). Similarly, the Chicago suburb of Oak Park incorporated in 1902 in part to avoid the threat from "a class of inhabitants that would favor the saloon" and "oppose Oak Park's well-known temperance policy" (Teaford, 1979, p. 18).

In contrast, a few communities incorporated to protect immoral pursuits. In 1903, a Los Angeles race track promoter incorporated his ranch as the city of Arcadia so that he could develop the Santa Anita raceway on the property with minimum interference from area police or bluenose do-gooders. Similarly, a Cleveland area devotee of horse racing imported temporary residents so that they could vote to incorporate his farm as the village of North Randall. It was a municipality custom-made as a site for horse racing, where local sportsmen could indulge in wagering under the jurisdiction of tolerant village authorities.

In other suburban communities, incorporation proponents pursued municipal status in order to benefit and protect manufacturing enterprises. During the 1890s, debt-burdened Hammond, Indiana, immediately east of Chicago, sought to annex the taxable property of the Standard Oil refinery in adjacent unincorporated Whiting. To thwart this predatory scheme and keep its tax bill to a minimum, Standard Oil encouraged the incorporation of its industrial enclave. The result was yet another municipality, a Standard Oil jurisdiction, off limits to Hammond's tax collectors. Likewise, in 1901, the Carnegie steel company aided in the incorporation of its Pittsburgh-area mill town of Munhall in order to avoid paying taxes for the support of services in the surrounding rural township. In 1895, the Cudahy Packing Company moved its operations to the company-created municipal corporation of Cudahy in order to evade regulations that nearby Milwaukee might impose on malodorous meatpacking houses. Eleven years later meatpackers incorporated National City, Illinois, across the Mississippi River from St. Louis, to escape the jurisdiction of potentially meddlesome township officials and to reduce the tax bill on their extensive slaughterhouses.

Suburban incorporation 1920s and 1930s

The acceleration of suburban growth during the emerging automobile age of the 1920s and 1930s produced hundreds of additional suburban incorporations. In Long Island's Nassau County, the number of municipalities rose from 20 to 65 between 1920 and 1940. St. Louis County was almost equally prolific with an increase from 15 in 1920 to 41 in 1940. Throughout the nation, 333 new municipalities appeared in metropolitan areas during the 1920s. The economic depression of the 1930s slowed the upward trend, but even that decade of hard times produced 158 additional suburban municipal incorporations (Hawley, 1959, p. 42).

Though the introduction of national prohibition in 1919 took the anti-saloon issue off the table, the motives for incorporation in the 1920s and 1930s did not differ greatly from those in earlier decades. Suburban residents and businesses wanted the improved services that corporate status could provide. In addition, they sought refuge from the higher taxes and regulations of neighboring communities. In 1921, oil was discovered on the site of Signal Hill, California, south of Los Angeles. Hundreds of oil derricks soon dotted the landscape, and in 1924, the oil town incorporated to protect its industry from annexation and regulation by nearby predatory municipalities and to ensure that local rule did not conflict with the needs of oil drillers. The city clerk explained:

> Because of the proximity of Signal Hill to several incorporated cities . . . and because of the impossibility of conforming such a gigantic industry to the restrictions and ordinances of the average metropolitan city, it was deemed wise . . . to incorporate Signal Hill as a separate municipality.
>
> *(Bigger and Kitchen, 1952, p. 91)*

Like Cudahy and National City, Signal Hill was a municipality tailor-made to protect the interests of a single industry.

An increasingly important motive for incorporation was the desire to preserve the community's existing or anticipated residential environment. In Nassau County, wealthy estate owners opted for municipal incorporation in order to prevent the development of small-lot residential subdivisions and avoid the tax levies necessary for urban improvements. As early as 1911, an estate owner secured incorporation of Saddle Rock, thereby relieving himself of the burden of paying for streets and sewers in nearby developments. Through incorporation he was able to escape higher taxes and preserve his semirural domain. In 1920, Saddle Rock had only 71 residents; 10 years later its population was 74, and in 1940, it had fallen to 69. Incorporation was, then, a means to perpetuate a privileged status quo. This Saddle Rock option appealed to other Nassau County gentry, accelerating the pace of incorporation in suburban Long Island. An account from the early 1940s identified Center Island, Cove Island, Head of the Harbor, Matinecock, North Hills, Old Brookville, and Old Westbury as protected estate municipalities off limits to dense development (Dobson, 1942).

Enhancing the importance of land use control as a motive for incorporation was the introduction of zoning ordinances in the post–World War I period. Comprehensive land use zoning was a new and significant power exercised by municipalities and one more reason for Americans concerned about property values to cast their ballots for incorporation. Since a municipality could not forcibly be annexed to an adjacent city, homeowners in the Detroit suburb of Huntington Woods defended themselves from the advances of nearby Royal Oak and Ferndale through incorporation. Not only did incorporation allow Huntington Woods to preserve its independence, it enabled the new municipality to forcefully exploit single-family zoning and preserve its character as a residential refuge for the upper-middle class. In St. Louis County, elite Ladue incorporated to avoid annexation by adjoining Clayton and through zoning maintain its low-density residential environment. The municipality's zoning commission made clear its intention "to protect and continue the spacious residential character" and boasted that Ladue was "one of the few communities in St. Louis County that are unspoiled by uses generally objectionable to desirable residential sections" (Teaford, 1997, pp. 17–18).

By the onset of World War II, thousands of suburban communities had opted for incorporation for a variety of reasons. Both industrial and residential communities resorted to defensive incorporation to avoid annexation to adjacent cities with higher tax burdens and land use controls detrimental to the distinct interests of the incorporating municipalities. Suburbanites sought to keep the big central city at arm's length, but incorporation also defended against land grabs by nearby outlying communities. Thus, Whiting escaped the governance of Hammond, and Ladue avoided absorption by Clayton. Some communities incorporated to repel sin in the form of alcohol, and others chose independence to embrace race track betting. Many wanted better streets, drainage, lighting, and schools. No matter the motive, the prevailing tradition of permissiveness allowed virtually every crossroads hamlet or residential subdivision to establish itself as a municipality.

Post-World War II incorporations

This openhanded attitude spawned an especially generous crop of new municipalities during the two decades following World War II. As new subdivisions sprouted up in former pastures, the suburban population soared. These new suburbanites demanded the services, protection, and self-rule offered by municipal government. The proliferation of new incorporations was most evident in St. Louis County, where the municipal count rose from 41 in 1940 to 84 in 1950 and 98 in 1960. Many of the freshly minted corporations were miniscule. In 1946 a subdivision of 11 acres incorporated as the village of Mackenzie. By 1951, St. Louis County included 26 municipalities with areas of less than 100 acres. In the especially fragmented north county region, nascent, and longer-standing municipalities fought mind-boggling territorial battles. In 1950, the municipalities of Bel-Nor, Hanley Hills, and Wellston each submitted papers for annexation of the same suburban tract. At the same time, residents of the territory in dispute filed an incorporation petition with the county court to become the independent municipality of Greendale, thereby endeavoring to escape rule by any of its warring neighbors. Meanwhile, through incorporation Pagedale thwarted the annexation attempts of Wellston and Hanley Hills. Commenting on this confusing scramble for territory and independence, Wellston's newspaper reported: "If everybody gets incorporated, a map maker who tries to make a map of each town in a different color is going to run out of colors" (Teaford, 1997, p. 67).

In Detroit's suburban Oakland County, the number of municipalities rose from 24 in 1940 to 38 in 1960, and nowhere were the annexation/incorporation battles harder fought than in South-field Township. Whereas some Southfield residents wanted to incorporate the entire township, others sought to divide it into protected municipal fragments. In 1950, advocates for an independent Lathrup Village raced to submit their incorporation petition at the county courthouse before the township-wide party could do so. The Lathrup Villagers won by 20 minutes, leaving the courthouse just as the laggard township-wide petitioners were entering. Since the county processed incorporation papers in order of submission, the Lathrup Villagers could hold their incorporation election first. Succeeding at the polls, Lathrup Village incorporation advocates secured municipal independence and freedom from the larger township.

The Southfield Township struggle, however, was not over. The adjacent suburb of Oak Park mounted an unsuccessful campaign to annex a tax-rich section of the township, and the city of Berkley also cast greedy glances at unincorporated Southfield's lucrative territory. Meanwhile, in 1953 and 1955, two semirural tracts in northern Southfield Township incorporated as the municipalities of Franklin and Bingham Farms. Through the zoning protection afforded by incorporation, wealthy, large-lot Franklin could preserve what its municipal charter referred to as "our simple, rural way of life" (Teaford, 1997, p. 62). Likewise, as its name implied, Bingham Farms conceived of itself as a simple rural enclave for those wealthy and privileged enough to avoid middle-class subdivisions and commercial development. By the end of the decade the balkanization of Southfield was complete. The incorporation of Beverly Hills, a municipality whose name conveyed its pretensions, added another jurisdiction to the northern third of the township, and the city of Southfield governed the southern two-thirds.

In Southern California the story of municipal proliferation was somewhat different. By the mid-twentieth century, Los Angeles County provided a full range of services to unincorporated areas, thus eliminating a significant motive for municipal status. Between 1930 and 1954, only one new municipality was created in the county. In 1952, however, the city of Long Beach embarked on a campaign to annex the unincorporated tract-house community of Lakewood. This aroused consternation among residents and interested parties who felt incorporation was the best defense against Long Beach aggression. The opening of a Long Beach rehabilitation facility for alcoholics adjacent to Lakewood angered many Lakewooders and strengthened their desire to take charge of local land use regulation through municipal incorporation. Private utility companies serving unincorporated Lakewood also opposed absorption into Long Beach, a city serviced by municipal-owned utilities. Annexation would most likely mean a serious loss of paying customers. Los Angeles County sweetened the incorporation alternative by offering to contract its services to the new municipality. In other words, the municipality of Lakewood would retain satisfactory county services and avoid having to hire an expensive army of city employees. Consequently, the tax burden would not rise. Lakewood would maintain independence from Long Beach at no significant extra expense or loss of services. In 1954 Lakewood voters chose this low-cost, county-serviced municipal option, which became known as the Lakewood Plan.

The Lakewood Plan proved so attractive that it resulted in a wave of additional incorporations. Between 1954 and 1960, 26 communities with a combined population of 500,000 incorporated in Los Angeles County. As in earlier decades the motives for municipal independence were diverse. Both the cities of Industry and Commerce incorporated as tax havens for factories and warehouses. Largely devoid of homeowners, the City of Industry only met California's minimum population requirement for incorporation by counting as residents the occupants of a local mental hospital. Dairy farmers incorporated the city of Dairy Valley to protect themselves from suburbanites who complained of cattle barns infringing on their subdivisions. This agricultural municipality was the self-proclaimed "city of 500 people and 60,000 cows" (Schiesl, 1982, p.

230). Other communities incorporated to perpetuate their exclusive lifestyle. Like their wealthy counterparts in Franklin, Rolling Hills, residents sought to maintain their privileged large-lot, low-density existence. No matter the motive, the Lakewood Plan had broad appeal. It offered all the benefits of incorporation without the downside of higher taxes.

By the late 1950s, however, there was a growing outcry against the rampant fragmentation of America's metropolitan regions. Critics wrote of the crazy-quilt map of municipalities with an emphasis on the crazy. The incorporation free-for-all seemed to defy rationality and enabled privileged residents and businesses to escape their social responsibility to the metropolitan region as a whole. Moreover, with scores of municipalities exercising independent authority within a single metropolitan area, regional cooperation was difficult, if not impossible. Incorporation seemed a tool to perpetuate inequity and a roadblock to concerted regional action. Policymakers increasingly rebelled against the permissiveness of the past and sought to curb the seemingly mindless fragmentation.

The result was some effort to bring order to the creation of local government units. Like many other states, Minnesota had experienced a boom in municipal creations during the post-World War II period. During a single decade, 45 new municipalities arose in five metropolitan counties. Nearly half of these nascent cities had fewer than a thousand residents at the time of incorporation; one was home to only 43 people. By the close of the 1950s, 130 municipalities cluttered the map of the Twin Cities metropolitan area. Commenting on this proliferation of governments, a state legislative report lamented that "multiplying villages, like rabbits, can out-distance all progress achieved by otherwise intelligent planning" ("History of Municipal Boundary Adjustments," 2016). To forestall further chaotic governmental fragmentation, in 1959, the state legislature created the Minnesota Municipal Board to exercise administrative review over all municipal incorporations and boundary changes. Aspiring municipalities had to present their case before the state board and meet the board's standards. As a consequence, the rate of municipal creation plummeted, with the board closing the door on the incorporation of miniscule fragments. Whereas from 1950 to 1959, there were 62 municipal corporations created with an average area of 7.6 square miles, during the first two decades of the board's existence there were a total of only 15 new municipalities averaging 30 square miles.

Inspired by Minnesota, some other states took action. Michigan followed the Minnesota model and established a state boundary commission with veto power over new incorporations. In 1963 the California legislature delegated administrative review of municipal incorporations and boundary adjustments to county Local Agency Formation Commissions (LAFCOs). There was to be a LAFCO in each county, and this agency was to pass judgment on whether an aspiring municipality merited incorporation. By creating county boards rather than a statewide review commission, California lawmakers acceded to demands for local control. Yet, as in Minnesota the goal was to impose some order on the mad rush for local self-rule.

More recent suburban incorporations

During the late twentieth and early twenty-first centuries, the perceived problem of proliferating municipal corporations seemed to be waning. Whereas 1,074 new municipalities incorporated nationwide in the 1950s, this number dropped to 338 during the 1980s, and to 148 from 2000 through 2009 (Rice et al., 2014, p. 141; Waldner et al., 2013, p. 63). This in part reflected a decline in population growth in the United States. The metropolitan population was increasing at a slower rate, so the need for new municipalities was diminishing. A less permissive attitude toward incorporation in states such as Minnesota also had an impact. Possibly significant was the rise of county governments as service providers to unincorporated suburban areas. In metropolitan areas throughout the nation, counties expanded their role and in effect became unincorporated cities, offering the services and protection associated with municipal governments.

Yet, county rule did not satisfy some suburbanites, and in certain metropolitan areas this dissatisfaction produced a new wave of incorporations. Repeating the familiar rhetoric of past incorporation battles, suburbanites lauded government close to the people and complained of the distant authority of remote rulers in the county courthouse. The desire for self-rule for each fragment of the metropolis was not dead. Each community sought control over its own destiny and the power to pursue its own self-interests.

This new wave of incorporations redrew the map of Miami-Dade County, Florida. Between 1960 and 1991, no new municipality had formed in Miami-Dade. Instead, a county bureaucracy of 28,000 people supported by a budget larger than the outlays of 12 states had provided services for the growing portion of the population that lived in unincorporated areas. In the early 1990s, however, the wealthy communities of Key Biscayne, Pinecrest, and Aventura incorporated. Resentful that Aventura was paying too much to the county and receiving too little in return, an incorporation proponent argued, "We're just not getting services. . . . We're not getting the bang for our buck. We want lower taxes and far increased services" (Waldner and Smith, 2015, p. 187). Another incorporation advocate complained of distant government: "We don't need to be going downtown to try to find a county commissioner to take care of our problems. We're looking to have people in our own community in office" (Waldner and Smith, 2015, p. 187). Between 1997 and 2005 Sunny Isles Beach, Miami Lakes, Palmetto Bay, Miami Gardens, Doral, and Cutler Ridge likewise opted for independence from the county. A Cutler Ridge supporter of incorporation offered an argument heard throughout the county when she contended, "Incorporation will provide residents with a local government consisting of members of the community, elected by the community" (Teaford, 2008, p. 130).

The fragmentation of Miami-Dade also empowered African Americans who resented the dominance of white Cubans in county affairs. Incorporated in 2003, Miami Gardens had 105,000 residents, 79 percent of whom were black. With an African American mayor and city council, the newly independent suburb could proudly proclaim itself "the heart of Black Miami" (Teaford, 2008, p. 60).

Meanwhile, in the Atlanta metropolitan area county officials faced a similar suburban revolt. Georgia was one of the few states in which the state legislature still authorized municipal incorporation referenda through special legislation. Throughout the late twentieth century, a Democrat-controlled legislature had blocked incorporation efforts in Atlanta's Republican north Fulton County suburbs. Municipal independence would erode the powers and tax revenues of the county's Democratic commissioners. When Republicans won a legislative majority in the early twenty-first century, however, the gates opened for new suburban municipalities. In 2005, Sandy Springs led the way, with 94 percent of those voting in the incorporation referendum approving city status. The well-to-do community of 90,000 residents could now pursue its own destiny. One Sandy Springs resident explained, "My major thing is, let's make the decisions here rather than downtown" in Atlanta's county building (Teaford, 2008, p. 133). In 2006, two additional north county communities, Milton and Johns Creek, incorporated, followed by south county Chattahoochee Hills in 2007. In all three communities, more than 80 percent of those casting ballots chose municipal status. The state legislator who sponsored the bill to allow an incorporation vote in Milton argued, "Fulton County, with 850,000 residents . . . cannot provide true local government and local representation no matter how hard it tries" (Teaford, 2008, p. 134).

Critics of this fragmentation of Fulton County claimed racial motives underlay the incorporation movements. Generally affluent, predominantly white communities were abdicating their responsibility for paying for services in poor, black areas of the county, and white north Fulton residents were escaping rule by African American county commissioners. Yet, black suburban-ites also opted for independence. In 2016, the nearly 100,000 predominantly African American residents of South Fulton approved incorporation for their community. Like white north Fulton suburbanites, advocates of incorporation in South Fulton emphasized the need "to make local government more accessible, accountable, and responsive to citizen needs and desires because of smaller size and more direct access to local officials" ("Why Incorporate?: What Are the Advan-tages of Incorporation?," 2016). In addition, the predominantly black city of Atlanta was seeking to annex most of South Fulton, and South Fulton residents seemed especially fearful that annexa-tion would mean absorption into the inferior Atlanta school district, a change that "would be devastating to our children" ("Why Incorporate?: What Are the Advantages of Incorporation?," 2016). Through incorporation South Fulton blacks could keep Atlanta at bay and assume control of their future.

On the west coast suburbanites were joining in the challenge to county rule. In Sacramento County, California, Citrus Heights incorporated in 1997, the first new municipality in the county in 51 years. Elk Grove and Rancho Cordova chose municipal status in 2000 and 2001, with residents expressing familiar complaints. An Elk Grove advocate of incorporation argued: "The time has come to abolish county governments . . . The counties have made bad decisions on land use planning and allowed leap-frogging development. They should get out the business" (Waldner and Smith, 2015, p. 191). In Rancho Cordova, it was reported that voters "frequently chafed at unpopular decisions made by haughty officials 'downtown'" (Waldner and Smith, 2015, p. 191).

In Seattle's King County, a new crop of cities likewise redrew the map of local government. Whereas no new municipalities were created in the county from 1962 through 1989, during the 1990s, ten communities opted for corporate status. Key to many incorporation efforts was a desire to wrest authority over land use planning from the county. After the county targeted Federal Way as a site for apartment complexes, the community incorporated in order "to clamp down on multi-family development and put a lid on growth" (Waldner and Smith, 2015, p. 198). A proponent of incorporation in Kenmore contended that "as a city, we can negotiate growth and density targets. . . . Although we will still have to accept some growth, we can better control how, when and where it goes and what it looks like" (Teaford, 2008, p. 137). Seattle-area resi-dents reiterated oft-heard complaints about the remoteness of county officials. "When I realized King County planners hadn't actually been here, I realized we needed more control," observed a proponent of incorporation in Woodinville (Teaford, 2008, p. 137).

As seen in Florida, Georgia, California, and Washington, suburban America remained a battle-field for municipal status. Though the number of new municipalities had abated, and some states had imposed procedures aimed at limiting metropolitan fragmentation, the desire for self-rule still motivated suburbanites to seek independence from the big central city, from neighboring subur-ban cities, and from county commissioners and planners. In suburbia, big was not synonymous with good, and the refrain about government by friends and neighbors was still heard. Devotion to grassroots governance was secure and persistent.

The rhetoric about self-rule, however, had long provided a cover for the selfish interests of a variety of suburban players. Industrialists had exploited municipal incorporation to avoid regula-tion and higher taxes. Municipal status had also served the interests of teetotalers and race track gamblers. It was a method by which wealthy estate owners could maintain a bucolic existence free from middle-class riff-raff. And it enabled middle-class homeowners to exclude apartment dwellers and unsightly commerce. Through incorporation, black suburbanites in South Fulton could keep their children from the bane of Atlanta schools, and African Americans in Miami Gardens could wield power in an increasingly Hispanic Miami-Dade County.

Because of the generally permissive tradition of municipal incorporation in the United States, suburban Americans could separate themselves from other residents in the metropolis, wield power, and pursue policies tailored to their desires. Self-rule may often have been barely distinguishable from selfishness, but the American tradition, for better or worse, had prescribed that each metropolitan fragment had the right to follow its own destiny.

Guide to further reading

Connor, M. (2013) "'Public Benefits from Public Choice': Producing Decentralization in Metropolitan Los Angeles, 1954–1973." *Journal of Urban History*, 39, pp. 79–100.

Connor, M. (2014) "Metropolitan Secession and the Space of Color-Blind Racism in Atlanta." *Journal of Urban Affairs*, 37, pp. 436–461.

Miller, G. (1981) *Cities by Contract: The Politics of Municipal Incorporation.* Boston, MA: MIT Press.

Teaford, J. (1979) *City and Suburb: The Political Fragmentation of Metropolitan America, 1850–1970.* Baltimore, MA: Johns Hopkins University Press.

Teaford, J. (1997) *Post-Suburbia: Government and Politics in the Edge Cities.* Baltimore, MA: Johns Hopkins University Press.

Waldner, L. and Smith, R. (2015) "The Great Defection: How New City Clusters Form to Escape County Governance." *Public Administration Quarterly*, 39, pp. 170–219.

References

Bigger, R. and Kitchen, J. (1952) *How the Cities Grew: A Century of Municipal Independence and Expansionism in Metropolitan Los Angeles.* Los Angeles: University of California.

Dobson, M. (ed.) (1942) *This Is Long Island: The Sunrise Homeland.* New York: Long Island Association.

Hawley, A. (1959) "The Incorporation Trend in Metropolitan Areas, 1900–1950." *Journal of the American Institute of Planners*, 25, pp. 41–45.

"History of Municipal Boundary Adjustments in Minnesota." Available at: www.mba.state.mn.us/History.html [Accessed 20 December 2016].

Rice, K., Waldner, L., and Smith, R. (2014) "Why New Cities Form: An Examination into Municipal Incorporation in the United States, 1950–2010." *Journal of Planning Literature*, 29, pp. 140–154.

Schiesl, M. (1982) "The Politics of Contracting: Los Angeles County and the Lakewood Plan, 1954–1962." *Huntington Library Quarterly*, 45, pp. 227–243.

Teaford, J. (1979) *City and Suburb: The Political Fragmentation of Metropolitan America: 1850–1970.* Baltimore, MA: Johns Hopkins University Press.

Teaford, J. (1997) *Post-Suburbia: Government and Politics in the Edge Cities.* Baltimore, MA: Johns Hopkins University Press.

Teaford, J. (2008) *The American Suburb: The Basics.* New York: Routledge.

Waldner, L., Rice, K., and Smith, R. (2013) "Temporal and Spatial Dimensions of Newly Incorporated Municipalities in the United States." *The Geographical Review*, 103, pp. 59–79.

Waldner, L. and Smith, R. (2015) "The Great Defection: How New City Clusters Form to Escape County Governance." *Public Administration Quarterly*, 39, pp. 170–219.

"Why Incorporate?: What Are the Advantages of Incorporation?" Available at: http://voteyescitysfulton.com/fact-sheets/?v=400b9db48c62 [Accessed 10 December 2016].

Williams, J. (1985) "The Invention of the Municipal Corporation: A Case Study in Legal Change." *The American University Law Review*, 34, pp. 369–438.

16

Immigrants in U.S. suburbs

Kyle Walker

Theoretical models of suburban immigrant settlement

Since the early twentieth century, theoretical models of immigrant settlement in U.S. metropolitan areas have been heavily influenced by the Chicago School of Sociology's scholarship (Park et al., 1925). The Chicago School scholars used their classic concentric zone model of urban form to explain the dynamics of immigrant settlement within the city, and they linked the social position of immigrants and the periodicity of immigration to immigrants' geographical locations. In his chapter "The Growth of the City: An Introduction to a Research Project," Burgess (1925, p. 56) discusses "slums" near to the central business district, which he describes as "crowded to overflowing with immigrant colonies – the Ghetto, Little Sicily, Greektown, Chinatown – fascinatingly combining old world heritages and American adaptations." He follows that the zone of "second immigrant settlement" beyond the immediate urban core represents an "escape" from the slum for second-generation immigrants, who "in turn look[s] to the 'Promised Land' beyond" (1925, p. 56) on the urban fringe. Importantly, in these passages Burgess links the *social location* of immigrants within American society to their *spatial location* in the metropolis, associating outer ring suburban settlements with higher social status.

These formulations are closely related to major theoretical models of urban immigrant settlements that persist in importance nearly a hundred years after the original publication of *The City*. One aspect of this is *ethnic enclave theory*, which links the clustering of immigrant populations in urban cores to ethnic economies that employ many of those immigrants (Light et al., 1994). Portes (1981, p. 291) further explains that the "basic characteristic" of ethnic enclaves "is that a significant proportion of the immigrant labor force works in enterprises owned by other immigrants." As opposed to the Chicago School's observations, however, more modern ethnic enclaves can provide opportunities for immigrants to thrive economically, albeit largely within a concentrated area of co-location with other immigrants and co-ethnics.

In Burgess's formulation, immigrants who leave the enclave look toward the "Promised Land" of fringe neighborhoods associated with elevated socioeconomic status. Massey (1985) formalized this idea into a theory of *spatial assimilation* linking the Chicago School ecological model with theories of social assimilation, suggesting that assimilation has both social and spatial components. For immigrants, the theory proposes that assimilation within the host society (in this case, the

United States) is associated with improvements in one's residential location, which often means settlement in neighborhoods where native-born residents predominate rather than immigrants.

A wide interdisciplinary literature has sought to evaluate the tenets of spatial assimilation theory, generally finding support for it. Such studies have frequently modeled socioeconomic differences of immigrants who live in urban cores as opposed to suburban fringes and found that an increase in socioeconomic status is associated with suburban residential location (Alba and Logan, 1991; Alba et al., 1999; Allen and Turner, 1996). However, as Hall (2009) acknowledges, demographic shifts within both cities and suburbs call into question the salience of the suburbs as an assimilative destination. Additionally, Wright et al. (2005) have critiqued the common association between residential co-location with whites and immigrant spatial assimilation.

These critiques dovetail with observations of enclave-like immigrant communities forming in suburban locations that nonetheless have high levels of socioeconomic status. Li (2009) terms these communities *ethnoburbs*. Ethnoburbs are immigrant communities in which at least one ethnic group is prominent, though it does not necessarily make up a majority of the population. In turn, the ethnoburb serves as both a socioeconomic and spatial middle ground between the ethnic enclave and spatial assimilation models. While ethnoburbs exhibit spatial clustering, like ethnic enclaves, they also show evidence of immigrant upward mobility and suburban residential attainment, as in the spatial assimilation model.

Other scholarship suggests that immigrant suburbanization may require neither the dissolution of immigrant spatial communities, as suggested in the spatial assimilation model, nor immigrant clustering, as suggested in the enclave and ethnoburb models. The model of *heterolocalism*, proposed by Zelinsky and Lee (1998), posits that coherent immigrant communities within metropolitan areas may nonetheless be residentially dispersed and in turn often located in the suburbs. Ethnic communities then coalesce around institutions such as churches, cultural centers, or festivals, and are connected either via telecommunications (e.g., the internet) or personal transportation.

These models of suburban immigrant settlement are often found in the contemporary American metropolis. The size and scale of large American metropolitan areas mean that no single theoretical model of suburban immigration can precisely explain the complexity of their geography in practice, especially given the widely disparate circumstances under which immigrants of different national origins and socioeconomic statuses migrate to the United States. However, aggregate demographic analysis across metros and regions of origin do suggest overarching trends in immigrant suburbanization, which are explored in the following section.

The geography of suburban immigration

The section that follows includes a series of data summaries and visualization of trends in immigrant suburbanization across the United States, with a focus on trends in the Chicago and Washington, DC metropolitan areas. Data in this section come from the National Historical Geographic Information System (Minnesota Population Center, 2011) and Brown University's Longitudinal Tract Database (Logan et al., 2014). Using tools provided by these projects, historical data are adjusted to 2010 census tract boundaries, allowing for the use of consistent boundaries over time.

Across the United States, the overarching trend of immigrant settlement in American metropolitan areas has been one of suburbanization. As illustrated in Figure 16.1, proportionally fewer and fewer immigrants in large U.S. metropolitan areas are living in census tracts near urban centers. The figure, broken down by region of origin, shows the percentage of the foreign-born population in the 50 largest U.S. metropolitan areas living in the 20 percent of census tracts nearest to their respective city halls, as measured by the 2000 census and the 2011–2015 American Community Survey.

Percent of immigrants living in nearest fifth of census tracts, 50 largest metropolitan areas

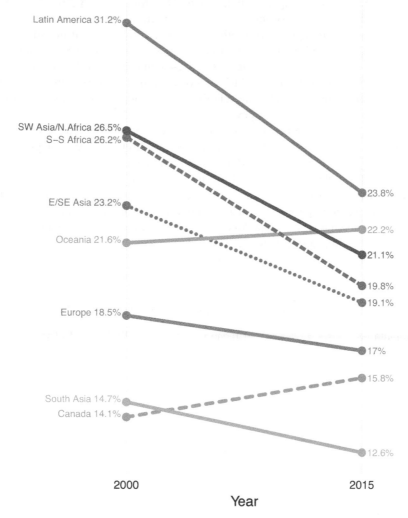

Figure 16.1 Percentage of immigrants living in the nearest fifth of census tracts for the 50 largest metropolitan areas

Source: Kyle Walker

Latin American immigrants – immigrants from Mexico, Central and South America, and the Caribbean – were the most urban of all the regional groups in 2000, with 31.2 percent living in the nearest fifth of census tracts to their respective urban cores. However, as measured in the 2011–2015 American Community Survey, this percentage fell significantly since 2000, with only 23.8 percent living in those neighborhoods. Several other regional groups suburbanized since 2000. In 2000, over 23 percent of East and Southeast Asian immigrants lived in the most urban metropolitan neighborhoods; this dropped to just over 19 percent in 2011–2015. Immigrants from South Asia, Europe, Sub-Saharan Africa, and Southwest Asia/North Africa also suburbanized during this period. The two regional groups that urbanized – in relative terms – during this period are immigrants from Oceania (largely Australia and New Zealand) and Canada.

The regional trends suggested in Figure 16.1 provide some potential evidence for demographic inversion among U.S. metropolitan immigrants. Data for 2000 and 2011–2015 show trends in immigrant suburbanization for almost all groups, with the notable exceptions of Canada and Oceania. Unlike many immigrants from other parts of the world, Canadian and Oceanic immigrants are perhaps more likely to be racialized as white in the United States and likely already speak English. In turn, the graphic suggests that these immigrants may be participating in trends characteristic of native-born whites, which are slowly becoming more populous in urban cores. In contrast, non-white immigrants are suburbanizing, which is consistent with overall trends in non-white metropolitan demography.

Latin American immigrants are perhaps the most prominent group in Figure 16.1, as they are the most urban of all the regional groups in both 2000 and 2011–2015, and the group that suburbanized the most during this period. Additionally, immigrants from Latin America numbered approximately 21.6 million in 2011–2015, making up over half of the total U.S. immigrant population. While dominated by immigrants from Mexico, this group includes several other prominent urban immigrant populations, such as Dominicans, Cubans, and Salvadorans.

Figure 16.2 breaks down trends in Latin American suburbanization by metropolitan area for nine large U.S. metros in all major regions of the country.

Percent of Latin American immigrants living in nearest fifth of census tracts

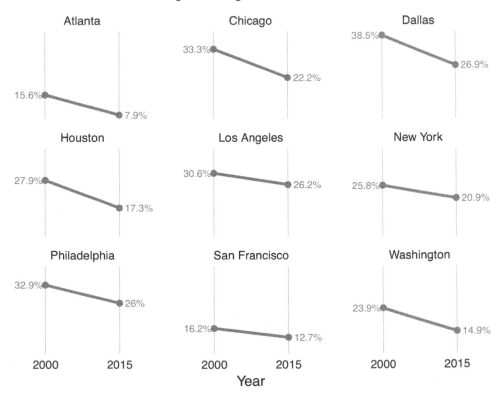

Figure 16.2 Percentage of Latin American immigrants living in the nearest fifth of census tracts

Source: Kyle Walker

Percent of East and SE Asian immigrants living in nearest fifth of census tracts

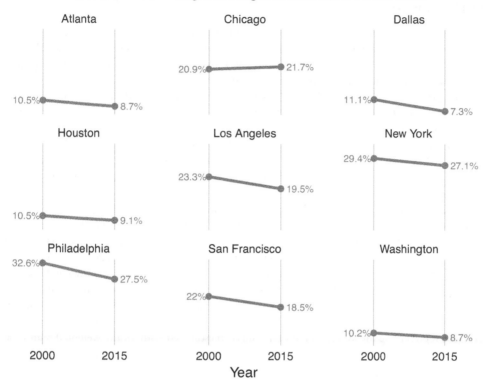

Figure 16.3 Percentage of Eastern and Southeastern Asian immigrants living in the nearest fifth of census tracts

Source: Kyle Walker

In each of the nine metropolitan areas, the percentage of Latin American immigrants living near the urban cores declined between 2000 and 2011–2015. The steepest declines were found in the Chicago, Dallas-Fort Worth, and Houston metropolitan areas, where each decline exceeded 10 percentage points. Notably, in metropolitan areas like Atlanta, San Francisco-Oakland, and Washington, DC, Latin American immigrants were already heavily suburban in 2000, with the closest percentages in the census tracts below 25 percent.

In 2011–2015, the second largest region of origin of immigrants to the United States was East and Southeast Asia, numbering just under eight million. Figure 16.3 illustrates shifts in the metropolitan location of immigrants from these regions for the same nine large metropolitan areas. The graphic reveals a similar process of suburbanization for East and Southeast Asian immigrants, though with a gentler slope than Latin Americans. Part of this reflects the already heavily suburban orientation of East and Southeast Asian immigrants in many large U.S. metros. This is particularly true of metropolitan areas in the South and Southeast; the percentages of East and Southeast Asian immigrants living in the most urban census tracts dropped below 10 percent in 2011–2015, but in each instance only represented a fall from just over 10 percent in 2000. Curiously, Chicago bucks the overall trend, as this regional group became more urban in its residential locations between 2000 and 2011–2015.

Trends in immigrant suburbanization in Chicago and Washington, DC

While the graphs in the previous section are useful in showing overall trends in immigrant suburbanization, they are limited in their ability to reveal within metropolitan trends. The following section focuses on immigration trends within the Chicago and Washington, DC metropolitan areas, tracing the geography of immigration within these metros back to 1970.

Chicago and Washington, DC were chosen as representatives of different metropolitan contexts with regard to immigration. The classic typology of metropolitan immigrant gateways is formulated in Singer (2004) and updated in Singer (2015). Chicago is one of four "Major-continuous gateways," where "the proportion of their foreign-born populations has exceeded the national average for every decade of the past century" (Singer, 2015, p. 1). Washington, DC, in contrast, is a "Post-World War II gateway," which is representative of metropolitan areas that had small immigrant populations prior to 1950 but now accompany major-continuous gateways as principal hubs for immigrant settlement in the United States.

The graphical tool used to accomplish this is *distance profile visualization*, which traces how the relative concentration of immigrants varies by distance from the urban core in a given metropolitan area. Applications of distance profile visualization include Estiri et al.'s (2015, 2018) studies of generational trends in U.S. metropolitan areas, and Walker (2018), which explores how neighborhood racial and ethnic diversity varies in U.S. metros. In the visualizations that follow, immigrant concentration is represented by *location quotients*, which indicate the extent to which census tract immigrant populations exhibit greater than or lower than average concentrations for a given metropolitan area, with one representing the average. Line values are then estimated with locally weighted regression.

In the Chicago metropolitan area, the distance profile visualization in Figure 16.4 reveals a steady process of immigrant suburbanization since 1970. For all five datasets – 1970, 1980,

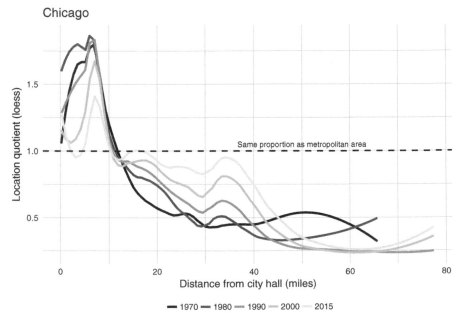

Figure 16.4 Location quotients of immigrants in suburban Chicago

Source: Kyle Walker

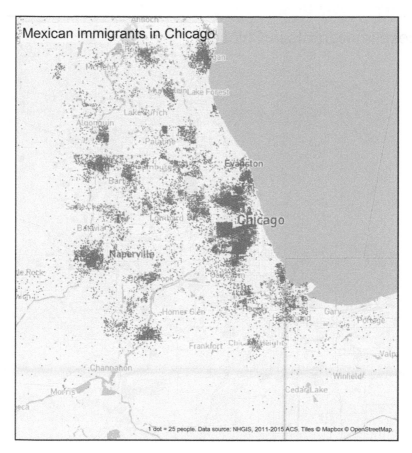

Figure 16.5 Mexican immigrants in suburban Chicago

Source: Kyle Walker

1990, 2000, and the 2011–2015 American Community Survey – the greatest concentration of immigrants in the Chicago metropolitan area is about 8 miles from the Chicago Loop. This reflects the continued settlement of immigrants on Chicago's North, West, and Southwest Sides. Notably, however, this urban immigrant concentration relative to the overall metropolitan area has declined since 1970 and risen distinctly throughout the suburbs, especially in areas between 30 and 40 miles from the urban core. These areas tend to be located around Chicago's satellite cities, which include Elgin, Aurora, Joliet, and Waukegan.

Within the Chicago metropolitan area, regions of origin vary significantly geographically. In the satellite cities proper, Mexican immigrants tend to predominate; clusters of Mexican-origin immigrants have since extended beyond these satellite cities, often into proximate suburbs. In suburban economic hubs like Naperville and Schaumburg, concentrations of South Asian immigrants are common. Immigrant geographies of inner-ring suburbs tend to reflect the demographics of proximate neighborhoods within the city of Chicago. Harwood Heights and Norridge, to the northwest of Chicago, have large Polish populations; Cicero and Berwyn, to the west of the city, house significant Mexican concentrations; and Skokie includes concentrations of East Asian and Middle Eastern immigrants. Examples of the geography of immigration in the Chicago metropolitan area are found in Figures 16.5 and 16.6, which are dot-density maps of Mexican and Chinese-origin immigrants.

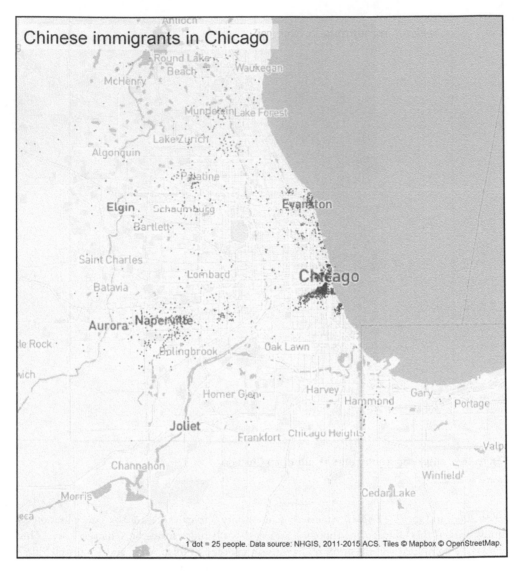

Figure 16.6 Chinese immigrants in suburban Chicago

Source: Kyle Walker

The map reiterates that the strongest concentrations of Mexicans in the Chicago metropolitan area are located in the urban core on the city's Northwest and Southwest sides, as well as Chicago's satellite cities 30 to 40 miles from the city center. However, the map also illustrates several secondary concentrations of Mexicans throughout the Chicago suburbs. Mexican immigrants have settled in inner suburbs adjacent to the city's West Side, such as Cicero, Berwyn, and Melrose Park, in addition to suburban communities further from the urban core, such as Des Plaines and Carpentersville (Badillo, 2004; Peña, 2008; Vicino, 2013).

Scholars have studied various aspects of Mexican suburbanization in the Chicago region. Badillo (2004) and Peña (2008) built communities and impacted the suburban cultural landscape through religion. In Cicero and Berwyn, for example, Mexican Catholics have inherited parishes historically linked to Eastern European communities, helping sustain these churches. Suburban settlement can present challenges for Mexican immigrants, however. Troche-Rodriguez (2009) writes of housing discrimination experienced by Latinos in the Chicago suburbs. This is illustrated in Vicino (2013), who details attempts by the village of Carpentersville to exclude undocumented immigrants through local policy, which will be covered in more depth at the end of this chapter.

Chinese immigrants, as evidenced in Figure 16.6, represent a very different pattern of suburban settlement than Mexicans. As opposed to settlement in Chicago's satellite cities, concentrations of Chinese immigrants are found in the upscale suburbs of Naperville and Evanston. Notable on the map, however, is the concentration of Chinese immigrants in the urban core to the south of downtown Chicago. As illustrated earlier in Figure 16.6, Chicago was the only one of the nine profiled metropolitan areas where the East and Southeast Asian immigrant population has grown more urban since 2000. This is in part due to the attractiveness of Chicago's Chinatown to an economically diverse Chinese population, and the influx of wealthy Chinese investors to high-demand Chicago neighborhoods like the Gold Coast (Spula, 2014; Eltagouri, 2016).

Overall, Washington, DC exhibits similar trends to Chicago in immigrant suburbanization, as evidenced in the distance profile visualization in Figure 16.7. In 1970, above-average concentrations of immigrants were found within 10 miles of the urban core. By 2011–2015, notable immigrant concentrations could be found about 12 and 22 miles from the urban core, the latter growing significantly during the time period under study. Further, by 2011–2015, the most

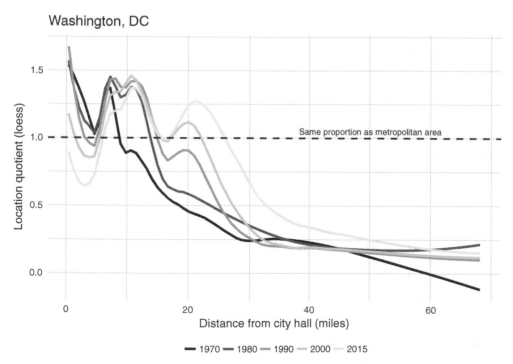

Figure 16.7 Location quotients for immigrants in suburban Washington, DC

Source: Kyle Walker

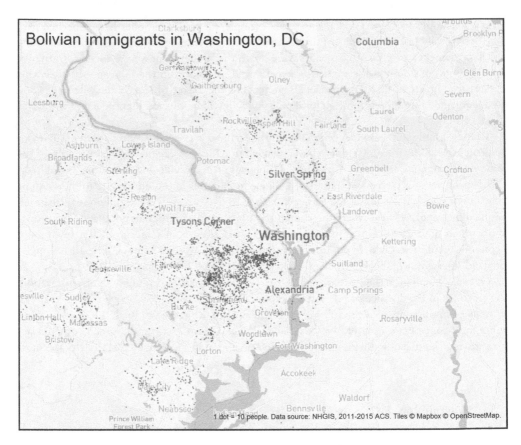

Figure 16.8   Bolivian immigrants in suburban Washington, DC

Source: Kyle Walker

urban census tracts in the DC metropolitan area exhibited a below-average concentration of immigrants for the metropolitan area as a whole.

Much of this reflects the rapid growth of immigrant populations in areas like Fairfax and Prince William County, Virginia, and Montgomery County, Maryland. As opposed to the Chicago metropolitan area, however, Mexican immigrants are far less prominent. In contrast, Latin American immigrants who have settled in areas like Prince William County often come from countries like El Salvador, Guatemala, and Bolivia. The Washington metropolitan area has attracted immigrants from many other regions of the world as well, notably Ethiopians who are concentrated both within the central city but also in the Montgomery County suburbs. Figures 16.8 and 16.9 display dot-density maps for Bolivians and Ethiopians in the Washington, DC metropolitan area.

As shown in Figure 16.8, Bolivians in the Washington, DC area are heavily suburban, with few concentrations found in the District of Columbia. Instead, Bolivians largely cluster to the west of the urban core in Virginia, dotting communities in Arlington and Fairfax County. Smaller clusters of Bolivians are found in Maryland's Montgomery County to the north of the city, and in Manassas and Prince William County on the exurban fringes of the metropolitan area. The suburban residential locations of Bolivians in the DC metropolitan area connect with broader

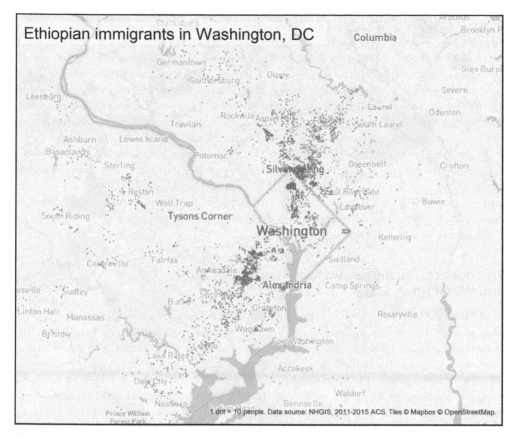

Figure 16.9   Ethiopian immigrants in suburban Washington, DC

Source: Kyle Walker

frameworks such as heterolocalism and transnationalism. Price et al. (2005) comment on the role of immigrant soccer leagues in the Washington region in tying together heterolocal Latino communities, with teams often organized by nation or even village of origin. Strunk (2014) further observes how Bolivian hometown associations (HTAs) in the DC suburbs – often linked to soccer leagues, as well as other cultural institutions – in turn facilitate fundraising projects for migrants' communities of origin in Bolivia.

Like Bolivian immigrants in the Washington, DC area, Ethiopian immigrants are also largely suburban, with major concentrations in Alexandria, Virginia, and Silver Spring, Maryland, as illustrated in Figure 16.9. However, unlike the Bolivian community, an Ethiopian enclave exists within the central city in the Adams Morgan and Shaw neighborhoods of Northwest Washington (Chacko, 2011). In these communities, Ethiopians have gained a foothold within American society by filling what Chacko (2016) terms "occupational niches." Such economic activities include driving taxicabs in Washington and neighboring suburbs as well as founding ethnic businesses that sell both Ethiopian food and goods, catering to the Ethiopian community (Chacko, 2003). Like Bolivians in the region, Ethiopians in Washington, DC are transnational; Chacko (2011) details both the economic and political links between the Ethiopian community in Washington, DC and the Ethiopian capital city of Addis Ababa.

These examples from Chicago and Washington, DC illustrate the complex settlement patterns of contemporary metropolitan immigrant populations, which cannot be characterized neatly with a single theoretical framework. While Ethiopians and Bolivians in the Washington, DC region are largely suburban, they both exhibit concentrations consistent with an ethnoburb model as well as heterolocal practices. In the Chicago region, Mexican immigrants settle throughout the metropolitan area, in urban enclaves as well as dispersed throughout the suburbs. Further, the trend in Chinese immigration to Chicago is one of urbanization, spearheaded by Chicago's thriving Chinatown and attractive real estate for wealthy Chinese immigrants.

Immigrants in the Chicago and Washington, DC regions have formed thriving communities that are integral parts of the suburbs in which they now reside. However, suburbs have responded in mixed ways to their presence. Whereas some suburbs have pursued initiatives to welcome immigrants to their communities, others have attracted national attention for their efforts to exclude immigrants. The following section explores the suburban politics of immigration in Chicago and Washington, DC.

The politics of suburban immigration in Chicago and Washington, DC

In the mid-2000s, hundreds of local governments and several state governments took up legislation to deal with issues related to immigration in their communities. In many cases, this "localization" of immigration policy was framed by localities as a way to fill a void left by federal inaction on the issue. The bulk of these policies were located in metropolitan regions, both due to U.S. demographics and to the fragmented nature of U.S. metropolitan areas, which may contain dozens if not hundreds of local governments. In turn, many metropolitan areas included some suburbs that adopted inclusive policies in addition to other suburbs that enacted exclusionary ones (Walker and Leitner, 2011). The Chicago and Washington, DC metropolitan areas were two such regions.

In the Washington, DC metropolitan area, the most prominent exclusionary response to immigration was found in Prince William County and Manassas, an independent city located within the county. Prince William County drew national attention for its 2007 resolution directing local police to inquire into the legal status of detainees and permitting the denial of public services to undocumented immigrants (Wilson et al., 2010). Around the same time, Manassas changed its municipal housing code to include a very restrictive definition of "family" to address issues of overcrowding, which was believed to target the local immigrant population (Walker, 2014). Local activists, such as the Help Save Manassas organization, and national organizations, such as the Federation for American Immigration Reform, supported both movements (Walker, 2014).

Other suburbs within the Washington, DC metropolitan area responded quite differently to the presence of immigrants. Takoma Park, Maryland, an inner suburb bordering the District of Columbia to the north, allows noncitizens – including undocumented immigrants – to vote in local elections. Additionally, Takoma Park has a long-standing status as a "sanctuary city" – first in response to the 1980s influx of refugees from Central America but later adapted to prohibit city officials or police from checking individuals' legal statuses (Walker, 2014). Additionally, immigrants and immigrant advocacy organizations within the Washington, DC suburbs have mobilized in response to restrictive immigration policies. Leitner and Strunk (2014) document such organizations challenging policies that target immigrants in the region. Examples include CASA de Maryland, based in the suburban Montgomery and Prince George's Counties, and Mexicans Without Borders, based in Prince William County, Virginia. Additionally, Strunk and Leitner (2013) study efforts in Arlington, Virginia to resist the federal Secure Communities and 287(g) programs, which were designed to involve law enforcement officials in immigration enforcement.

Like Washington, DC, the Chicago metropolitan area includes a wide variety of local responses to immigrants. While Chicago is a "sanctuary city" where local officials do not check immigration status, policy positions throughout the suburbs are more varied. The village of Carpentersville pursued a notable example of an exclusionary policy in 2006, when village trustees proposed an Illegal Immigration Relief Act ordinance (IIRA) (Vicino, 2013). This policy, based on that of Hazleton, Pennsylvania, was designed to prohibit employers from hiring undocumented immigrants and required landlords to check tenants' legal status before leasing property (Vicino, 2013; Walker, 2015).

In contrast to Carpentersville, in 2008 activists in the inner-ring suburb of Evanston, to the north of Chicago, pushed for the consideration of a local sanctuary policy and pro-immigration resolutions. In fact, activists designed this policy in part to respond to anti-immigration policies pursued in Carpentersville and other communities around the country (Walker, 2015). While the sanctuary policy failed to pass the Evanston City Council, the Council did pass a pro-immigration resolution, and the Evanston Police Department incorporated the sanctuary policy through a General Order (Walker, 2014).

These suburban immigration policy movements in Chicago and Washington, DC have a direct relationship to shifting suburban demographics, especially in the context of the theories of urban demography outlined earlier in this chapter. Hanlon and Vicino (2016) write of "new metropolitan realities" in which many suburbs are no longer characterized by the upward mobility implied by the Chicago model but are rather experiencing processes of decline. As Walker (2014) explores, a clash between normative expectations of immigrant assimilation in suburbs and these new demographic realities can in some cases spark exclusionary policy. However, in some instances the fragmented jurisdictional nature of American suburbia also creates opportunities for local inclusive policy, as evidenced in Takoma Park.

Conclusion

Like the American suburb itself, the dynamics of contemporary immigrant settlement in the suburban United States are multifaceted. While theoretical models explaining immigrant settlement patterns in metropolitan areas remain salient, no single model adequately explains how immigrants live within American cities and suburbs. Instead, models such as spatial assimilation, ethnic enclave theory, and heterolocalism figure in varying degrees between and within immigrant communities as they integrate and adapt to American suburbia. Such integration is not always easy, as evidenced by the exclusionary policies targeting immigrants in the suburban United States and the movements to contest these policies. However, demographic trends suggest that the classic story of immigration to the metropolitan United States will increasingly be one that unfolds in the suburbs.

Guide to further reading

Singer, A., Hardwick, S. W., and Brettell, C.B. (eds.) (2008) *Twenty-First Century Gateways: Immigrant Incorporation in Suburban America*. Washington, DC: Brookings.

Varsanyi, M. (ed.) (2010) *Taking Local Control: Immigration Policy Activism in U.S. Cities and States*. Stanford, CA: Stanford University Press.

Vicino, T. J. (2013) *Suburban Crossroads: The Fight for Local Control of Immigration Policy*. Lanham, MD: Lexington Books.

References

Alba, R.D. and Logan, J.R. (1991) "Variations on Two Themes: Racial and Ethnic Patterns in the Attainment of Suburban Residence." *Demography*, 28, pp. 431–453.

Alba, R.D., Logan, J.R., Stults, B.J., Marzan, G., and Zhang, W. (1999) "Immigrant Groups in the Suburbs: A Reexamination of Suburbanization and Spatial Assimilation." *American Sociological Review*, 64, pp. 446–460.

Allen, J.P. and Turner, E. (1996) "Spatial Patterns of Immigrant Assimilation." *Professional Geographer*, 48(2), pp. 140–155.

Badillo, D.A. (2004) "Mexicanos and Suburban Parish Communities: Religion, Space, and Identity in Contemporary Chicago." *Journal of Urban History*, 31(1), pp. 23–46.

Burgess, E.W. (1925) "The Growth of the City: An Introduction to a Research Project." In: Park, R.E., Burgess, E.W. and McKenzie, R.D. *The City*. Chicago, IL: University of Chicago Press, pp. 47–62.

Chacko, E. (2003) "Ethiopian Ethos and the Making of Ethnic Places in the Washington Metropolitan Area." *Journal of Cultural Geography*, 20(2), pp. 21–42.

Chacko, E. (2011) "Translocality in Washington, D.C. and Addis Ababa: Spaces and Linkages of the Ethiopian Diaspora in Two Capital Cities." In: Brickell, K. and Datta, A. (eds.) *Translocal Geographies: Spaces, Places, Connections*. New York: Routledge, pp. 163–178.

Chacko, E. (2016) "Ethiopian Taxicab Drivers: Forming an Occupational Niche in the US Capital." *African and Black Diaspora: An International Journal*, 9(2), pp. 200–213.

Eltagouri, M. (2016) "Here's Why Chicago's Chinatown Is Booming, Even as Others across the U.S. Fade." *Chicago Tribune*, May 13. Available at: www.chicagotribune.com/news/local/ct-chicago-chinatown-growth-met-20160513-story.html.

Estiri, H. and Krause, A. (2018) "A Cohort Location Model of Household Sorting in US Metropolitan Regions." *Urban Studies*, 55(1), pp. 71–90.

Estiri, H., Krause, A., and Heris, M.P. (2015) "'Phasic' Metropolitan Settlers: A Phase-Based Model for the Distribution of Households in US Metropolitan Regions." *Urban Geography*, 36(5), pp. 777–794.

Hall, M. (2009) "Interstate Migration, Spatial Assimilation, and the Incorporation of US Immigrants." *Population, Space, and Place*, 15, pp. 55–77.

Hanlon, B. and Vicino, T.J. (2016) "Local Immigration Legislation in Two Suburbs: An Examination of Immigration Policies in Farmers Branch, Texas, and Carpentersville, Illinois." In: Anacker, K.B. (ed.) *The New American Suburb: Poverty, Race and the Economic Crisis*. New York: Routledge, pp. 113–132.

Leitner, H. and Strunk, C. (2014) "Assembling Insurgent Citizenship: Challenging Exclusionary Local Immigration Policies in the Washington D.C. Metro Area." *Urban Geography*, 35(7), pp. 943–964.

Li, W. (2009) *Ethnoburb: The New Ethnic Community in Urban America*. Honolulu, HI: University of Hawaii Press.

Light, I., Sabagh, G., Bozorgmehr, M., and Der-Martirosian, C. (1994) "Beyond the Ethnic Enclave Economy." *Social Problems*, 41(1), pp. 65–80.

Logan, J.R., Xu, Z., and Stults, B.J. (2014) "Interpolating U.S. Decennial Census Tract Data from as Early as 1970 to 2010: A Longitudinal Tract Database." *The Professional Geographer*, 66(3), pp. 274–283.

Massey, D. (1985) "Ethnic Residential Segregation: A Theoretical and Empirical Synthesis." *Sociology and Social Research*, 69, pp. 315–350.

Minnesota Population Center (2011) *National Historical Geographic Information System: Version 2.0*. Minneapolis, MN: University of Minnesota Press.

Park, R.E., Burgess, E.W., and McKenzie, R.D. (1925) *The City*. Chicago, IL: University of Chicago Press.

Peña, E. (2008) "Beyond Mexico: Guadeloupian Sacred Space Production and Mobilization in a Chicago Suburb." *American Quarterly*, 60(3), pp. 721–747.

Portes, A. (1981) "Modes of Structural Incorporation and Present Theories of Labor Immigration." In: Kritz, M., Kelley, C.B. and Tomasi, S. (eds.) *Global Trends in Migration*. New York: Center for Migration Studies, pp. 279–297.

Price, M., Cheung, I., Friedman, S., and Singer, A. (2005) "The World Settles In: Washington, DC as an Immigrant Gateway." *Urban Geography*, 26, pp. 61–83.

Singer, A. (2004) "The Rise of New Immigrant Gateways." In: *Metropolitan Policy Program*. Washington, DC: Brookings Institution Press.

Singer, A. (2015) "A Typology of Immigrant Gateways, 2014." In: *Metropolitan Policy Program*. Washington, DC: Brookings Institution Press.

Spula, I. (2014) "Why Chinese Millionaires Are Investing in Chicago Real Estate." *Chicago Magazine*, December 22. Available at: www.chicagomag.com/Chicago-Magazine/January-2015/The-New-China-Pipeline/.

Strunk, C. (2014) "'We Are Always Thinking of Our Community': Bolivian Hometown Associations, Networks of Reciprocity, and Indigeneity." *Washington, D.C. Journal of Ethnic and Migration Studies*, 40(11), pp. 1697–1715.

Strunk, C. and Leitner, H. (2013) "Resisting Federal-Local Immigration Enforcement Partnerships: Redefining 'Secure Communities' and Public Safety." *Territory, Politics, Governance*, 1(1), pp. 62–85.

Troche-Rodriguez, M. (2009) "Latinos and Their Housing Experiences in Metropolitan Chicago: Challenges and Recommendations." *Harvard Journal of Hispanic Policy*, 21, pp. 17–33.

Vicino, T.J. (2013) *Suburban Crossroads: The Fight for Local Control of Immigration Policy.* Lanham, MD: Lexington Books.

Walker, K.E. and Leitner, H. (2011) "The Variegated Landscape of Local Immigration Policies in the United States." *Urban Geography*, 32(2), pp. 156–178.

Walker, K.E. (2014) "Immigration, Local Policy, and National Identity in the Suburban United States." *Urban Geography*, 35(4), pp. 508–529.

Walker, K.E. (2015) "The Spatiality of Local Immigration Policy in the United States." *Tijdschrift voor economische en sociale geografie*, 106(4), pp. 486–498.

Walker, K.E. (2018) "Locating Neighborhood Diversity in the American Metropolis." *Urban Studies*, 55(1), pp. 116–132.

Wilson, J.H., Singer, A., and DeRenzis, B. (2010) "Growing Pains: Local Response to Recent Immigrant Settlement in Suburban Washington, DC." In: Varsanyi, M. (ed.) *Taking Local Control: Immigration Policy Activism in U.S. Cities and States.* Stanford, CA: Stanford University Press, pp. 193–215.

Wright, R., Ellis, M., and Parks, V. (2005) "Re-Placing Whiteness in Spatial Assimilation Research." *City & Community*, 4(2), pp. 111–135.

Zelinsky, W. and Lee, B.A. (1998) "Heterolocalism: An Alternative Model of the Sociospatial Behavior of Immigrant Ethnic Communities." *International Journal of Population Geography*, 4, pp. 281–298.

17

Poverty in U.S. suburbs

Katrin B. Anacker

Introduction

Over the past several decades the Great Risk Shift has individualized economic risk in the United States (i.e., it has partially shifted risk from corporations and the federal, state, and local governments to households) (Hacker, 2008). Over the past decade, the Great Recession, which started in December 2007 and technically ended in June 2009, has resulted in precarious household balance sheets and possible poverty for some households, especially those that are Black/African American and Hispanic/Latino. Reasons may include an increase in workers coping with a new, lower-paying job, involuntary part-time work, underemployment, or foreclosure (Barr, 2012; Morduch and Schneider, 2017; National Bureau of Economic Research, n.d.; Noah, 2012; Reich, 2013, 2015; Shapiro, 2017). Other factors may have been unplanned medical expenses (which may have been partially addressed by the Affordable Care Act), relationship dissolution, or a spouse's death (Sanders, 2011; Warren, 2014, 2017).

Poverty is typically clustered and concentrated in certain neighborhoods (Madden 2003a, 2003b). The number of high poverty neighborhoods (i.e., neighborhoods with a poverty rate of 40 percent or higher) increased from 1970 to 1990 and then decreased from 1990 to 2000 (Jargowsky, 2003). However, somewhat recent analyses show another increase in the number of high poverty neighborhoods, to 2,822 in 2005/2009 from 2,075 in 2000, or a 36 percent change, based on the 2000 Census and the 2005/2009 American Community Survey (Kneebone and Nadeau, 2015). The Great Risk Shift, the Great Recession, and structural and individual factors may have either caused or contributed to a poverty rate of 12.7 percent in 2016 (Abramsky, 2013; Coates, 2015; Desmond, 2016; Gest, 2016; Murray, 2013; Semega et al., 2017; Vance, 2016).

Poverty typically results in concentrated disadvantage at the individual as well as the community level. At the neighborhood level, poverty may translate to high crime and unemployment rates, a large proportion of vacant buildings, low-performing schools, fewer employers, lackluster public and private services, a low homeownership rate, and real estate's failure to appreciate, among many other factors (Edelman, 2013; Ehrenreich, 2001; Harrington, 2012; Isenberg, 2016). Poverty may also translate into economic mobility challenges. For example, Chetty et al. (2016) found that absolute income mobility, measured as the proportion of children who earn more than their parents, has decreased from 92 percent for children born in 1940 to about 50 percent for children born in 1984. They conclude that increasing absolute income mobility requires a greater distribution of gross domestic product growth rates rather than just high gross domestic product growth rates.

This chapter focuses on poverty in U.S. suburbs and revisits the policy blind spot hypothesis (Puentes and Orfield, 2002), arguing that the suggested major budget cuts to the Community Development Block Grant (CDGB) program and the Home Investment Partnership Program (HOME), two place-based programs, through the Tax Cuts and Jobs Act and the scheduled end of the Neighborhood Stabilization Program (NSP) will result in suburbs entering a *true* policy blind spot (Congress.gov, 2017). This chapter is structured as follows. First, it discusses differences between suburban and inner-city poverty. Second, it focuses on suburban poverty, and the suburban social welfare infrastructure in particular. Third, it discusses the policy blind spot hypothesis, including place-based as well as people-based policies benefiting suburbs as well as suburbanites. It ends with a conclusion.

Suburban versus inner-city poverty

In the past, poverty was discussed in terms of the differences between inner cities and suburbs. Until somewhat recently, the vast majority of the literature focused on poverty in inner cities (Desmond, 2016), small towns (Vance, 2016), and rural areas (Abramsky, 2013; Wilkerson, 2010). Currently, more Americans live below the poverty line in suburbs than in inner cities (Kneebone and Berube, 2013).

While some aspects of suburban poverty may be similar to inner-city poverty, other aspects are different (Kneebone and Berube, 2013). For example, there are similarities in the proportion of poor people who are under 18 years, work part-time or during a part of the year, do not work, have a bachelor's degree or higher, are male and head a household (regardless of whether they have children), and are female single householders (Kneebone and Berube, 2013).

There are also differences. Compared to inner cities, suburbs have a higher proportion of poor people who work full-time year-round, have a high school degree, some college, or an associate's degree, live in households with at least one worker, are homeowners, and are married (regardless of whether they have children) (Kneebone and Berube, 2013). These findings show that a full-time job, a high school or college degree, homeownership, and marriage do not necessarily protect people from poverty, consistent with the Great Risk shift along with the growing insecurity among U.S. households. Suburbs have a higher proportion of poor people over the age of 65 and are non-Hispanic white (Kneebone and Berube, 2013). There may be diversity among suburban neighborhoods (Holliday and Dwyer, 2009). While the establishment of Social Security in 1935 decreased the proportion of poor seniors in the long run, it did not eradicate poverty among seniors altogether. Indeed, past discussions about poverty have focused on people of color, ignoring non-Hispanic whites, as current discussions show (Isenberg, 2016; Gest 2016; Vance, 2016).

Suburban poverty

In the 1980s and 1990s, discussions focused on changes in terms of demographics, socioeconomics, and housing gradually acknowledged that suburbs were no longer the home of native-born, home-owning, high-income, highly educated non-Hispanic whites (Baldassare, 1986; Jackson, 1985; Kunstler, 1993; Orfield, 1997; Rusk, 1999). In the 2000s, scholars used the terms "suburban decline" or "suburban diversity," possibly indicating that they were not ready to acknowledge that there was suburban poverty, which may have sounded like an oxymoron to some (Gallagher, 2013; Hanlon, 2010; Hudnut, 2003; Kotkin, 2001; Lucy and Phillips, 2000, 2006; Orfield, 2002;

Vicino, 2008). In the 2010s, poverty was explicitly acknowledged and discussed in the context of suburbs (Anacker, 2015; Kneebone and Berube, 2013; Kneebone and Nadeau, 2015; Nicolaides, 2002; Roth and Allard, 2015; Wiese, 2004).

In the recent past, poverty was discussed in terms of the differences between and among suburbs. Scholars have created typologies in order to understand suburban poverty better (Peters, 2009). For example, Orfield (2002) classified *challenged* suburban municipalities in 25 metropolitan regions according to tax capacity and costs into the following three types: at-risk, segregated (348 out of 4,711 municipalities; 7.39 percent); at-risk, older (391 out of 4,711 municipalities; 8.30 percent); and at-risk, low-density (1,104 out of 4,711 municipalities; 23.43 percent). Madden (2003b) found that poverty has increasingly concentrated in suburbs that are part of older metropolitan areas, especially in the Northeast and Midwest. Mikelbank (2004) analyzed national population, place, economic, and government census data and found the following two *challenged* suburban types: Black (179 out of 3,567 places; 5.02 percent) and struggling (828 out of 3,567 places; 23.21 percent).

Murphy and Wallace (2010) analyzed the availability (or lack thereof) of organizational resources available for the poor, focusing on three U.S. metropolitan areas. They differentiated among three types of organizations: first, hardship organizations, which help clients meet their daily needs, such as shelters, food pantries, and legal services; second, employment organizations, which provide job placements; and third, educational organizations, which provide trade school and computer training. Based on logistic regression analyzes, they found that poor suburban neighborhoods are more likely to be organizationally deprived than poor urban neighborhoods, especially with regard to hardship organizations.

Murphy (2010) conducted in-depth interviews with nonprofit, antipoverty organizations in eight suburbs in the Philadelphia and Pittsburgh metropolitan areas and found three types of poor suburbs. First, symbiotic suburbs, those in close proximity to the inner-city and that mirror poor urban neighborhoods, have a majority of census tracts with a poverty rate greater than 20 percent and a high crime rate, a high proportion of households with low educational attainment, a low labor force participation rate, and are of color or headed by a single parent. Second, skeletal suburbs, once home to Rustbelt firms that closed several years or decades ago, have a majority of census tracts with a poverty rate greater than 20 percent and a high crime rate, a high proportion of households with low educational attainment, a low labor force participation rate, a low number of commercial establishments and amenities, and a deteriorating housing stock. Third, overshadowed suburbs are relatively affluent but have segregated pockets of invisible poverty.

Kneebone and Nadeau (2015) identified nationwide 223 high poverty suburban census tracts in 2000 and 344 tracts in 2005/2009, a 54.3 percent change. Of these census tracts, 79 and 114 (44.3 percent change) were high-density; 100 and 156 (56 percent change) were mature suburban; 36 and 58 (61 percent change) were emerging suburban; and 8 and 16 (100 percent change) were exurban in 2000 and 2010, respectively. Kneebone and Nadeau (2015) also found that the increase was highest in mature suburbs in terms of number, and highest in exurbs in terms of proportions, indicating an increase in high poverty neighborhoods even further away from the city center, extending the growth models suggested by the Chicago School several decades ago (Burgess, 1925; Harris and Ullman, 1945; Hoyt, 1939).

Interestingly, some scholars have discussed the lagging state of mind of many suburbanites who "have had the mistaken belief that an impenetrable wall stands between them and the problems and threats in the city" (Bier, 1991, p. 48). Denton and Gibbons (2013, p. 15; see also Murphy, 2010) discussed "hidden frames, . . . individual conceptions of what a suburb is based on one's age, location, and social status" and suspect that they "may conjure different varied conceptions of suburbs, their problems, and their assets."

Being poor and residing in a poor neighborhood, regardless of location, is in itself challenging. However, being poor and residing in a poor suburban neighborhood may be different from being poor and residing in an inner-city neighborhood. Compared to some poor inner-city neighborhoods, some poor suburban neighborhoods may still have lower crime and unemployment rates, fewer vacant buildings, higher performing schools, more employers, better public and private services, a higher homeownership rate, and more robust real estate appreciation. Where poor neighborhoods differ is in terms of their social welfare infrastructure, thus resulting in additional challenges in navigating poverty.

Suburban social welfare infrastructure

The United States has a social welfare system, funded at the federal, state, and local levels, which provides benefits for eligible individuals. Federal benefits may be administered directly by federal agencies, or indirectly through state and local agencies or nonprofit social service providers. State and local benefits may be administered directly by state agencies or indirectly through local agencies or nonprofit social service providers. However, over the past several decades, the state's role has gradually decreased due to devolution, while the role of many nonprofits and philanthropies has grown due to increased need after the Great Recession (Twombly, 2001; Wolch, 1996).

Several scholars have studied social service providers. With regard to organizational structure and activities, Roth and Allard (2015) studied social service providers in the suburbs of the Chicago, Illinois; Los Angeles, California; and the Washington, DC metropolitan areas, based on an analysis of and expert interviews with 100 diverse nonprofit social service organizations. They found that there is great variety in terms of four aspects. First, there was variety in terms of poverty rates and in their change from 2000 to 2005/2009 among the different suburbs. Second, there was variety in terms of the budget size of the interviewed suburban service providers: 74 percent of the interviewees were secular nonprofit and 26 percent were religious nonprofit organizations. Third, there was variety in terms of offered services: 45.7 percent of the analyzed providers had a budget larger than $1 million, 28.7 percent had a budget between $200,000 and $1 million, 20.2 percent had a budget between $50,000 and $200,000, and 5.3 percent had a budget less than $50,000. Fourth, there was variety in terms of services offered: 50 percent of the providers offered six or more different services while 28.6 percent offered two to five different services. More specifically, most providers focused on food assistance or meals (68.4 percent); family or individual counseling (57.1 percent); job training, search, or placement (54.6 percent); youth programs (54.1 percent); clothing or household items (53.1 percent); assistance finding affordable housing (49 percent); emergency cash or utility assistance (48 percent); assistance paying rent (43.9 percent); assistance with GED, ESL, or high school completion (32.7 percent); temporary shelter or housing programs (31.6 percent); senior programs (30.6 percent); outpatient mental health assistance (27.6 percent); programs for ex-offenders (21.4 percent); outpatient substance abuse programs (19.4 percent); and physical and/or developmental disability programs (14.3 percent).

Mitchell-Brown (2013) studied three community development corporations (CDCs) in mature suburbs in the Cincinnati, Ohio metropolitan area, conducting expert and stakeholder interviews in addition to analyzing and visiting the neighborhoods in which the three CDCs were located. CDCs are nonprofit community-based organizations that provide job training, commercial revitalization, and affordable housing for eligible low- and moderate-income residents in challenged neighborhoods, until recently typically in inner cities. She found that the three studied suburban CDCs focused on promoting economic development through the construction or rehabilitation of affordable housing units, possibly filling a gap that municipalities are unable or unwilling to address, rather than social development, community leadership, and empowerment. Indeed, the CDCs developed new infill housing and single-family housing rehabilitation, trained potential homeowners, and developed community partnerships with local governments, civic/community groups, faith-based institutions, and other CDCs (Mitchell-Brown, 2013).

With regard to the spatial distribution and service delivery of social service providers, Wolch and Geiger (1983) analyzed the distribution of social service providers in Los Angeles County and found that they locate in stable, growing, mature low- and middle-income inner-ring suburbs that have relatively high service needs. Joassart-Marcelli and Wolch (2003) investigated the spatial distribution and service delivery of service providers in California's Los Angeles, Orange, Riverside, San Bernardino, and Ventura Counties, based on 1996 Internal Revenue Service and 1990 census data. They found that older middle- and high-income inner suburbs had a relatively high proportion of antipoverty nonprofits and a high level of expenditures, while very low-income, old industrial inner-ring suburbs had a relatively low proportion of nonprofits, indicating a spatial mismatch (Joassart-Marcelli and Wolch, 2003).

Hendrick and Mossberger (2009) analyzed the delivery of social services and the capacity of 114 suburban townships and municipalities in the Chicago metropolitan area, based on two surveys between March and September 2008 and an analysis of the 2000 Census. They found uneven service delivery across the metropolitan area, including townships that tax and spend less or more on social services due to high resources or high needs. They also found that townships, which receive little intergovernmental assistance, depend heavily on property taxes, although in terms of raising property tax rates they do have certain constraints, such as a lack of home rule status and local voters unwilling to vote for tax increases. In sum, the political and economic capacity to raise revenue for service delivery has been weak and will most likely remain weak. Therefore, there could be better coordination of social service provisions within and among townships and counties in the metropolitan area (Hendrick and Mossberger, 2009).

With regard to copying service providers' strategies, Roth and Allard (2015) found that interviewees prioritized clients by degree of need (50 percent), referred a greater number of clients out (45.6 percent), reduced full-time and/or part-time staff (41.7 percent), reduced overhead costs (41.3 percent), expanded (the) waitlist(s) (35.6 percent), reduced available services (33.3 percent), reduced the number of clients served (27.5 percent), reduced salaries (16.1 percent), reduced their hours of operation (14.1 percent), merged with another organization (8.6 percent), and anticipated closing down (4.6 percent) (Roth and Allard, 2015).

Scholars have also discussed the similarities and differences between urban and suburban service providers. In terms of similarities, they operate in an environment with increased overall need yet steady or decreased overall support, forcing them to do more with less (Brown, 1980). In terms of differences, urban service providers receive more grants and gifts and have higher expenditures and assets per capita. They also may serve a higher number of people per office, potentially resulting in lower quality of service but greater efficiency compared to suburban service providers (Allard, 2009; Allard et al., 2003; Murphy, 2010; Roth and Allard, 2015; Small and McDermott, 2006; Wolpert, 1993). Over time, the number of suburban residents in need has increased, but the number of service providers' offices in the suburbs has remained steady or declined, resulting in a spatial mismatch in social service provision (Hendrick and Mossberger, 2009; Mitchell-Brown, 2013; Roth and Allard, 2015). The relatively low number of suburban service providers may result from the lack of history of social service provision in the first place, a limited sense of community and civic culture, and possibly suburban nonprofits providing select local amenities such as arts, educational, and health services (Joassart-Marcelli and Wolch, 2003).

Public policy implications

More than a decade ago, Puentes and Orfield (2002) stated, "first suburbs [in the Midwest] are caught in a policy blind spot. Unlike inner cities, they are not poor enough to qualify for many federal . . . reinvestment programs and not large enough to receive federal . . . funds directly" (Puentes and Orfield, 2002, p. 3). Puentes and Orfield (2002) define a policy blind spot using a two-tiered criterion. The first criterion relates to relative location with specific characteristics (i.e., a first suburb that is an inner-ring community just outside an inner-city that has older housing, an aging infrastructure, some struggling neighborhoods and commercial areas, but also growing job centers). The second criterion relates to household income eligibility that benefits low- and moderate-income households.

Interestingly, there are three recent federally established programs benefiting suburbs and suburbanites in need. Thus, one may wonder about the hypothesis that first suburbs, at least in the Midwest, are in a policy blind spot. The aforementioned programs are the Community Development Block Grant (CDBG), established in 1974; the Home Investment Partnership Program (HOME), established in 1990; and the Neighborhood Stabilization Program (NSP), established in 2008, among other state and local programs (U.S. Department of Housing and Urban Development, n.d.b). These three programs will be discussed as follows.

First, the primary goal of the CDBG Program is to improve low-income communities by providing funding for decent housing, a suitable living environment, and expanded economic opportunities in communities with a high proportion of low- and moderate-income persons (U.S. Department of Housing and Urban Development, n.d.b). Moderate-income households have a household income below 80 percent of the area median income (AMI), whereas low-income households have a household income below 50 percent of the AMI (U.S. Department of Housing and Urban Development, n.d.b). Recipients of the funds are grantees (i.e., states, counties, and cities) (U.S. Department of Housing and Urban Development, n.d.b). Entitlement communities (i.e., principal cities of Metropolitan Statistical Areas [MSAs], cities in MSAs with at least 50,000 people, or counties with at least 200,000 people) receive 70 percent of the annual CDBG funding, based on their own programs and priorities (U.S. Department of Housing and Urban Development, n.d.a). Non-entitlement communities (i.e., smaller towns and rural communities) receive 30 percent of the annual CDBG funding on a competitive basis, passed along by their respective states (U.S. Department of Housing and Urban Development, n.d.b). The grant amounts are based on the higher score of two aspects: first, overcrowded housing, population, and poverty; or second, age of housing, population growth lag, and poverty (U.S. Department of Housing and Urban Development, n.d.b).

Eligible activities are classified in the following categories:

- activities related to housing (including homeownership assistance, rental, and homeowner rehabilitation activities, housing services in connection with the HOME program, and lead-based paint testing and abatement);
- other real estate or property activities (including acquisition, disposition, clearance and demolition, code enforcement, and historic preservation);
- public facilities (including infrastructure, special needs facilities, and community facilities);
- activities related to economic development (including microenterprise assistance, commercial rehabilitation, and special economic development activities);
- activities related to public services (including job training and employment services, healthcare and substance abuse services, child care, crime prevention, and fair housing counseling);

- assistance to Community-Based Development Organizations (including neighborhood revitalization, community economic development, and energy conservation); and
- other types of activities.

(U.S. Department of Housing and Urban Development, n.d.b)

In sum, while suburbs with at least 50,000 people, of whom a high proportion are moderate- and low-income, may be potentially eligible for the CDBG Program, only a fraction of suburbs benefit from it, as most suburbs fall below this population threshold (Anacker, 2009).

Second, the primary goal of the HOME Program, based on the National Affordable Housing Act of 1990, is to increase the supply of affordable housing for eligible low- and very low-income households (U.S. Department of Housing and Urban Development, n.d.b). Low-income households have a household income below 50 percent of the Area Median Income (AMI), whereas very low-income households have a household income between 30 and 50 percent of the AMI (U.S. Department of Housing and Urban Development, n.d.b). The HOME Program has the secondary goal of supporting the development and sustainability of Community Housing Development Organizations (U.S. Department of Housing and Urban Development, n.d.b). The funds' recipients are participating jurisdictions (i.e., states, counties, cities, and consortia) (U.S. Department of Housing and Urban Development, n.d.b). Applicants must have resided within a community for at least one year and be either first-time homebuyers, displaced householders, or applicants who have not owned homes for three years (U.S. Department of Housing and Urban Development, n.d.b). Eligible activities are as follows:

- homeowner repair, rehabilitation, or reconstruction of owner-occupied homes;
- homebuyer acquisition, rehabilitation, or new construction of owner-occupied homes;
- investor acquisition, rehabilitation, or new construction of rental housing;
- assistance for rent, security deposits, and (under certain conditions) utility deposits for/by tenants.

(U.S. Department of Housing and Urban Development, n.d.b)

HOME imposes a maximum purchase price of 95 percent of the median purchase price in the MSA for owner-occupied units, which must meet state and local codes and ordinances as well as local building codes (U.S. Department of Housing and Urban Development, n.d.b). HOME also imposes rent and occupancy requirements over the length of an affordability period, depending on both the amount of assistance per unit or buyer and the activity (U.S. Department of Housing and Urban Development, n.d.b). The affordability period is five (10, 15) years for an assistance amount of less than $15,000 (between $15,000 – $40,000, more than $40,000, respectively), 15 years for refinancing rental housing, and 20 years for new construction of rental housing (U.S. Department of Housing and Urban Development, n.d.b). Regardless of the period, only income-eligible households are allowed to occupy a HOME-assisted unit (U.S. Department of Housing and Urban Development, n.d.b).

In sum, while suburbs and suburbanites may potentially benefit from the HOME program, only a certain number will. Some owner-occupiers may not be able or willing to repair, rehabilitate, or reconstruct their homes because they do not have the resources, may predict that they will not be able to age in their community, or may be concerned about finding a buyer who will pay an acceptable amount of money. Some suburbs are (almost) completely built up and built out, making new construction of owner-occupied homes difficult. Some investors may not be interested in acquiring, rehabilitating, or constructing rental housing due to not-in-my-backyard (NIMBY) concerns. Also, some homebuyers may not have resided in their community for a year (Anacker, 2009).

Third, the Neighborhood Stabilization Program (NSP), established in 2008, provides emergency assistance to stabilize communities with a high proportion of abandoned and foreclosed homes. The first round of NSP funding provided $3.92 billion to 307 state and local governments based on a formula. The second round of funding was $1.93 billion to 56 states, local governments, nonprofits, and consortia of nonprofit entities on a competitive basis. Finally, the third round of NSP funding provided $1 billion to 270 state and local governments, again based on a formula. Eligible activities are

- establishing financing mechanisms for purchase and redevelopment of foreclosed homes and residential properties;
- purchasing and rehabilitating homes and residential properties abandoned or foreclosed;
- establishing land banks for foreclosed homes;
- demolishing blighted structures; and
- redeveloping demolished or vacant properties.

(U.S. Department of Housing and Urban Development, n.d.c)

NSP grantees must use at least 25 percent of the funds appropriated to households whose incomes do not exceed 50 percent of the area median income. To the author's knowledge, there is currently little systematic research on the proportion and types of suburbs that have benefited from NSP funds.

I argue that the impending major budget cuts to the CDBG and HOME programs and the scheduled end of the NSP will result in suburbs entering a *true* policy blind spot. In March 2017 the Trump Administration considered $6 billion in cuts to the FY 2018 budget of the U.S. Department of Housing and Urban Development (HUD). These cuts would impact the CDBG and the HOME Investment Partnerships (HOME) programs, among many others (Congress.gov, 2017; Daniel, 2017a, 2017b; National Low Income Housing Coalition, 2017). Most grantees have drawn down their NSP funds or will do so shortly (U.S. Department of Housing and Urban Development, n.d.d). In summary, the three major place-based programs that benefit suburbs might soon be history.

Nevertheless, there are still people-based programs benefiting suburbanites, along with people who reside in inner cities as well as rural areas. These are cash or in-kind public assistance programs, such as Temporary Assistance for Needy Families (TANF; administered by the Department of Health and Human Services), the Supplemental Nutrition Assistance Program (SNAP; administered by the U.S. Department of Agriculture), Unemployment Insurance (UI; administered by the U.S. Department of Labor), and the Earned Income Tax Credit (EITC; administered by the Internal Revenue Service) (Roth and Allard, 2015). Unlike many place-based programs discussed earlier, these people-based programs might be safe from budget cuts in the near and possible distant future.

Conclusion

In the coming years, income and wealth inequality in mature suburbs will most likely increase despite the fact that government support is not likely to follow suit (Stiglitz, 2012). Suburban stakeholders will be forced to do more with less, possibly by increasing collaboration to improve effectiveness and efficiency (Hendrick and Mossberger, 2009). The question remains if (and if so, how) philanthropy will respond, given the lagging gap between the perceptions and realities of suburban poverty (Adams, 2014).

In terms of neighborhoods in the mature suburbs, there will be differences between those that have investment potential and those that do not (Anacker and Morrow-Jones, 2008; Bollens,

1988). Neighborhoods that fall into the former group may be close to public transit (especially rail) and may have a housing stock with an aesthetically pleasing architecture, such as Colonial Revival, which might result in either teardowns or in rehabilitation activities (Charles, 2013, 2014). Neighborhoods that fall into the latter group may be poorly connected to public transit and may have an affordable housing stock that may house low-income residents.

More than 150 years ago, John Stuart Mill (1848, p. 969, quoted in Joassart-Marcelli and Wolch, 2003) argued that "charity almost always does too much or too little: it lavishes its bounty in one place, and leaves people to starve in another." Over the past centuries, many inner-city institutions have benefited from philanthropic activities. Over the past two decades or so, the suburbs have demonstrated their own need but philanthropic activities have been scarce. Given the new landscape of federal funding, future efforts, possibly conducted by Metropolitan Planning Organizations, could focus on developing a regional fair share plan of hubs that provide social services (Wolch, 1996).

Guide to further reading

Anacker, K.B. (ed.) (2015) *The New American Suburb: Poverty, Race, and the Economic Crisis.* Burlington, VT: Ashgate.

Hanlon, B. (2010) *Once the American Dream: Inner-Ring Suburbs of the Metropolitan United States.* Philadelphia, PA: Temple University Press.

Hanlon, B., Short, J.R., and Vicino, T.J. (2010) *Cities and Suburbs: New Metropolitan Realities in the U.S.* New York: Routledge.

Heiman, R. (2015) *Driving after Class: Anxious Times in an American Suburb.* Oakland, CA: University of California Press.

Kneebone, E. and Berube, A. (2013) *Confronting Suburban Poverty in America.* Washington, DC: Brookings Institution Press.

Lung-Amam, W.S. (2017) *Trespassers? Asian American and the Battle for Suburbia.* Oakland, CA: University of California Press.

Nicolaides, B.M. and Wiese, A. (eds.) (2016) *The Suburb Reader.* New York: Routledge.

Niedt, C. (ed.) (2013) *Social Justice in Diverse Suburbs: History, Politics, and Prospects.* Philadelphia, PA: Temple University Press.

Vicino, T.J. (2008) *Transforming Race and Class in Suburbia: Decline in Metropolitan Baltimore.* New York: Palgrave Macmillan.

Vicino, T.J. (2012) *Suburban Crossroads: The Fight for Local Control of Immigration Policy.* Lanham, MD: Lexington Books.

References

Abramsky, S. (2013) *The American Way of Poverty: How the Other Half Still Lives.* New York: Nation Books.

Adams, C.T. (2014) *From the Outside In: Suburban Elites, Third-Sector Organizations, and the Reshaping of Philadelphia.* Philadelphia, PA: Temple University Press.

Allard, S.W. (2009) *Out of Reach: Place, Poverty, and the New American Welfare State.* New Haven, CT: Yale University Press.

Allard, S.W., Tolman, R.M., and Rosen, D. (2003) "Proximity to Service Providers and Service Utilization among Welfare Recipients: The Interaction of Place and Race." *Journal of Policy Analysis and Management,* 22(4), pp. 599–613.

Anacker, K.B. (2009) *Analyzing Mature Suburbs in Ohio through Property Values.* Saarbrücken, Germany: Verlag Dr. Müller.

Anacker, K.B. (2015) "Analyzing Census Tract Foreclosure Risk Rates in Mature and Developing Suburbs in the United States." *Urban Geography,* 36(8), pp. 1221–1240.

Anacker, K.B. and Morrow-Jones, H.A. (2008) "Mature Suburbs, Property Values, and Decline in the Midwest? The Case of Cuyahoga County." *Housing Policy Debate,* 19(3), pp. 519–549.

Baldassare, M. (1986) *Trouble in Paradise: The Suburban Transformation in America.* New York: Columbia University Press.

Barr, M.S. (2012) *No Slack: The Financial Lives of Low-Income Americans.* Washington, DC: Brookings Institution Press.

Bier, T.E. (1991) "Public Policy Against Itself: Investments That Help Bring Cleveland (and Eventually Suburbs) Down." In Schorr, A. (ed.) *Cleveland Development: A Dissenting View.* Cleveland, OH: David Press, pp. 43–52.

Bollens, S.A. (1988) "Municipal Decline and Inequality in American Suburban Rings, 1960–1980." *Regional Studies,* 22(4), pp. 277–285.

Brown, M.A. (1980) "Do Central Cities and Suburbs Have Similar Dimensions of Need?" *Professional Geographer,* 32(4), pp. 400–411.

Burgess, E.W. (1925) "The Growth of the City: An Introduction to a Research Project." In: Park, R.E., Burgess, E.W. and McKenzie, R.D. (eds.) *The City.* Chicago, IL: University of Chicago Press, pp. 47–61.

Charles, S.L. (2013) "Understanding the Determinants of Single-Family Residential Redevelopment in the Inner-Ring Suburbs of Chicago." *Urban Studies,* 50(8), pp. 1505–1522.

Charles, S.L. (2014) "The Spatio-Temporal Pattern of Housing Development in Suburban Chicago, 2000–2010." *Urban Studies,* 51(12), pp. 2646–2664.

Chetty, R., Grusky, D., Hell, M., Hendren, N., Manduca, R., and Narang, J. (2016) "The Facing American Dream: Trends in Absolute Income Mobility Since 1940." National Bureau of Economic Research Working Paper No. 22910. National Bureau of Economic Research, Cambridge, MA.

Coates, T.-N. (2015) *Between the World and Me.* New York: Spiegel and Grau.

Congress.gov. (2017) *H.R.1: Tax Cuts and Jobs Act: 115th Congress.* Washington, DC: U.S. Congress.

Daniel, D. (2017a) "HUD Weighs CDBG and HOME Program Eliminations in Fiscal Year 2018 Budget." National Association of Counties (NACO) Blog, March 10.

Daniel, D. (2017b) "Commissioner Highlights Importance of CDBG and HOME at Capitol Hill Roundtable Discussion." National Association of Counties (NACO) Blog, May 30.

Denton, N. and Gibbons, J.R. (2013) "Twenty-First Century Suburban Demography: Increasing Diversity Yet Lingering Exclusion." In: Niedt, C. (ed.) *Social Justice in Diverse Suburbs: History, Politics, and Prospects.* Philadelphia, PA: Temple University Press, pp. 13–30.

Desmond, M. (2016) *Evicted: Poverty and Profit in the American City.* New York: Crown Publishers.

Edelman, P. (2013) *So Rich, So Poor: Why It's So Hard to End Poverty in America.* New York: The New Press.

Ehrenreich, B. (2001) *Nickel and Dimed: On (Not) Getting By in America.* New York: Henry Hold and Company.

Gallagher, L. (2013) *The End of the Suburbs: Where the American Dream Is Moving.* New York: Portfolio.

Gest, J. (2016) *The New Minority: White Working Class Politics in an Age of Immigration and Inequality.* New York: Oxford University Press.

Hacker, J.S. (2008) *The Great Risk Shift: The New Economic Insecurity and the Decline of the American Dream.* New York: Oxford University Press.

Hanlon, B. (2010) *Once the American Dream: Inner-Ring Suburbs of the Metropolitan United States.* Philadelphia, PA: Temple University Press.

Harrington, M. (2012) *The Other America: Poverty in the United States.* New York: Scribner.

Harris, C. and Ullman, E.L. (1945) "The Nature of Cities." *Annals of the Academy of Political and Social Sciences,* 242, pp. 7–17.

Hendrick, R. and Mossberger, K. (2009) *Uneven Capacity and Delivery of Human Services in the Chicago Suburbs: The Role of Townships and Municipalities.* Chicago, IL: University of Illinois at Chicago.

Holliday, A.L. and Dwyer, R.E. (2009) "Suburban Neighborhood Poverty in U.S. Metropolitan Areas in 2000." *City and Community,* 8(2), pp. 155–176.

Hoyt, H. (1939) *The Structure and Growth of Residential Neighborhoods of Residential Neighborhoods in American Cities.* Washington, DC: U.S. Government Printing Office.

Hudnut, W.H. (2003) *Halfway to Everywhere: A Portrait of America's First-Tier Suburbs.* Washington, DC: Urban Land Institute.

Isenberg, N. (2016) *White Trash: The 400-Year Untold History of Class in America.* New York: Penguin Press.

Jackson, K.T. (1985) *Crabgrass Frontier: The Suburbanization of the United States.* New York: Oxford University Press.

Jargowsky, P.A. (2003) *Stunning Progress, Hidden Problems: The Dramatic Decline of Concentrated Poverty in the 1990s.* Washington, DC: Brookings Institution Press.

Joassart-Marcelli, P. and Wolch, J.R. (2003) "The Intrametropolitan Geography of Poverty and the Nonprofit Sector in Southern California." *Nonprofit and Voluntary Sector Quarterly,* 32(1), pp. 70–96.

Kneebone, E. and Berube, A. (2013) *Confronting Suburban Poverty in America.* Washington, DC: Brookings Institution Press.

Kneebone, E. and Nadeau, C.A. (2015) "The Resurgence of Concentrated Poverty in America: Metropolitan Trends in the 2000s." In: Anacker, K.B. (ed.) *The New American Suburb: Poverty, Race and the Economic Crisis.* Burlington, VT: Ashgate, pp. 15–38.

Kotkin, J. (2001) *Older Suburbs: Crabgrass Slums or New Urban Frontier?* Los Angeles, CA: Reason Public Policy Institute.

Kunstler, J.H. (1993) *The Geography of Nowhere: The Rise and Decline of America's Man-Made Landscape.* New York: Touchstone.

Lucy, W.H. and Phillips, D.L. (2006) *Tomorrow's Cities, Tomorrow's Suburbs.* Chicago, IL: Planners Press.

Lucy, W.H. and Phillips, D.L. (2000) *Confronting Suburban Decline: Strategic Planning or Metropolitan Renewal.* Washington, DC: Island Press.

Madden, J.F. (2003a) "The Changing Spatial Concentration of Income and Poverty among Suburbs of Large US Metropolitan Areas." *Urban Studies*, 40(3), pp. 481–503.

Madden, J.F. (2003b) "Has the Concentration of Income and Poverty among Suburbs of Large US Metropolitan Areas Changed over Time?" *Papers in Regional Science*, 82, pp. 249–275.

Mikelbank, B.A. (2004) "A Typology of U.S. Suburban Places." *Housing Policy Debate*, 15(4), pp. 935–964.

Mill, J.S. (1848) *Principles of Political Economy.* London: John W. Parker.

Mitchell-Brown, J. (2013) "First Suburbs and Nonprofit Housing: How Do Urban CDCs Develop Affordable Housing in Suburban Communities?" In: Niedt, C. (ed.) *Social Justice in Diverse Suburbs.* Philadelphia, PA: Temple University Press.

Morduch, J. and Schneider, R. (2017) *The Financial Diaries: How American Families Cope in a World of Uncertainty.* Princeton, NJ: Princeton University Press.

Murphy, A.K. (2010) "The Symbolic Dilemmas of Suburban Poverty: Challenges and Opportunities Posed by Variations in the Contours of Suburban Poverty." *Sociological Forum*, 25(3), pp. 541–569.

Murphy, A.K. and Wallace, D. (2010) "Opportunities for Making Ends Meet and Upward Mobility: Differences in Organizational Deprivation across Urban and Suburban Poor Neighborhoods." *Social Science Quarterly*, 91(5), pp. 1164–1186.

Murray, C. (2013) *Coming Apart: The State of White America, 1960–2010.* New York: Crown Forum.

National Bureau of Economic Research (n.d.) *U.S. Business Cycle Expansions and Contractions.* Cambridge, MA: National Bureau of Economic Research.

National Low Income Housing Coalition (2017) *Trump FY18 Budget Calls for Massive Cuts to Affordable Housing Programs.* Washington, DC: National Low Income Housing Coalition.

Nicolaides, B. (2002) *My Blue Heaven: Life and Politics in the Working-Class Suburbs of Los Angeles, 1920–1965.* Chicago, IL: University of Chicago Press.

Noah, T. (2012) *The Great Divergence: America's Growing Inequality Crisis and What We Can Do about It.* New York: Bloomsbury Press.

Orfield, M. (1997) *Metropolitics: A Regional Agenda for Community and Stability.* Washington, DC: Brookings Institution Press.

Orfield, M. (2002) *American Metropolitics: The New Suburban Reality.* Washington, DC: Brookings Institution Press.

Peters, D.J. (2009) "Typology of American Poverty." *International Regional Science Review*, 32(1), pp. 19–39.

Puentes, R. and Orfield, M. (2002) *Valuing America's First Suburbs: A Policy Agenda for Older Suburbs in the Midwest.* Washington, DC: Brookings Institution Press.

Reich, R.B. (2015) *Saving Capitalism: For the Many, Not the Few.* New York: Alfred A. Knopf.

Reich, R.B. (2013) *Aftershock: The Next Economy and America's Future.* New York: Vintage Books.

Roth, B.J. and Allard, S.W. (2015) "The Response of the Nonprofit Safety Net to Rising Suburban Poverty." In: Anacker, K.B. (ed.) *The New American Suburb: Poverty, Race, and the Economic Crisis.* Burlington, VT: Ashgate, pp. 247–284.

Rusk, D. (1999) *Inside Game/Outside Game: Winning Strategies for Saving Urban America.* Washington, DC: Brookings Institution Press.

Sanders, B. (2011) *The Speech: A Historic Filibuster on Corporate Greed and the Decline of Our Middle Class.* New York: Nation Books.

Semega, J.L., Fontenot, K.R., and Kollar, M.A. (2017) *Income and Poverty in the United States: 2016.* Washington, DC: U.S. Bureau of the Census.

Shapiro, T.M. (2017) *Toxic Inequality: How America's Wealth Gap Destroys Mobility, Deepens the Racial Divide, and Threatens Our Future.* New York: Basic Books.

Small, M.L. and McDermott, M. (2006) "The Presence of Organizational Resources in Poor Urban Neighborhoods: An Analysis of Average and Contextual Effects." *Social Forces*, 84(3), pp. 1697–1724.

Stiglitz, J.E. (2012) *The Price of Inequality: How Today's Divided Society Endangers Our Future*. New York: W.W. Norton.

Twombly, E.C. (2001) *Human Service Nonprofits in Metropolitan Areas during Devolution and Welfare Reform*. Washington, DC: The Urban Institute.

U.S. Department of Housing and Urban Development (n.d.a) *CDBG Entitlement Program Eligibility Requirements*. Washington, DC: U.S. Department of Housing and Urban Development.

U.S. Department of Housing and Urban Development (n.d.b) *HOME and CDBG: Working Together to Create Affordable Housing*. Washington, DC: U.S. Department of Housing and Urban Development.

U.S. Department of Housing and Urban Development (n.d.c) *NSP Eligibility Requirements*. Washington, DC: U.S. Department of Housing and Urban Development.

U.S. Department of Housing and Urban Development (n.d.d) *NSP Closeout Guidance*. Washington, DC: U.S. Department of Housing and Urban Development.

Vance, J.D. (2016) *Hillbilly Elegy: A Memoir of a Family and Culture in Crisis*. New York: HarperCollins.

Vicino, T.J. (2008) *Transforming Race and Class in Suburbia: Decline in Metropolitan Baltimore*. New York: Palgrave Macmillan.

Warren, E. (2014) *A Fighting Chance*. New York: Metropolitan Books.

Warren, E. (2017) *This Fight Is Our Fight: The Battle to Save America's Middle Class*. New York: Metropolitan Books.

Wiese, A. (2004) *Places of Their Own: African American Suburbanization in the Twentieth Century*. Chicago, IL: University of Chicago Press.

Wilkerson, I. (2010) *The Warmth of Other Suns: The Epic Story of America's Great Migration*. New York: Vintage Books.

Wolch, J.R. (1996) "Community-Based Human Service Delivery." *Housing Policy Debate*, 7(4), pp. 649–671.

Wolch, J.R. and Geiger, R.K. (1983) "The Distribution of Urban Voluntary Resources: An Exploratory Analysis." *Environment and Planning A*, 15, pp. 1067–1082.

Wolpert, J. (1993) "Decentralization and Equity in Public and Nonprofit Sectors." *Nonprofit and Voluntary Sector Quarterly*, 22(4), pp. 281–296.

18

From Sanford to Ferguson

Race, poverty, and protest in the American suburb

Willow Lung-Amam and Alex Schafran

Introduction

"We must decide that poverty is now an imbedded suburban problem and we must therefore reorganize city and suburb to address this reality."
                                    – *Sandra Moore (2015) "A Vision for Ferguson and Beyond"*

On February 26, 2012, Trayvon Martin, an unarmed black teenager, was shot and killed by a neighborhood watch member in Sanford, Florida, a gated community roughly 20 miles from downtown Orlando. Two and a half years later, Michael Brown, another unarmed black teenager, was shot six times and killed by a white police officer in Ferguson, Missouri, an inner-ring suburb of St. Louis. In the sweltering Missouri summer heat, Brown's body lay in the street for more than four hours while his neighbors and family gathered before being removed by police. These two incidents sparked waves of protests that lasted months, sometimes turned violent, and catalyzed social movements and policy reforms at the local, state, and federal level. In the wake of Martin's killing, Black Lives Matter was born; following that of Brown, it became an international movement. Martin's hoodie and Brown's hands in the air became infamous symbols of violent, racialized, and predatory policing practices recognized around the world.

Media coverage of the shootings and the unrest they inspired broadcast a somewhat unfamiliar image of American suburbs onto the front pages of newspapers, televisions, and social media sites. As crowds filled the streets of Sanford and Ferguson proclaiming "Justice for Trayvon" and "Black Lives Matter," they showed suburbia at the forefront of contemporary struggles for civil rights and racial and economic justice. In doing so, they exposed a wrinkle in scholarly and popular notions of suburban racial politics and political movements. Unlike in the early twentieth century, when African Americans and other marginalized groups largely fought for access to suburban communities, the protests in Sanford and Ferguson emerged over the unequal treatment of African Americans already present in diverse and struggling suburbs.

Many of the conditions that gave rise to the troubled race relations in Ferguson and Sanford were strikingly similar. While incorporated as nearly all-white, middle-class suburbs, by the time of the killings both were majority non-white communities with high unemployment rates and median household incomes below the national average. The 2011–2015 American Community Survey (ACS) shows both had nearly identical and shockingly high poverty rates – 22 percent for

Table 18.1 A tale of two suburbs: demographic profile of Ferguson, Missouri, and Sanford, Florida

	Ferguson, MO	Sanford, FL
Population	21,120	56,170
Median Age	35.7	33.5
Median Household Income	42,738	38,273
% in Poverty	22.1	21.8
% under 18 in Poverty	36.2	33.5
Unemployment Rate	10.7	12.4
% Non-Hispanic White	29.8	45.2
% Non-Hispanic Black	66.1	25.6
% Hispanic	1.4	23
% Homeowner	57.3	53.3
% Vacant Units	14.8	26.1
% Moved into Unit after 2010	36.7	45.8
% Renters Paying More than 35% of Income on Rent	49	52.8

Source: 2011–2015 American Community Survey

individuals (see Table 18.1). More than one-third of children lived in poverty. Housing vacancy rates were high, yet roughly half of all renters spent more than 35 percent of their income on rent. As in many lower-income communities, mobility rates were high, with more than a third of households having moved into their homes after 2010. Both were racially diverse but economically challenged communities.

In other ways, however, these two suburbs could not be more different. Sanford was a booming new Sunbelt suburb in which more than 30 percent of the housing units were built after 2000. Ferguson, in contrast, was a classic inner-ring Rustbelt suburb largely built out by the early post-war period; more than half its homes were built by 1960, and less than 2 percent after 2000. Ferguson also had a very familiar history of metropolitan segregation. In the post-war period, it grew as a result of white middle-class flight from St. Louis. As suburbs opened up in the late 1960s, black migration out of St. Louis precipitated the flight of established white residents, and Ferguson saw increasing poverty rates and a declining municipal tax base as the concentration of people of color increased. Sanford, on the other hand, was a historic commercial center that, like many Sunbelt towns, became a sprawling suburb of its greater metropolitan area after the 1960s housing boom (Lang and LeFurgy, 2007). The Retreat at Twin Lakes, where Martin was killed, was a newly built gated community that symbolized a more exclusive version of the suburban dream, but it was also hit hard by the 2008 foreclosure crisis and its aftermath (Green, 2012).

In their similarities and differences, Ferguson and Sanford signify the way metropolitan regions around the United States have fragmented into diverse and dispersed pockets of poverty and privilege. The tragic deaths of Trayvon Martin and Michael Brown showed these communities as key sites for understanding contemporary processes shaping metropolitan inequality and racial politics. While radically different types of suburbs, Ferguson and Sanford were both subject to forces of decline and disinvestment that followed African Americans as they moved around the metropolis and the metropolis moved around them. In analyzing what led to the shootings and the subsequent unrest, we argue that the suburban landscape was not simply the backdrop to the events. The spatial structure of these communities and the processes that produced them, foreground the ways in which young African American men were positioned and the rallying cries of protesters. They underscore the intricate ties between *what happened* and *where it happened*.

Shifting suburban racial politics

In news coverage of the protests following the deaths of Martin and Brown, and the acquittal and failure to indict their shooters, respectively, Sanford and Ferguson appeared as surprising backdrops to stories about "urban" unrest. In both communities, prominent civil rights leaders Jesse Jackson and Al Sharpton marched alongside residents and activists through wide streets surrounded by cars and strip malls rather than tall buildings and row houses. In Sanford, protestors paraded down the rustic, tree-lined Lake Avenue. In Ferguson, the convenience store that Martin had left only minutes before being shot was located in a single-story commercial center surrounded by a sea of parking that became the staging ground for multiple protests. The grassy knoll where neighbors witnessed the shooting became the site of several candlelight vigils.

While the suburbs of European cities like Paris are well-known sites for uprisings over the conditions of socially and economically marginalized groups, American suburbs are not. From the rebellions that engulfed U.S. cities following the deaths of Martin Luther King in the 1960s to Rodney King in the 1990s, protests, acts of civil disobedience, rioting, and other expressions of resistance to racial inequality have generally played out on the streets of major urban centers. Even given documented civil rights protests in classic post-war suburbs like Levittown (Harris, 2010), suburban politics remain strongly associated with the consolidation of social and fiscal conservatism that bolstered contemporary right-wing religious and political movements (Kruse, 2005).

The protests in Sanford and Ferguson underscore the emergence of a different form of suburban politics around race and a different role for suburbia in racial politics. While revanchist politics are common to many suburban communities, many are also home to a range of progressive political movements, from those centered on racial justice to immigrant rights (Vicino, 2012; Lung-Amam, 2017). In the last few decades, social justice movements have sprouted up in inner-ring suburbs and far-flung exurbs (Niedt, 2013; Schafran et al., 2013). Heavily impacted by economic restructuring and foreclosure, efforts to retrofit and redevelop suburbs have spurred a politics of development (Sweeney and Hanlon, 2016) and equitable development (Lung-Amam et al., 2014) once only imaginable in urban neighborhoods.

Contemporary suburban political movements reflect the growing diversity of suburbanites and the frustration, felt by many, that the suburbs have not lived up to the promise they once held for many who have fought so hard to get there (Pfeiffer, 2012). Contrary to popular stereotypes, by 2010 the majority of poor people and people of color living in the largest metropolitan areas of the United States lived in suburbia (Kneebone and Berube, 2013). Suburban racial and economic diversity has become common knowledge, reported on by the media, academics, and policy wonks, often in hyperbolic and even derogatory terms (Schafran, 2013a). Far less reported are the causes and consequences of these trends, especially for African Americans. In fact, as African Americans have settled in suburbia, their metropolitan regions have restructured around them, making room for them in new places but continuing to disinvest in them, reproducing patterns of uneven development and new forms of segregation (Schafran, 2013b). In recent years, the struggles of these communities became evident in their high foreclosure rates (Anacker, 2015) and declining infrastructure, particularly in inner-ring suburbs (Vicino, 2008; Hanlon, 2010). Capacity-strapped social service providers and municipalities are serving more poor people but with few of public policy's traditional tools at hand (Kneebone and Berube, 2013; Roth and Allard, 2015).

In analyzing the historical and contemporary processes that produced Ferguson and Sanford as different kinds of suburbs with linked fates, we illuminate ways in which race has foregrounded their disinvestment and the struggles of communities of color within them. African Americans have survived amidst the constant threat of not only police violence but also various forms of the state-led structural violence that followed them from the city to the suburbs.

Ferguson

As protesters confronting militarily armed police on the streets of Ferguson were subjected to curfews, round-up arrests, tear gas, and rubber bullets, questions emerged about where such tense community-police relations came from and why they had landed in this inner-ring suburb, only about 10 miles north of downtown St. Louis. Some scholars and commentators pointed out that African Americans' frustrations did not begin with the killing of Michael Brown (Rothstein, 2014; Oliveri, 2015). African Americans were subject to violence at the hands of city councils, the state and federal government, banks, real estate and insurance agents, and their white suburban neighbors for generations. These other forms of violence shaped the St. Louis metropolitan area into a highly segregated and unequal landscape that, by the time of Brown's killing, left Ferguson struggling against forces of disinvestment.

The foundations of contemporary metropolitan segregation in the St. Louis region were set in 1875 by a provision in the Missouri Constitution allowing St. Louis, then one the most populous and least segregated cities in the nation, to separate from its surrounding rural hinterlands and govern itself by home rule (Gordon, 2009). As historian Colin Gordon (2009) documented in *Mapping Decline: St. Louis and the Fate of the American City*, as African Americans flooded into the city from southern states during the Great Migration and the run-up to World War I, this law allowed whites, who were abandoning the city at some of the most dramatic rates in the nation, to establish independent municipalities. At the same time, it denied St. Louis the power to annex new areas as its population grew. By 2014, St. Louis County had 91 separate municipalities, including the 6-square mile town of Ferguson, and was among the most racially segregated and fragmented metropolitan areas in the country.

Incorporated in 1894, Ferguson, like many of its neighboring suburbs, used a host of private as well as local, state, and federal policies and practices to craft itself as an exclusive, white, middle-class community. David Rothstein (2014) and others have documented the range of tools that shaped Ferguson's social and spatial character over the course of the twentieth century (Oliveri, 2015). These included racial and exclusionary zoning, discriminatory local and federal lending and insurance practices, racial deed restrictions and neighborhood associations, individual and collective acts of violence, racial steering, block busting, and other real estate practices.

Through the mid-1970s, such tools effectively sealed off the St. Louis suburbs from African Americans. The few who were able settle in the suburbs did so largely in communities like Kinloch, the first incorporated majority African American community in the state. Like black suburbs elsewhere, many Kinloch residents worked as nannies and housekeepers in adjacent communities like Ferguson, but they were otherwise kept separated by dead-end and blocked streets as well as threats of violence and intimidation (Wright, 2000; Wiese, 2004). The vast majority of African Americans lived in segregated St. Louis neighborhoods that fell into steep decline as jobs, white residents, and the city's middle-class tax base decamped to the suburbs. Many industrial jobs left the region altogether in a pattern of deindustrialization that hit Rustbelt cities such as St. Louis particularly hard. Highways ripped through black neighborhoods to speed the downtown commute for white suburbanites, while urban renewal cleared "blighted" areas for white-collar office towers, leaving African Americans more concentrated in poor neighborhoods and segregated public housing projects. The 2,870-unit Pruitt-Igoe complex became an infamous symbol of the deplorable conditions of segregated public housing when it was razed to the ground in dramatic fashion in 1972.

Displaced by urban renewal and facing rapidly deteriorating conditions in inner-city neighborhoods, African American suburbanization took root in the mid-1970s, as legal restrictions preventing racially discriminatory real estate practices waned. Most settled in older, inner-ring suburbs with affordable housing options in northern St. Louis County, adjacent to the city's nearly all-black north side neighborhoods. Along this "black belt" was Ferguson, a suburb initially zoned for single-family homes. During World War II, however, it allowed some multi-family construction on its eastern edge, including the Canfield Green Apartments where Michael Brown lived (Rothstein, 2014). In Ferguson, African Americans leaving inner-city St. Louis joined many former Kinloch residents, who had survived multiple urban renewal battles in the past, only to be displaced by a proposal for the Lambert International Airport in the 1980s that never materialized (Oliveri, 2015).

In a pattern that has affected black suburbs around the United States, as African Americans moved into Ferguson, many white residents left, and neighborhoods like Michael Brown's began to resemble those that African Americans had left behind in St. Louis. While in 1970 Ferguson was 99 percent white, by 2010, 69 percent of its 21,000 residents were black. By the time of Brown's shooting, poverty and unemployment rates were high and growing, both nearly doubling between 2000 and 2012 to 25 percent and 13 percent, respectively (Kneebone, 2014). North County was the epicenter of the region's foreclosure crises, as subprime loans that had preyed on African Americans went south. From their peak in 2007, home prices in Ferguson were down 37 percent by August 2014 (Nicklaus, 2014). The average value of real estate was nearly a third of the county's (Mollenkopf and Swanstrom, 2015). The high school where Brown had graduated only months before his death was among the lowest-performing schools in Missouri, with a nearly all-black student body that, alongside the rest of the school district, was operating without state accreditation (Oliveri, 2015).

African Americans in Ferguson faced far worse prospects than whites, many of whom lived in the city's few remaining middle-class neighborhoods, leading many commentators to refer to not one but "two Fergusons." According to the 2010–2014 ACS, around the time of the shooting African Americans' median household income was about $23,000 less than whites ($34,103 compared to $53,283). In contrast to St. Louis, where African Americans had successfully built powerful political coalitions and leaders, black political power was lacking in Ferguson. The city's mayor was white, as were five of its six city council members and six of its seven school board members. The police chief was white and only three of the 56 city police officers were black.

Though less well off than their white neighbors, African Americans in Ferguson bore the brunt of municipal finance through predatory fees and fines. With state laws that capped increases in local tax rates without approval by citywide referendums, small incorporated cities were often on the hook for financing critical services such as police and fire departments. Levying excessive fees and fines, particularly by over-policing poor communities of color, had become a common fiscal strategy for many Missouri cities. In Ferguson, though African Americans made up only 67 percent of residents, they accounted for 85 percent of all traffic stops, 90 percent of citations, and 93 percent of arrests between 2012 and 2014. Multiple citations were often levied for a single traffic stop, and residents were often saddled with excessive court fees over minor infractions (U.S. Department of Justice, 2015). In 2015, municipal fines and fees accounted for nearly a quarter of Ferguson's general funds, while property taxes (disproportionately paid by white homeowners and businesses) accounted for just fewer than 12 percent (U.S. Department of Justice, 2015; Johnson, 2015).

The violence that erupted on the streets of Ferguson emerged from a collective political struggle for the visibility, rights, and respect long denied African Americans in the greater St. Louis area. As Marc Lamont Hill wrote, Ferguson and the Canfield Green Apartments were spaces of "civic vulnerability," and the injustices within them "just as insidious as poll taxes and Jim Crow" (2016, p. 28). The events that sparked the Ferguson uprising stoked the ire of many not because of their exceptionalism but because they were so commonplace and persistent. Indeed, such state-led, structural violence had followed St. Louis's African American community from the inner-city to the inner suburbs, connecting this Midwestern Rustbelt town to the gated suburbs of the American Sunbelt.

Sanford

While the majority of Sanford's built environment is new, the city itself, incorporated in 1877, is not much younger than Ferguson. Sanford's troubling racial history, however, dates back to its founding by the wealthy Connecticut scion and diplomat Henry Shelton Sanford. As Lincoln's ambassador to Belgium and an ardent supporter of King Leopold, Sanford encouraged the United States to recognize Leopold's claim to the Congo, a site that would see some of the worst atrocities of the colonial era. He also advocated sending recently freed slaves to the Congo to avoid political struggles in the United States (Hochschild, 1999). The racial legacy left by Sanford would haunt the community for more than a century, as it grew from a small port town into a booming Sunbelt suburb.

Like so many American suburbs, Sanford was a city in its own right before it was engulfed by the greater Orlando area. Located along the St. John's River, the city was an ideal location for exporting vegetables and other goods to population centers farther north, making it a residential and commercial hub in the early twentieth century. The 1910 census placed Sanford's population just a few hundred shy of Orlando's (3,570 compared to 3,894). By then, Sanford was also engaged in the region's competitive growth politics, which were spurred by rapid metropolitan growth and efforts amongst municipalities to secure their tax bases. In 1911, when faced with a proposal by neighboring Sanford Heights to incorporate, Sanford leaders instead went on the offensive. They appealed to the Florida state legislature to dissolve the city and reincorporate Sanford with the neighboring city of Goldsboro. Goldsboro was founded by the Freedman's Bureau in 1891 and was one of Florida's oldest black townships (Simpson, 2012). With reincorporation, Sanford increased its footprint but the city of Goldboro was literally wiped off the map. The annexation of Goldboro stands in contrast to the experience of many suburban black townships that have remained unincorporated or self-incorporated because of opposition from neighboring white communities (Wiese, 2004). Even its streets were renamed. In a particularly pernicious renaming, Clark Street, named after the African American merchant who helped found Goldsboro, became Lake Avenue, for Sanford Mayor Forrest Lake, the man who, in the words of Seminole County historian Altermese Smith Bentley, "engineered the town's demise" (Bentley, 2000, p. 57).

The ugly racial legacy that started with Henry Shelton Sanford and was epitomized by the Goldboro annexation continued after World War II. In 1946, Jackie Robinson attempted to make his integrated baseball debut in Sanford as part of a minor league team, but he was prevented from doing so by local threats and the county sheriff (Lamb, 2004; Simpson, 2012). The city's lone swimming pool was filled in during Jim Crow to avoid racial integration (Lee, 2013). Such racial tensions were still evident in 1997, when an official apology from Sanford's mayor for the Jackie Robinson incident drew criticism by some residents (Lamb, 2004).

Sanford's population boom (and further annexations) came, in large part, in the last decades of the twentieth and first decades of the twenty-first century. During the same period, the Orlando metropolitan area became the twenty-sixth largest in the United States. It was also an important center for the aerospace industry and had already become a global tourism destination after

Disney World opened in 1971. Sanford grew rapidly to over 56,000 residents, though its population constituted less than 3 percent of the region. This period coincided with increasing racial and economic diversity in Sanford, but not spatial integration. The Retreat at Twin Lakes, where Martin was killed, was comprised mostly of newly built gated homes. Goldboro, however, had become stereotyped as "inner-city" Sanford and struggled with issues of disinvestment that sat in stark contrast with the rest of the city. While census tract 206, which includes Twin Lakes, was 13 percent African American in the 2011–2015 ACS, tract 204.01, the heart of Goldboro, was 78 percent African American.

Like Ferguson, by the time of Martin's tragic shooting, Sanford was a deeply divided community with pockets of both privilege and poverty. However, the latter tended to be overlooked by those both inside and outside Sanford; when one critic claimed part of the problem Martin faced in Twin Lakes was the purposeful lack of sidewalks purportedly meant to increase the community's exclusive feel (Youngerman, 2012), few also noted that many neighborhoods in Goldsboro lacked sidewalks as well. For Goldboro, however, the lack of sidewalks was symptomatic of the city's chronic underinvestment in its infrastructure. In an exposé about Goldsboro more than a year after Martin's death, journalist Trymaine Lee (2013) told the tale of the "two Sanfords." While the city's historic waterfront had seen major reinvestment over the past few decades, Goldsboro had a creaky sewer system and many abandoned sites, including several shuttered public housing complexes (Lee, 2013).

Such disparities worsened long-standing tensions over racialized police violence in Goldsboro and the larger Sanford area. By the time of Martin's death, Goldsboro was still simmering over the police shooting of an unarmed black man, Eugene Scott, two years prior. Yet, it was not until Martin was killed in the gated, wealthier, and whiter part of Sanford that much attention was paid to police violence in the city. The protests following Martin's shooting highlighted Goldboro as key to understanding the frustrations felt by African Americans over the serial acts of violence perpetrated on their community. It was Goldsboro, not the Retreat at Twin Lakes, where protests began and grew. Allen Chapel A.M.E. Church, founded in 1893 before Goldboro was incorporated into Sanford, became "ground zero" for protesters, civil rights groups, and organizers (Lee, 2013). The rallying cry "Justice for Trayvon" took on new meaning when posed against the backdrop of this historically African American community suffering from decades of disinvestment and loss.

While there are key differences between the two Sanfords and the two Fergusons, the internal divisions within these cities-cum-suburbs provided important context for the protests that emerged within them. And like Ferguson's neighboring city of Kinloch, Goldsboro served as a reminder that suburban poverty has grown not only because of poverty moving from central cities into suburbs, but also from poverty rising among long-established and long-neglected communities within suburbs. Racialized urban space, and the structural violence and protest it engenders, exist both inside and outside city limits, in big cities and in the smaller cities that surround them.

After uprising: in search of suburban social justice

The protests in Sanford and Ferguson forced the eyes of the nation to struggling suburbs and the conditions of their communities of color. They demonstrated that the problems ran deeper than a few bad apples on the police force or vigilante homeowners. The events in both communities were a continuation of the structural violence suffered by African Americans since the cities were founded – processes that produced areas of poverty and disinvestment in suburbia, just as they had in the inner-city.

In searching for justice in both communities, protesters found some relief. In Sanford, the white police chief who served during the Martin shooting was fired and replaced by a long-time African American officer who had honed his skills in community policing on Chicago's west side and made repairing race and community relations in the city a public priority. In Ferguson, the uprising helped African Americans secure three seats on the city council in the elections shortly after Brown's shooting, and the hiring of a new municipal judge and interim city manager, both of who were black. An independent Ferguson Commission was set up by Governor Nixon to examine race relations, failing schools, and other social and economic issues in the city. Community policing efforts got underway under an African American interim police chief's supervision, and the city made commitments to hiring more black police officers, additional training, and the use of body cameras for its officers. Perhaps most significantly, a new state law was passed that put a 20 percent limit on how much cities can collect from traffic fines and municipal court fees, and the Ferguson court system reduced its fees and jail time for minor offenses. The city, private foundations, and employers also increased funding for job training, education, and economic development efforts (Davey, 2015).

Despite these changes, in neither suburb have the protests fundamentally shifted the calculus for African Americans. Ferguson, for instance, still lacks many of the social service agencies needed to address the problems associated with rising poverty and inadequate public transportation (Dreier and Swanstrom, 2014). African Americans in St. Louis County still face vast disparities in exposure to poor neighborhoods and schools compared to whites, and they have unemployment rates that are roughly three times higher (Nicklaus, 2014). The region has some the lowest rates of intergenerational economic mobility among U.S. metropolitan areas (Chetty et al., 2014). Facing a $2.8 million budget shortfall, city residents turned down a proposed property tax increase in 2016 that would have helped to fund the deficit and many of the reforms promised by municipal leaders in a settlement agreement with the U.S. Department of Justice. As Peter Dreier and Todd Swanstorm (2014) concluded during the uprising, "Ferguson is simply too small and too poor to address the underlying racial and economic disparities that are fueling the current protests." Similar disparities continue to exist between Goldboro and other neighborhoods in Sanford and the greater Orlando area.

The uprisings in Ferguson and Sanford marked a significant moment in suburban political history. In both places, protests helped to uncover the ever-shifting roots of metropolitan inequality, including the conditions in which communities of color were living in suburbia and the forces that had and continue to perpetuate uneven patterns of development. Without the uprisings, many scholars, public policymakers, or activists might never have heard of Sanford, Ferguson, Trayvon Martin, or Michael Brown. As disadvantaged communities move or are pushed further to the urban periphery, there is an acute danger of their struggles being exacerbated by, and yet made more invisible by, their suburban location. The unrest widened the gaze for those committed to finding new possibilities to America's deepening social and economic divides. Whether in inner-ring suburbs of older industrial regions or the farther-out exurbs of Sunbelt metropoles, suburbia is more than ever where marginalized groups are staking their claim to a more just and equitable piece of the American Dream. In cul-de-sacs, parking lots, and shopping malls, protesters showed the need for scholarship to open up and interrogate suburban spaces to understand how the metropolis is not simply shaped by social inequalities, but also shapes them.

In holding up Ferguson and Sanford as symbols of increasing metropolitan inequality, we are also reminded of what is often missing in the narrative – the ever-more exclusive, generally white suburbs. There are many places in suburban America where young black men like Brown and Martin face little danger because they are still so unlikely to be there in the first place. As more and more media, academic, and political attention is paid to struggling racialized suburbs, vigilance is also needed to ensure that the story of metropolitan inequality is written not only from the places where protest appears, but also where it does not.

Guide to further reading

Anacker, K.B. (ed.) (2015) *The New American Suburb: Race, Poverty, and the Economic Crisis*. Farnham: Ashgate.

Gordon, C. (2009) *Mapping Decline: St. Louis and the Fate of the American City*. Philadelphia, PA: University of Pennsylvania Press.

Kneebone, E. and Berube, A. (2013) *Confronting Suburban Poverty*. Washington, DC: Brookings Institution Press.

Lang, R.E. and LeFurgy, J.B. (2007) *Boomburbs: The Rise of America's Accidental Cities*. Washington, DC: Brookings Institution Press.

Niedt, C. (2013) *Social Justice in Diverse Suburbs*. Philadelphia, PA: Temple University Press.

Rothstein, R. (2014) *The Making of Ferguson: Public Policies at the Root of its Troubles*. Washington, DC: Economic Policy Institute. Available at: www.epi.org/files/2014/making-of-ferguson-final.pdf [Accessed 6 June 2017].

Wiese, A. (2004) *Places of Their Own: African American Suburbanization in the Twentieth Century*. Chicago, IL: University of Chicago Press.

References

Anacker, K.B. (ed.) (2015) *The New American Suburb: Race, Poverty, and the Economic Crisis*. Farnham: Ashgate.

Bentley, A.S. (2000) *Seminole County*. Charleston, SC: Arcadia Publishing.

Chetty, R., Hendren, N., Kline, P., and Saez, E. (2014) "Where Is the Land of Opportunity? The Geography of Intergenerational Mobility in the United States." *The Quarterly Journal of Economics*, 129(4), pp. 1553–1623.

Davey, M. (2015) "A Year Later, Ferguson Sees Change, But Asks If It's Real." *New York Times*. Available at: www.nytimes.com/2015/08/06/us/in-year-since-searing-death-ferguson-sees-uneven-recovery.html [Accessed 7 June 2017].

Dreier, P. and Swanstrom, T. (2014) "Suburban Ghettos Like Ferguson Are Ticking Time Bombs." *Washington Post*. Available at: www.washingtonpost.com/posteverything/wp/2014/08/21/suburban-ghettos-like-ferguson-are-ticking-time-bombs/?utm_term=.64ec273f90f1 [Accessed 7 June 2017].

Gordon, C. (2009) *Mapping Decline: St. Louis and the Fate of the American City*. Philadelphia, PA: University of Pennsylvania Press.

Green, A. (2012) "Zimmerman's Twin Lakes Community Was on Edge before Trayvon Shooting." *Daily Beast*. Available at: www.thedailybeast.com/zimmermans-twin-lakes-community-was-on-edge-before-trayvon-shooting [Accessed 7 June 2017].

Hanlon, B. (2010) *Once the American Dream: Inner-Ring Suburbs of the Metropolitan United States*. Philadelphia, PA: Temple University Press.

Harris, D.S. (2010) *Second Suburb: Levittown*. Pittsburgh, PA: University of Pittsburgh Press.

Hill, M.L. (2016) *Nobody: Casualties of America's War on the Vulnerable, from Ferguson to Flint and Beyond*. New York: Simon and Schuster.

Hochschild, A. (1999) *King Leopold's Ghost: A Story of Greed, Terror, and Heroism in Colonial Africa*. Vancouver: Houghton Mifflin Harcourt.

Johnson, W. (2015) "Ferguson's Fortune 500 Company." *The Atlantic*. Available at: www.theatlantic.com/politics/archive/2015/04/fergusons-fortune-500-company/390492/ [Accessed 6 June 2017].

Kneebone, E. (2014) "Ferguson, Mo: Emblematic of Growing Suburban Poverty." [Blog]. Available at: www.brookings.edu/blog/the-avenue/2014/08/15/ferguson-mo-emblematic-of-growing-suburban-poverty/ [Accessed 6 June 2017].

Kneebone, E. and Berube, A. (2013) *Confronting Suburban Poverty*. Washington, DC: Brookings Institution Press.

Kruse, K.M. (2005) *White Flight: Atlanta and the Making of Modern Conservatism*. Princeton, NJ: Princeton University Press.

Lamb, C. (2004) *Blackout: The Untold Story of Jackie Robinson's First Spring Training*. Lincoln, NE: University of Nebraska Press.

Lang, R.E. and LeFurgy, J.B. (2007) *Boomburbs: The Rise of America's Accidental Cities.* Washington, DC: Brookings Institution Press.

Lee, T. (2013) "In Black Sanford: A Place to Gather and Wait for a Verdict." *MSNBC.* Available at: www. msnbc.com/msnbc/black-sanford-place-gather-and-wait [Accessed 6 June 2017].

Lung-Amam, W. (2017) *Trespassers?: Asian Americans and the Battle for Suburbia.* Berkeley, CA: University of California Press.

Lung-Amam, W., Pendall, R., Scott, M., and Knaap, E. (2014) "Equitable Transit-Oriented Development in Diverse Suburbs: Promise and Challenges." In: *Transit, Development and Forme Urbaine: Washington et Paris Symposium.* College Park, MD: National Center for Smart Growth Research and Education. Available at: http://smartgrowth.umd.edu/assets/documents/research/dcparis_equitable_tod_10.04.14_(2).pdf [Accessed 6 June 2017].

Mollenkopf, J. and Swanstrom, T. (2015) "The Ferguson Moment: Race and Place." [Blog]. Available at: http://furmancenter.org/research/iri/essay/the-ferguson-moment-race-and-place [Accessed 6 June 2017].

Moore, S.M. (2015) "A Vision for Ferguson and Beyond." *St. Louis Post-Dispatch,* p. A15.

Nicklaus, D. (2014) "Frustration in North County Has Deep Economic Roots." *St. Louis Post-Dispatch,* p. E1.

Niedt, C. (2013) *Social Justice in Diverse Suburbs.* Philadelphia, PA: Temple University Press.

Oliveri, R.C. (2015) "Setting the Stage for Ferguson: Housing Discrimination and Segregation in St. Louis." *Missouri Law Review,* 80, pp. 1053–1075.

Pfeiffer, D. (2012) "African Americans' Search for 'More for Less' and 'Peace of Mind' on the Exurban Frontier." *Urban Geography,* 33(1), pp. 64–90.

Roth, B.J. and Allard, S.W. (2015) "The Response of the Nonprofit Safety Net to Rising Suburban Poverty." In: Anacker, K.B. (ed.) *The New American Suburb: Poverty, Race, and the Economic Crisis.* Farnham: Ashgate, pp. 247–284.

Rothstein, R. (2014) *The Making of Ferguson: Public Policies at the Root of Its Troubles.* Washington, DC: Economic Policy Institute. Available at: www.epi.org/files/2014/making-of-ferguson-final.pdf [Accessed 6 June 2017].

Schafran, A. (2013a) "Discourse and Dystopia, American Style: The Rise of 'Slumburbia' in a Time of Crisis." *City,* 17(2), pp. 130–148.

Schafran, A. (2013b) "Origins of an Urban Crisis: The Restructuring of the San Francisco Bay Area and the Geography of Foreclosure." *International Journal of Urban and Regional Research,* 37(2), pp. 663–688.

Schafran, A., Lopez, O.S., and Gin, J.L. (2013) "Politics and Possibility on the Metropolitan Edge: The Scale of Social Movement Space in Exurbia." *Environment and Planning A,* 45(12), pp. 2833–2851.

Simpson, M. (2012) "Racial Tension Runs through Sanford's Roots." *National Public Radio,* March 22.

Sweeney, G. and Hanlon, B. (2016) "From Suburb to Post-Suburb: The Politics of Retrofit in the Inner Suburbs of Upper Arlington, Ohio." *Journal of Urban Affairs,* 39(2), pp. 241–259.

U. S. Department of Justice, Civil Rights Division (2015) *Investigation of the Ferguson Police Department.* Washington, DC: U.S. Department of Justice.

Vicino, T. (2008) *Transforming Race and Class in Suburbia: Decline in Metropolitan Baltimore.* New York: Palgrave Macmillan.

Vicino, T. (2012) *Suburban Crossroads: The Fight for Local Control of Immigration Policy.* Lanham, MD: Lexington Books.

Wiese, A. (2004) *Places of Their Own: African American Suburbanization in the Twentieth Century.* Chicago, IL: University of Chicago Press.

Wright, J.A. (2000) *Kinloch: Missouri's First Black City.* Charleston, SC: Arcadia Publishing.

Youngerman, Z. (2012) "Did Bad Neighborhood Design Doom Trayvon Martin?" *Boston Globe, Opinion.* Available at: www.bostonglobe.com/opinion/2012/04/06/did-bad-neighborhood-design-doom-trayvon-martin/8TSIJBEdBla6NBb1z1VHOO/story.html [Accessed 7 June 2017].

19

Stigma and the U.S. suburbs

Whitney Airgood-Obrycki and Cody R. Price

Introduction

Stigma is a growing concern among suburban communities. Historically associated solely with urban environments in the United States, stigma has spread outward geographically because of the decentralization of poverty, long-term suburban disinvestment, and major catastrophic events and crises in suburban space. Suburbs, once mythically viewed as white, wealthy havens separate from the social ills of the city, are increasingly susceptible to negative perceptions and stereotypes. The rise of suburban stigma is studied here through the lens of Goffman's (1963) seminal work on stigma as well as Wacquant's (2007) more recent work on territorial stigma. This chapter highlights the historical and social construction of suburban stigma, which was produced in the urban context and decentralized over time. Stigma is not a fixed condition, but something that results from a chain of events within a certain society or historical period (Foster, 2009, 2012; Nédélec, 2017). Thus, the process of territorial stigmatization, combining both place and people, theorizes the "contamination" of a suburb because of its population and vice versa.

From social stigma to territorial stigma

Stigma is a social as well as a physical construct. In his book, *Stigma: Notes on the Management of Spoiled Identity* (1963, p. 1), Erving Goffman theorizes three types of stigmas. These include "abominations of the body," "blemishes of the individual character," and "tribal stigma" (Goffman, 1963, p. 14). His analyses focus on individuals and their disabilities and/or physical abnormalities, and he explains that these individual blemishes can be transmitted throughout family lineages (Goffman, 1963, p. 14). Within this new thinking about family, definitions of stigma also expanded. Stigmatized individuals no longer needed to have a physical sign or abnormality. Rather, he claimed that certain traits and attributes such as race, social class, or physical disability function as an indication of "undesired differentness," causing them to be stigmatized.

Groups of stigmatized individuals that become spatially isolated create a place subject to stigmatization. The concept of territorial stigma, referred to as a "blemish of place" (Wacquant, 2007), came about from Wacquant's comparative work of Chicago's black belt and the Parisian *banlieue* (Wacquant, 1996) and was popularized in his later work (Wacquant, 2007; Wacquant et al., 2014). Territorial stigma emerges as stereotypes, stigma, and symbolic power (Bourdieu, 1991) of a space come together (Wacquant, 2007; Wacquant et al., 2014). Territorial stigma can be used to understand the marginality of spaces from both above and below:

> Bourdieu works from above, following the flow of efficient representations from symbolic authorities such as state, science, church, the law, and journalism, down to their repercussions upon institutional operations, social practices, and the self; Goffman works from below, tracing the effects of procedures of sense-making and techniques of 'management of spoiled identity' across encounters and their aggregations into organizations. They can thus be wedded to advance our grasp of the ways in which noxious representations of space are produced, diffused, and harnessed in the field of power, by bureaucratic and commercial agencies, as well as in everyday life in ways that alter social identity, strategy, and structure.
>
> *(Wacquant et al., 2014, pp. 1272–1273)*

Places can also become subject to stigmatization due to negative media discourse or events that reinforce symbolic power. Historical events and representations of those events in the media stand at the core of the stigmatization process because they can emit images or stereotypes of that place or program. For example, the 1963 bombing of the 16th Street Baptist Church, an act of white supremacist terrorism, caused Birmingham, Alabama to be stigmatized as a racist city (Foster, 2009, 2012; Nédélec, 2017). Additionally, the 1970s publicized demolition of Pruitt-Igoe ignited the stigma of public housing (U.S. Department of Housing and Urban Development, 1974). As Foster (2009, p. 18) points out, "once created, place-based stereotypes gain and retain popularity as journalists refer to the sensational events of the past, although they have no bearing on current activities." Thus, negative attention and portrayal in the media helps perpetuate and drive public opinion and stereotypes even if these negative portrayals of place are no longer true. As Lippmann (1922, 2012) states,

> The subtlest and most pervasive of all influences are those which create and maintain the repertory of stereotypes. We are told about the world before we see it. We imagine most things before we experience them. And those preconceptions, unless education has made us acutely aware, govern deeply the whole process of perception.
>
> *(Lippmann, 2012, p. 49)*

Thus, as individuals consumed with the ever-growing media coverage in today's world, we tend to take shortcuts to process mass quantities of information – resulting in the use of stereotypes. This process helps us understand why people perceive a place a certain way and what they may expect from it, even if it is negative. As a result, a stigmatized place enters into a perpetual cycle where the "normative expectations," theorized by Goffman, are believed to be the norm, even if they are not accurate of the place's history (Nédélec, 2017). This is evident in Nédélec's (2017) work where he examined the city of Las Vegas and found that the individual level and social stigma of its residents have led to the entire city being stigmatized and branded as Sin City.

The urban roots of suburban territorial stigma

Territorial stigma is the product of stigma and prejudices against the poor and racial and ethnic minorities. When stigmatized populations become spatially concentrated, private capital retracts and the place itself becomes stigmatized. Racial and class segregation has been the primary mechanism for urban territorial stigmatization in the U.S. context. Policies, planning practices, and economic systems throughout U.S. history have spatially concentrated stigmatized groups in urban neighborhoods. We discuss the forces that have created territorial stigmatization in urban communities and the impact that territorial stigma has had on reinforcing stigmas of race and class.

Urban segregation deepened from 1900–1940 as the United States industrialized and blacks relocated from rural areas to cities. Massey and Denton (1993) described the process of urban segregation that laid the groundwork for urban territorial stigma. Industrialization required unskilled labor in centralized manufacturing locations. The managerial middle-class class was differentiated from the unskilled laborers. Because black workers were relegated to unskilled labor positions, class and race were highly correlated. Housing types and locations bifurcated along these divisions. Residential segregation became exacerbated to a level that had not present since the Civil War. Industrialization and social stigmas provided the basis for future urban territorial stigmas.

At the same time, housing policy perpetuated segregation and accentuated urban territorial stigmatization. The Federal Housing Administration (FHA) fueled the suburban boom from the 1930s to the 1970s. The FHA, Veterans Association (VA), and private mortgage providers promoted white suburbanization by adopting the Homeowners Loan Corporation (HOLC) appraisal standards, codifying the practice of redlining (Jackson, 1985). The HOLC standards were a signal of the territorial and social stigmatization that existed because urban neighborhoods and those that were predominantly black or showing signs of racial transition were deemed risky. However, the standards also reinforced territorial stigmas. FHA and VA mortgage insurance programs enabled middle-class whites to flee cities, leaving disinvested urban neighborhoods with high concentrations of low-income, racial minority residents (Denton, 2006). The FHA also promoted black, urban segregation by encouraging the use of racially restrictive covenants, preventing the sale of homes to certain populations. Housing policies catalyzed suburban white flight and urban disinvestment. Such discriminatory practices, grounded in territorial and social stigmas, ensured that racial and ethnic minorities were spatially concentrated in certain neighborhoods.

The public housing program further cemented the spatial concentration of low-income black residents in urban centers, contributing to urban stigmatization. White opposition to creating housing agencies in suburban municipalities and white political power ensured that public housing projects were sited in urban neighborhoods (Massey and Denton, 1993). As white households were increasingly able to access affordable mortgages and suburban neighborhoods, public housing became predominantly low-income and black. Problems associated with the underfunding and architecture of public housing projects heightened the stigmatization of black and impoverished urban neighborhoods (von Hoffman, 1996).

While the 1968 Fair Housing Act sought to legally end discriminatory housing practices and policies, discrimination fortified patterns of segregation and the spatial concentration of racial minorities (Denton, 2006). Steering by real estate agents reinforced patterns of segregation; white homebuyers were more frequently advised to select homes in whiter, wealthier neighborhoods than their black counterparts (Galster and Godfrey, 2005). Additionally, white hostility in suburban neighborhoods prevented potential black residents from moving outside of urban or predominantly black suburban neighborhoods (Keating, 1994). Exclusionary zoning, functioning as a discriminatory mechanism since the inception of suburbia, persistently increased racial and economic segregation (Rothwell and Massey, 2009; Keating, 1994).

As low-income residents and racial minorities spatially concentrated in urban neighborhoods, capital investment evaporated and the impacts of economic restructuring were heightened. Private capital decentralized in the post-war period, funding suburban housing, school districts, and employment centers. Urban disinvestment, segregation, and a lack of employment opportunities coalesced to create an "underclass" population. As social problems deepened in disinvested urban neighborhoods, territorial stigma reached its peak. The "underclass" discourse centered around deep, concentrated, racialized urban poverty that produces a so-called subculture (Jargowsky and Yang, 2006). Without work, role models, or attachment to mainstream institutions, the "underclass" was marked by weak labor force attachment, welfare dependency, criminal behavior, drug use, single parent households, and educational attrition (Ricketts and Sawhill, 1988; McLanahan and Garfinkel, 1989; Wilson, 1987). Associated with neighborhood disorder, the "underclass" produced territorial stigmas from social stigmas (Sampson and Raudenbush, 2005).

Segregation and urban neighborhood stigmatization served as a mechanism for maintaining class and racial inequalities (Keene and Padilla, 2010; Bass, 2001). Further inequity and stigmatization has occurred through predatory lending and increased policing. Policing of black residents and neighborhoods has been justified based on social prejudices and territorial stigmas. Policing is a means of social control over minorities (Bass, 2001). Black residents are inequitably policed, as demonstrated in the high rates of incarceration and police brutality of people of color. Tensions between racial groups and frustrations about the treatment of black citizens have erupted into protest and race riots (Perez et al., 2003). Riots then perpetuated racial prejudices and urban stigmas. They were used to further justify policing and promoted white flight to suburban neighborhoods (Keating, 1994). Policing of black residents continues to be a substantial issue impacting and stigmatizing urban neighborhoods. The death of Freddie Gray and subsequent protests in Baltimore is just one example of the poor relationship between black residents and police power; the injustices against racial minorities continue to mark urban places.

The spatiality of racial injustices has also been documented in the predatory lending activities and resulting foreclosure crisis that deeply impacted urban neighborhoods. Lenders targeted racial minorities through directed advertising, recruiting at minority churches, and hiring minority store managers or loan officers in urban locations (Johnson, 2010). Black and Latino residents disproportionately received subprime loans, and urban neighborhoods had higher foreclosure rates on average (Hall et al., 2015). Residential segregation enabled subprime lending and heightened the impact of the foreclosure crisis (Rugh and Massey, 2010). Widespread foreclosures exacerbated vacancy, abandonment, and absentee ownership in urban neighborhoods, forcing already stigmatized places to face these additional challenges. The problems associated with the foreclosure crisis have endured. Black neighborhoods have not recovered as quickly from the foreclosure crisis as compared to white neighborhoods (Raymond et al., 2016) and are likely to bear the territorial stigma of foreclosure over a longer period of time.

Urban territorial stigma stemming from deep racial prejudice has existed throughout the history of the United States. Racial prejudice and territorial stigma informed a range of policies and practices that further cemented urban stigma. Thus, a perpetual production and reproduction of spatial stigma has shaped urban neighborhoods across the United States.

Suburban change and the spread of stigma

While there is a long history of urban territorial stigma, changes in the demographics and characteristics of suburban neighborhoods have more recently opened possibilities for suburban territorial stigma. The spread of an "urban demographic" into suburban neighborhoods has produced a similar trajectory of territorial stigma. The spatial dispersion of poverty, welfare usage, and racial change in certain suburban neighborhoods are interpreted as signals of decline. The perception of decline highlights the linkage between urban characteristics and territorial stigma. As suburbs increasingly look like urban neighborhoods, the stigmas associated with urban space have followed. Aging inner-ring suburbs, bypassed by policy and adjacent to city neighborhoods, have been particularly susceptible to formerly urban problems and territorial stigmatization. Gentrification in the city, subsidized housing programs that disperse poverty, the foreclosure crisis, and continual outer suburban growth have extended the geography of stigmatized groups and spatial stigma.

The suburbs have shifted dramatically from the stereotypical image of white, wealthy neighborhoods where residents own their own homes. Since the 1970s, scholars have documented changes in suburban characteristics; these changes are often perceived to be negative symptoms of declining suburban status. Decreasing income and increasing rates of poverty have been the primary indicators of decline. Lucy and Phillips (2006) found that 20 percent of suburbs in their study sample experienced greater income declines between 1960 and 2000 than their corresponding central cities. From 1980 to 2000, suburban poverty increased at a greater rate than city poverty, a trend that has continued into the most recent decades (Kneebone and Berube, 2013). Inner suburbs across the country have similarly seen incomes decline (Hanlon, 2010; Lucy and Phillips 2000; Bollens, 1988). Suburban neighborhoods are now home to more impoverished households than urban neighborhoods.

In addition to increasing poverty, suburban neighborhoods are seeing other population changes. Inner suburbs are experiencing population loss and aging (Leigh and Lee, 2005; Short et al., 2007; Keating and Bier, 2008; Vicino, 2008). As baby boomers age in place and pass away, these populations are not being fully replaced with younger households (Short et al., 2007). When younger residents do move in, they are increasingly non-traditional, female-headed households and may lack college degrees (Lee and Leigh, 2005, 2007); these households potentially have lower earning power. Population loss, population aging, and increasing poverty threatens the longevity of suburban communities because they reduce the available tax base on which so many suburban municipalities rely. Suburbs facing these challenges lack the revenue to update infrastructure and to invest in social services or public amenities.

Suburban neighborhoods are also more frequently home to racial minorities and immigrants than in past decades. Racial transition in suburban neighborhoods began as early as the 1960s. By the 1970s, the annual rate of black suburbanization exceeded that of white suburbanization (Logan and Schneider, 1984; Schwartz, 1980; Stahura, 1983). In recent decades, the proportion of black residents in suburbia has continued to rise. Racial minority households now account for 35 percent of suburban residents, and more than half of all minority households reside in suburban neighborhoods (Frey, 2011). Additionally, immigrants are now bypassing cities and moving straight to the suburbs (Katz et al., 2010). Suburban neighborhoods are currently home to more immigrant households than cities (Singer, 2013). In inner-ring suburbs alone, the foreign-born population has increased 262 percent since 1980 (Dunham-Jones and Williamson, 2011).

The demographic changes in suburbs have occurred fairly rapidly, and the prejudices against these groups have followed. Some suburbs have responded to racial change by implementing revanchist strategies (Niedt, 2006). Others have developed exclusionary policies aimed at immigrants (Hanlon and Vicino, 2015). The racial tensions that exist in urban neighborhoods also continued. The police shooting of teenager Michael Brown in Ferguson, Missouri set off a wave of protest. These events highlight the deteriorated racial relationships in suburbs that run parallel to urban circumstances and the resulting territorial stigma.

Changes in suburban housing have accompanied demographic shifts, reshaping the image of suburbia and making it more similar to urban neighborhoods. Suburban residents typically oppose the housing changes that have occurred, including the building of multi-family units (Obrinsky and Stein, 2007), the conversion of homeowner units to rental, and the inclusion of affordable or subsidized housing opportunities (Scally, 2012; Goetz, 2008; Hudnut, 2003). Opposition is rooted in fears that this housing will decrease property values, change the character of the neighborhood, and attract low-income, racially diverse households (Dear, 1991; Koebel et al., 2004; Nguyen et al., 2013, Pendall, 1999; Stein, 1992; Tighe, 2010). Variations in suburban housing conditions have resulted because of disinvestment, the foreclosure crisis, and urbanized redevelopment.

Suburban housing changes include shifts to rentals, decreased property values, high vacancy, and an influx of subsidized renter households. Suburban neighborhoods have growing proportions of rental housing, both through new construction and redevelopment. Thirty-nine percent of suburban tracts now have homeownership rates below 70 percent (Beck Pooley, 2015). The foreclosure crisis produced larger numbers of vacant, abandoned, and bank-owned properties in suburbs than ever before (Federal Reserve Bank of Boston, 2010). Low housing markets and vacancy in declining suburbs have reduced property values. In less desirable suburban neighborhoods, property values appreciated at a slower rate (Anacker and Morrow-Jones, 2008). Increasing rentals and depressed housing markets have made it profitable and attractive for suburban landlords to accept federal Housing Choice Vouchers. Anecdotally, housing voucher holders are increasingly finding rental housing in inner suburban jurisdictions (Hudnut, 2003). The Low-Income Housing Tax Credit program has also recently supported the expansion of subsidized housing into suburban communities. Certain changes in suburban housing are associated with neighborhood decline and racial or class transition, promoting the spatial stigmatization of suburbs that have undergone such changes.

The stigmas surrounding demographic and housing transformation have matured through decades of urban-suburban dichotomy. As suburban neighborhoods look more like urban neighborhoods, the urban stigmas associated with race and class follow. The urbanization of suburban neighborhoods produces similar forms of territorial stigma.

Urban stigma in the suburbs

Suburban stigma has arisen from the urbanization of suburban space. In addition to the spatialization of social stigmas, three types of stigma have emerged in both urban and suburban neighborhoods. The stigmas arise from long-term disinvestment, disrupting events, and crises. The perception of stigma is evident in the discourse around certain suburbs and the incorporation of urban terms.

East Cleveland, Ohio is one example of a suburb that underwent racial change, struggled with deep disinvestment, and became stigmatized in public discourse. The suburb adjacent to Cleveland was home to several industries in the first half of the twentieth century and boasted a Rockefeller residence. East Cleveland experienced a massive population change in the 1960s, becoming a predominantly black suburb. In the decades that followed, industry and investment left the suburb. In recent years, there has been discussion of East Cleveland merging back into Cleveland as a way to address the severe fiscal and infrastructural distress that the suburb faces. With a declining population and high rates of crime, East Cleveland has been categorized as a "dying suburb" (Renn, 2017). A resident interviewed on the local news remarked, "It used to be one of Cuyahoga County's best communities. Now, this is the ghetto. Looks like the ghetto, feels like the ghetto, seems like the ghetto" (Reid, 2013). It is certainly no coincidence that terms for stigmatized urban neighborhoods have appeared in the suburbs that faced similar processes of long-term disinvestment.

Disrupting events are increasingly present in suburban neighborhoods as well. While race riots and police shootings were previously the sole domain of inner cities, the recent events in Ferguson, Missouri, illustrate how disrupting events in the suburbs can leave a lasting stigma. Protests erupted after a police officer killed Michael Brown, an unarmed black teenager in

Ferguson; the officer was not indicted for the murder, sparking further protests and unrest similar to the Rodney King protests in Los Angeles. The police shooting of Michael Brown and the resulting protests have shaped perceptions of the suburb. Ferguson is now associated with racial inequality and police brutality, and the suburb is financially feeling the effects of the aftermath. The perceived stigma of Ferguson led to fewer home sales following the protests and many residents decided to move to other areas, which furthers the cycle of disinvestment and unrest (Bouscaren, 2014).

Finally, crisis has played a central role in stigmatizing suburban neighborhoods. The national foreclosure crisis cast stigma on many types of suburban communities across the country. Suburbs already facing decline found themselves with a greater supply of vacant and abandoned homes with decreasing property values. But even newer exurban communities were not immune from the stigma of foreclosure. The rise of the discourse around the suburban slum emerged with Leinenberger's 2008 article in *The Atlantic*, in which he argued that newly built suburban neighborhoods with high rates of foreclosure were the new slums. He supported his assertion with claims that renters "of dubious character" now occupy these suburban homes and that the neighborhoods show signs of physical distress. Egan's 2010 op-ed in the *New York Times* doubled-down on the slumburbia narrative, further propelling it into the public discourse.

While labels like ghetto and slums in the suburban context may seem harmless, Schafran (2013) points to the toxicity of discourse in reinforcing stigma. He notes that the portrayal of suburbs, like the portrayal of cities, can perpetuate problems and give them a spatial association. Rather than highlighting the roots of and solutions to these problems, the shifting discourse about suburban space reproduces stigmas and prejudices.

Conclusion

While some suburbs already face stigma, there are more that could become stigmatized in the future. In this chapter, we examined the historical and social construction of suburban stigma and how this has followed a similar trajectory of urban stigma. In our examination of suburban stigma, we documented factors such as poverty, long-term disinvestment, crises, and catastrophic events that have led to the decentralization of urban stigmas. Places that become stigmatized will most likely stay stigmatized. Territorial stigma triggers a perpetual cycle of discourse, and perception continues to shape a place even if it differs from reality. Thus, the spread of territorial stigmas can impact suburban communities for years to come.

For most suburbs, it is not too late. The spread of territorial stigma can be prevented. As Schafran (2013) notes, discourse is crucial. In future suburban research, scholars should be conscientious of the language they use to describe suburban neighborhood change and the conditions of specific suburbs. Planners, policymakers, and researchers should also strive to break down social stigmas and prejudices when possible to decrease the reproduction of territorial stigma. The occurrence of suburban territorial stigma is an understudied problem. Future research could examine stigmatized suburbs and those on the verge of stigmatization to identify strategies that are used to reverse and prevent the cycle of territorial stigma.

Guide to further reading

Dikeç, M. (2018) *Urban Rage: The Revolt of the Excluded.* New Haven, CT: Yale University Press.

United States Department of Justice Civil Rights Division. (2015) *The Ferguson Report: Department of Justice Investigation of the Ferguson Police Department Paperback.* Washington, DC.

Wacquant, L. (2008) *Urban Outcasts: A Comparative Sociology of Advanced Marginality.* Cambridge: Polity Press.

References

Anacker, K.B. and Morrow-Jones, H.A. (2008) "Mature Suburbs, Property Values, and Decline in the Midwest? The Case of Cuyahoga County." *Housing Policy Debate*, 19, pp. 519–552.

Bass, S. (2001) "Policing Space, Policing Race: Social Control Imperatives and Police Discretionary Decisions." *Social Justice*, 28, pp. 156–176.

Beck Pooley, K. (2015) "Debunking the 'Cookie-Cutter' Myth for Suburban Places and Suburban Poverty: Analyzing Their Variety and Recent Trends." In: Anacker, K.B. (ed.) *The New American Suburb: Poverty, Race and the Economic Crisis*. New York, NY: Ashgate, pp. 39–78.

Bollens, S.A. (1988) "Municipal Decline and Inequality in American Suburban Rings, 1960–1980." *Regional Studies*, 22, pp. 277–285.

Bourdieu, P. (1991) *Language and Symbolic Power*. Malden, MA: Polity Press.

Bouscaren, D. (2014) "Unrest in Ferguson May Speed Up Decline of Real Estate." *National Public Radio*, October 20.

Dear, M. J. (1991) *Gaining Community Acceptance*. Princeton, NJ: The Robert Wood Johnson Foundation.

Denton, N.A. (2006) "Segregation and Discrimination in Housing." In: Bratt, R. Stone, M. and Hartman, C. (eds.) *A Right to Housing: Foundation for a New Social Agenda*. Philadelphia, PA: Temple University Press, pp. 61–81.

Dunham-Jones, E. and Williamson, J. (2011) *Retrofitting Suburbia: Urban Design Solutions for Redesigning Suburbs*. Hoboken, NJ: John Wiley and Sons, Inc.

Egan, T. (2010) "Slumburbia." *The New York Times*, February 10.

Federal Reserve Bank of Boston (2010) *REO & Vacant Properties: Strategies for Neighborhood Stabilization*. Boston, MA: Federal Reserve Banks of Boston and Cleveland and the Federal Reserve Board.

Foster, J. (2009) *Stigma Cities: Dystopian Urban Identities in the United States and South in the Twentieth Century*, PhD diss. Las Vegas: University of Nevada at Las Vegas.

Foster, J. (2012) "Stigma Cities: Birmingham, Alabama and Las Vegas, Nevada in the National Media, 1945–2000." *Psi Sigma Siren*, 3, Article 3.

Frey, W.H. (2011) *Melting Pot Cities and Suburbs: Racial and Ethnic Change in Metro America in the 2000s*. Washington, DC: Brookings Institution Press.

Galster, G. and Godfrey, E. (2005) "By Words and Deeds: Racial Steering by Real Estate Agents in the US in 2000." *Journal of the American Planning Association*, 71, pp. 251–268.

Goetz, E.G. (2008) "Words Matter: The Importance of Issue Framing and the Case of Affordable Housing." *Journal of the American Planning Association*, 74, pp. 222–229.

Goffman, E. (1963) *Stigma: Notes on the Management of Spoiled Identity*. Englewood Cliffs, NJ: Prentice Hall.

Hall, M., Crowder, K., and Spring, A. (2015) "Neighborhood Foreclosures, Racial/Ethnic Transitions, and Residential Segregation." *American Sociological Review*, 80, pp. 526–549.

Hanlon, B. (2010) *Once the American dream: Inner-Ring Suburbs of the Metropolitan United States*. Philadelphia, PA: Temple University Press.

Hanlon, B. and Vicino, T.J. (2015) "Local Immigration Legislation in Two Suburbs: An Examination of Immigration Policies in Farmers Branch, Texas, and Carpentersville, Illinois." In: Anacker, K.B. (ed.) *The New American Suburb: Poverty, Race and the Economic Crisis*. New York, NY: Ashgate, pp. 39–78.

Hudnut, W. (2003) *Halfway to Everywhere: A Portrait of America's First-Tier Suburbs*. Washington, DC: Urban Land Institute.

Jackson, K.T. (1985) *Crabgrass Frontier: The Suburbanization of the United States*. New York: Oxford University Press.

Jargowsky, P.A. and Yang, R. (2006) "The 'Underclass' Revisited: A Social Problem in Decline." *Journal of Urban Affairs*, 28, pp. 55–70.

Johnson, C. (2010) "The Magic of Groups Identity: How Predatory Lenders Use Minorities to Target Communities of Color." *Georgetown Journal on Poverty Law & Policy*, 17, p. 165.

Katz, M.B., Creighton, M.J., Amsterdam, D., and Chowkwanyun, M. (2010) "Immigration and the New Metropolitan Geography." *Journal of Urban Affairs*, 32, pp. 523–547.

Keating, W.D. (1994) *The Suburban Racial Dilemma: Housing and Neighborhoods*. Philadelphia, PA: Temple University Press.

Keating, W.D. and Bier, T. (2008) "Greater Cleveland's First Suburbs Consortium: Fighting Sprawl and Suburban Decline." *Housing Policy Debate*, 19, pp. 457–477.

Keene, D.E. and Padilla, M.B. (2010) "Race, Class and the Stigma of Place: Moving to 'Opportunity' in Eastern Iowa." *Health & Place*, 16, pp. 1216–1223.

Kneebone, E. and Berube, A. (2013) *Confronting Suburban Poverty in America.* Washington, DC: Brookings Institution Press.

Koebel, C.T., Lang, R.E., and Danielsen, K.A. (2004) *Community Acceptance of Affordable Housing.* Washington, DC: National Association of Realtors.

Lee, S. and Leigh, N.G. (2005) "The Role of Inner Ring Suburbs in Metropolitan Smart Growth Strategies." *Journal of Planning Literature,* 19, pp. 330–346.

Lee, S. and Leigh, N.G. (2007) "Intrametropolitan Spatial Differentiation and Decline of Inner-Ring Suburbs: A Comparison of Four US Metropolitan Areas." *Journal of Planning Education and Research,* 27, pp. 146–164.

Leigh, N.G. and Lee, S. (2005) "Philadelphia's Space in between: Inner-Ring Suburb Evolution." *Opolis,* 1, pp. 13–32.

Leinenberger, C.B. (2008) "The Next Slum?" *The Atlantic,* March.

Lippmann, W. (2012 [1922]) *Public Opinion.* New York: Harcourt, Brace and Co. and Mineola: Dover Publications.

Logan, J.R. and Schneider, M. (1984) "Racial Segregation and Racial Change in American Suburbs, 1970–1980." *American Journal of Sociology,* 89, pp. 874–888.

Lucy, W.H. and Phillips, D.L. (2000) *Confronting Suburban Decline: Strategic Planning for Metropolitan Renewal.* Covelo, CA: Island Press.

Lucy, W.H. and Phillips, D.L. (2006) *Tomorrow's Cities, Tomorrow's Suburbs.* Chicago, IL: American Planning Association.

Massey, D.S. and Denton, N.A. (1993) *American Apartheid: Segregation and the Making of the Underclass.* Cambridge, MA: Harvard University Press.

McLanahan, S. and Garfinkel, I. (1989) "Single Mothers, the Underclass, and Social Policy." *The Annals of the American Academy of Political and Social Science,* 501, pp. 92–104.

Nédélec, P. (2017) "The Stigmatization of Las Vegas and Its Inhabitants: The Other Side of the Coin." In: Kirkness, P. and Tijé-Dra, A. (eds.) *Negative Neigbhourhood Reputation and Place Attachment: The Production and Contestation of Territorial Stigma.* New York: Routledge, pp. 9–28.

Nguyen, M.T., Basolo, V., and Tiwari, A. (2013) "Opposition to Affordable Housing in the USA: Debate Framing and the Responses of Local Actors." *Housing, Theory and Society,* 30, pp. 107–130.

Niedt, C. (2006) "Gentrification and the Grassroots: Popular Support in the Revanchist Suburb." *Journal of Urban Affairs,* 28, pp. 99–120.

Obrinsky, M. and Stein, D. (2007) "Overcoming Opposition to Multifamily Rental Housing." National Multi Housing Council (NMHC) White Paper, Washington, DC.

Pendall, R. (1999) "Opposition to Housing NIMBY and beyond." *Urban Affairs Review,* 35, pp. 112–136.

Perez, A.D., Berg, K.M., and Myers, D.J. (2003) "Police and Riots, 1967–1969." *Journal of Black Studies,* 34, pp. 153–182.

Raymond, E., Wang, K., and Immergluck, D. (2016) "Race and Uneven Recovery: Neighborhood Home Value Trajectories in Atlanta before and after the Housing Crisis." *Housing Studies,* 31, pp. 324–339.

Reid, M. (2013) "Homes to Be Demolished in East Cleveland." *Fox 8 Cleveland,* July 25. Available at: http://fox8.com/2013/07/25/homes-to-be-demolished-in-east-cleveland/.

Renn, A. (2017) "How to Save a Dying Suburb." *Citylab,* September 19.

Ricketts, E.R. and Sawhill, I.V. (1988) "Defining and Measuring the Underclass." *Journal of Policy Analysis and Management,* 7, pp. 316–325.

Rothwell, J. and Massey, D.S. (2009) "The Effect of Density Zoning on Racial Segregation in US Urban Areas." *Urban Affairs Review,* 44, pp. 779–806.

Rugh, J.S. and Massey, D.S. (2010) "Racial Segregation and the American Foreclosure Crisis." *American Sociological Review,* 75, pp. 629–651.

Sampson, R.J. and Raudenbush, S.W. (2005) "Neighborhood Stigma and the Perception of Disorder." *Focus,* 24, pp. 7–11.

Scally, C.P. (2012) "The Nuances of NIMBY: Context and Perceptions of Affordable Rental Housing Development." *Urban Affairs Review,* 49, pp. 718–747.

Schafran, A. (2013) "Discourse and Dystopia, American Style: The Rise of 'Slumburbia' in a Time of Crisis." *City,* 17, pp. 130–148.

Schwartz, B. (1980) "The Suburban Landscape: New Variations on an Old Theme." *Contemporary Sociology,* 9, pp. 640–650.

Short, J.R., Hanlon, B., and Vicino, T.J. (2007) "The Decline of Inner Suburbs: The New Suburban Gothic in the United States." *Geography Compass,* 1, pp. 641–656.

Singer, A. (2013) "Contemporary Immigrant Gateways in Historical Perspective." *Daedalus*, 142, pp. 76–91.

Stahura, J.M. (1983) "Determinants of Change in the Distribution of Blacks across Suburbs." *The Sociological Quarterly*, 24, pp. 421–433.

Stein, D. (1992) *Winning Community Support for Land Use Projects*. Bethesda, MD: Urban Land Institute.

Tighe, J.R. (2010) "Public Opinion and Affordable Housing: A Review of the Literature." *Journal of Planning Literature*, 25, pp. 3–17.

U.S. Department of Housing and Urban Development (1974) *Final Environmental Impact Statement*. St. Louis: Area Office. September.

Vicino, T.J. (2008) "The Quest to Confront Suburban Decline: Political Realities and Lessons." *Urban Affairs Review*, 43, pp. 553–581.

von Hoffman, A. (1996) "High Ambitions: The Past and Future of American Low-Income Housing Policy." *Housing Policy Debate*, 7, pp. 423–446.

Wacquant, L. (1996) "Red Belt, Black Belt: Racial Division, Class Inequality and the State in French Urban Periphery and the American Ghetto." In: Mingiane, E. (ed.) *Urban Poverty and the Underclass*. Oxford: Blackwell Publishing, pp. 234–274.

Wacquant, L. (2007) "Territorial Stigmatization in the Age of Advanced Marginality." *Thesis Eleven*, November, 91, pp. 66–77.

Wacquant, L., Slater, T., and Perira, V.B. (2014) "Territorial Stigmatization in Action." *Environment and Planning A*, 46, pp. 1270–1280.

Wilson, W.J. (1987) *The Truly Disadvantaged: The Inner City, the Underclass, and Public Policy*. Chicago, IL: University of Chicago Press.

Part IV

Planning, public policy, and reshaping the suburbs

20
Metropolitan governance in Paris

Theresa Enright

Introduction

Historically, Paris's urban development has been characterized by an intensive concentration of decision-making, productivity, amenities, services, and wealth. The iconic core of the monocentric city is highly valued by public and private institutions, while the *banlieue* or suburban areas are frequently ignored or maligned. Clear material and symbolic divides exist between Paris and its suburbs. Due to this structure, conflicts between the needs and interests of the city of Paris proper and its much larger regional territories have existed for centuries (Fourcaut et al., 2007). The spectacular urban uprisings of 2005 signaled the extreme unevenness of Paris's metropolitan development and the challenges of achieving an integrated and inclusive twenty-first century metropolis (Offner and Gilli, 2010).

While the political and social realities of a prosperous city walled off from its deprived peripheries should not be underestimated, processes of globalization and state decentralization have significantly eroded these dichotomous patterns. Indeed, today the city of Paris represents only a small fraction of the urbanized territory in the Île-de-France region, and the suburbs host diverse and essential networks of global financial, social, and cultural power. While Paris retains important political and tertiary economic functions (national ministries, publishing houses, universities, headquarters of major enterprises and banking institutions, and commercial spaces), in terms of demographics, employment, and productivity, the suburbs are "greater" than the urban core. La Défense, for example, located in the western *petit couronne* (inner suburbs), is France's principal business center and one of Europe's most powerful financial districts. In addition, new poles of the economy, such as Plaine-Saint-Denis in the north, have effectively captured highly competitive investments. Burgeoning research and technology clusters in Saclay, Évry, and Aubervilliers, and growing universities in Nanterre, Créteil, Villetaneuse, and Orsay, also suggest a centrifugal shift in informational and intellectual activity outside the walls of Paris. Even the largest commercial centers are located in the inner suburbs and not in Paris proper. The vast new Grand Paris Express "supermetro" (set to be operational in 2025) represents a €30 billion investment in the suburbs, and a recognition by the state and stakeholders of the metropolitan region's importance and vibrancy. Greater Paris is, for all intents and purposes, a large, complex, and multipolar space with an international presence.

Paris is functionally integrated with its *banlieue*, and there is now widespread recognition of the existence of an urban agglomeration beyond the walls of the *périphérique*. Yet, neither the attempted construction of a polycentric urban identity (Enright, 2014) nor the creation of a transit-connected region (Enright, 2013) has pointed toward a single scenario for how the metropolis should be managed and governed. The political space of the metropolis lags behind its speculative and technical correlates such that the experiential reality of Paris is "out of step with its political and administrative reality" (Sarkozy, 2010, p. 43). To complete the transformation from a monocentric city to a thriving multipolar urban region, Paris requires the construction of new governance arrangements and related transformations in urban authority, legitimacy, management, decision-making, citizenship, and democratic life.

This chapter considers the juridical and administrative supports on which Greater Paris is being founded. It begins with an overview of the often-deep-seated struggles between classes, territorial factions, political coalitions, and governmental agencies over Paris's political role and how a metropolitan polity should be built. It then traces some shifts in institutional landscapes and authority relationships in the greater Paris region leading up to the construction of the new metropolitan government, the Métropole du Grand Paris, in 2016. With a focus on enduring historical rationalities of the state and new techniques of metropolitanization, it seeks to answer the question, through what mechanisms and processes are the political institutions of the Parisian metropolis being built?

The challenges of metropolitan governance

In large urban regions around the world today, there is increasingly a divide between administrative and jurisdictional borders and other kinds of functional organizations. That is, the political boundaries rarely align with "urbanized" settlements, and even less so with economic relationships, labor and commodity markets, residential units, collective identities, infrastructure networks, private and public service catchments, or natural ecosystems. Moreover, with increased mobility in all areas of life, and residents and stakeholders increasingly defined by their multiple and overlapping relationships to places, what constitutes a political community as such is in a constant state of flux. The territorial base of contemporary local governments has thus been overturned; authority itself is multiple, complex, and difficult to pin down; and local officials are unable to autonomously control the spaces over which they preside.

With increasing global interdependence and national devolution, the presumed responsibilities of local collectivities are frequently at odds with the constitutional and legal spaces in which they are expected to operate. With increased responsibilities but diminished capacities and resources, large cities frequently lack the key political instruments to control their foundations. Local power rests in complex collaborative networks of public and private authorities at multiple scales. This is captured in the shift from municipal *government* to metropolitan *governance*, marking the widening realm of influential political actors "beyond the city" and "beyond the state."

Intensified connectivity, interdependencies, and externalities (both negative and positive) between administrative units pose problems for public service delivery and spatial and economic planning. For Michael Storper (2014), the permanent divide between functional and administrative territories results in a "principal-actor mismatch" whereby responsible institutions are misaligned with the problems they must solve. As a result, in large metropolises there is a continual cycle in which new public agencies and institutions are created, only to quickly become ill adapted to the changing realities they must address. Moreover, insofar as different policy sectors and planning endeavors will inevitably have their own geographic terms, it is nearly impossible to coordinate all of the service delivery, infrastructure, and policy-making on the metropolitan scale within a single fixed territorial unit.

The global city-region and its unwieldy tangle of flexible institutions have thus generated many new issues of governability (Jouve and Lefèvre, 2002; Kantor et al., 2012; Le Galès and Vitale, 2013). While the devolution of power means that the viability of cities is now dependent on new forms of inter-administrative cooperation and public-private partnerships, the fragmentation of authority also causes conflict. Vertically, this has resulted in problems of multi-level governance over jurisdictional authority, administrative competency, fiscal responsibility, and political legitimacy. Horizontally, it has led to cross-border competition between localities for power and resources, and between public and private coalitions over accountability, risk-bearing, and the distribution of surplus.

The matrix of political relations that constitutes the Paris metropolitan region is notoriously complex, and the administrative landscape of the Île-de-France is infamous for its multitudinous, competing, and overlapping governmental and nongovernmental structures. Described colloquially as a *"mille-feuille,"* a pastry consisting of a "thousand layers" of dough, governance in the capital region of France is multi-layered and fragile. This configuration has been criticized for impeding policy formation and implementation, and for making the functions of regional governance – the organization of production, the provision of social services, the regulation of activity – inefficient, ineffective, and undemocratic. According to local officials,

> The urban area of Paris is faced with all the challenges of a twenty-first-century metropolis, but like the majority of worldwide metropolises, still has to cope with the tools of the twentieth century government and with the administrative limits of the nineteenth century.
>
> *(Mairie de Paris, 2013)*

The political metropolis, as an entity with strong legitimacy, autonomy from other levels of government, wide-ranging jurisdiction, and appropriate territorial borders, is not yet present (Lefèvre, 1998, p. 2004).

In the last 15 years, however, there has been a series of local and national initiatives to restructure administrative territories, to streamline services, to implement interlocal regulatory policies, and to promote formal and informal collaborations, with the goal of metropolitanization. A farsighted institution at the metropolitan scale is a widely shared desire of diverse actors within the Île-de-France – indeed, "the metropolis" has taken on new life as a metanarrative of urbanization in Paris today – yet there remain significant conflicts over what form that institution should take, who should make up its ranks, and what the scope of its mandate should be.

The politics of metropolitanization

There are four main axes along which the conflicts over contemporary transformations in metropolitan governance are currently taking place: central and territorial authorities, Paris and the provinces, Paris and its *banlieue*, and left-right ideologies. Each of these represents a long-standing terrain of antagonism and negotiation over the shape of the metropolis and its techniques of authority. Indeed, to understand the eventual creation of the Métropole du Grand Paris, it is necessary to first sketch the context out of which it emerged.

Theresa Enright

Central and territorial authorities

The first dimension of metropolitan transformation concerns the political tension within French republicanism between a unitary and universal authority, and more disparate and diverse territorial arrangements. Debates between Jacobinism (a system founded on a highly ordered and powerful central command) and Girondism (in which power is decentralized into diffuse factions) have existed since the time of the Revolution. France is known for prioritizing the former but is equally defined by the latter. Under the Fifth Republic the central government has retained a prominent role in organizing national affairs, coordinating everything from universal educational curricula, to nativist cultural policy, to territorial and industrial production. Since the 1970s, however, there have also been significant devolutions of authority to subnational levels of government that have threatened the centralization of national power.

The Defferre Laws of 1982 (known as Act I of decentralization) began the far-reaching transformations of intergovernmental relations, and the movement of substantial decision-making power and policy responsibilities to subnational levels of government. Most important, the laws established the three official tiers (regional, departmental, municipal [communal]) of subnational government and their competencies. They also established new norms for how these institutions would interact. The three levels of government are arranged horizontally based on the principles of shared functions and non-subsidiarity. Many important policy issues (e.g., economic development, spatial planning) are under the control of several levels at once, and while each level may have the lead responsibility over particular issues, there is no domination exercised by one level over those under it.

Decentralization changed the character of local governance and its scope of operations. On the one hand, it confirmed the autonomy of local governments to direct their own affairs. On the other hand, with new responsibilities at the local level but inadequate capacity, new contractual arrangements, collaboration, and public-private arrangements proliferated throughout the 1980s and 1990s (Bernier, 1991; Cerny, 1989). Thus, while the state itself became oriented to regional and municipal scales (Brenner, 2004; Keating, 1983), local governments also began to alter their spatial policies, implementing new infrastructure and amenities, and pursuing place-based rebranding campaigns to attract economic investment and drive up financial assets.

In the wake of *dirigisme* emerged new regulatory arrangements of "negotiation, partnership, voluntary participation and flexibility" (Lefèvre, 1998, p. 18). Local councils in the Paris region had significant responsibilities but were too small and had insufficient resources to address the devastating urban impacts of deindustrialization such as unemployment, lack of affordable housing, and growing socio-spatial marginality. At the same time, the region of Île-de-France was better situated to deal with these broader urban policy geographies, but its powers were weak and its competencies minimal. Cross-border coalitions became necessary. In many sectors of local service delivery, from healthcare and education to welfare, policing, and basic infrastructural services (e.g., water, sewage), there are ongoing vertical and horizontal conflicts over who is responsible for issues, how power should be shared, and how the costs and benefits of policies should be distributed.

Even though significant planning, land use, and economic development capacities were transferred to the region in the Raffarin reforms of the 2000s (Act II of decentralization), the state reasserted its power in 2007 with the launch of the Grand Paris initiative. An umbrella term for new design, planning, and policy orientations, this centrally organized program saw the state take back the reins of power over Paris from more local authorities. The Grand Paris Act of 2010, for example, harnessed the energy of metropolitan reform. It established the Society for Grand Paris and endowed it with unprecedented powers in terms of land use and transportation – two essential areas of policy that were unsettled at the regional and local levels. In the early days of Grand Paris, the new Ministry of Development for the Capital Region and the Ministry of Cities assured the state's presence. Alongside these reforms, President Sarkozy's plan for metropolitan governance, outlined in the Balladur Report, was met with virulent criticism, from the left in the Île-de-France (which had just recently solidified its hold on power) and from local officials across the political spectrum, who were wary of losing autonomy and being structured out of existence.

Paris and the provinces

Decentralization is compounded by the second axis of governance reform, the asymmetries between Paris and the provinces resulting from the capital region's preeminence. The national "macrocephaly" refers to Paris's exceptional weight with respect to population, economic productivity, infrastructure, and political control in relation to the rest of the country. This condition results in two main contradictory discourses of the capital: Paris as national flagship and Paris as a drain on national resources.

The former is the belief that the Paris is the bellwether of national success (or failure) and that the city should be prioritized in any national growth agenda. While typically a national narrative, this is also the mantra of many local elites, for whom the grandeur of Paris is justification for exceptional resources and investment demands. The latter depiction is best expressed through Jean-François Gravier's (1947) famous thesis, "Paris and the French desert." Gravier denounces the extreme concentration of political and economic activity within the Paris region, and condemns the capital's monopoly of national resources as pathological. According to this perspective, which largely guided Keynesian distributive policies, the growth of Paris needs to be limited and territorial investment widespread to ensure more balanced national planning.

Against the constitutional principle of nonhierarchization among local collectivities, Paris clearly has more clout and receives more national attention in terms of investment than other communes and departments on the same formal footing. In 2007 the state confirmed the importance of Paris by identifying several key National Interest Operations (OINs) in the region (half of these national priority sites are in the Île-de-France) and by giving new support to powerful pro-growth development institutions to guide their creation. Most notably, however, the state's emphasis on Paris's national importance was marked by the launch of Grand Paris. Not only does Grand Paris seek special metropolitan status for Paris to strengthen its development capacities and to attract priority investments, but also the Grand Paris Act (2010) names the Parisian metropolis as *the* flagship territory of national development.

Elites outside of Paris, especially those in other large metropolitan regions, resent the special treatment of the capital. Local representatives of all political stripes, in fact, claim the opposite, that the state has not adequately prioritized the Île-de-France and that in order for the metropolitan region to flourish, even more targeted investments are needed.

Urban-suburban divides

The third axis of institutional reform concerns the over-determined relationship between Paris and its suburbs. Because of the history of regional centralization and their antagonistic historical development, it is difficult to achieve a shared agenda between the city of Paris and the surrounding suburbs. Mistrust and ignorance on both sides are legacies of an asymmetrical history, and conflicts persist over where to locate developments, who should pay for the externalities of growth, and who should decide on regional priorities. Officials in the city of Paris are hesitant to relinquish their long-held power in metropolitan power structures, while suburban representatives are skeptical of regional relations repeating historical patterns of annexation and exploitation.

Yet, one of the central developments driving the political processes of metropolitanization is a shift in the balance of power toward the periphery. Over the past decade there has been a "revenge of the suburbs" (Gilli and Gonguet, 2015), bolstered by changes in the demographic and economic influence of the *banlieue* and the assertion of a uniquely suburban power bloc. While obstacles to metropolitan government remain, and interterritorial disputes between Paris and its suburbs or between the suburbs themselves (especially the *petit* and *grand couronne*, inner and outer territories) are not likely to be resolved, metropolitanization has seen a more concerted effort by local officials to organize and cooperate across boundaries, and to transcend parochial concerns in pursuit of common goals and projects.

This dynamic also builds on patterns of intermunicipal cooperation developed over the past 30 years. Initially intercommunal bodies were created for service provisions, such as water distribution and waste collection, but these institutions, especially in the form of public authorities for intercommunal cooperation with fiscal capacities, have been clamoring to direct larger initiatives, such as regional economic development, social welfare, and physical infrastructure, including public transportation.

The period from 1999 to 2007 was marked by a general intensification of local coordination. In 2000 one the most powerful local intermunicipal institutions, Plaine Commune, was formed with five municipalities of the northern Parisian suburbs. Arising out of a long history of working-class organizing and shared socialist and communist leadership, Plaine Commune has become a coherent stage for territorial planning with significant influence over urban planning, transportation, economic development, urban policy, and facilities. Organizations such as Plaine Commune thus have a significant geopolitical function: to weigh in on debates on governance in the face of the state, the region, the departments, and the city of Paris, and to push for a more confederate style of coordination at the scale of the metropolis (Subra, 2012).

The emergence of intermunicipal partnerships between Paris and its cross-border suburbs was another key development in local affairs. In particular, Bertrand Delanoë, elected as mayor of Paris in March 2001, played a key role in improving relations between Paris and the *banlieue*. Delanoë and his team launched a proactive "process of atonement" (Mansat, 2012, p. 12) to establish better regional relations. Through this leadership, Paris showed its commitment to cooperation agreements, common projects with neighboring municipalities, and metropolis-wide policies. Delanoë was also instrumental in the creation of the powerful joint authority Paris Métropole, which aimed to "put a definitive end to the traditional relationship between Paris and the suburbs" (Mansat, 2012, p. 13). Paris Métropole brings together more than 100 local authorities to cooperate in pursuit of shared interests. While Paris Métropole is deeply legitimate and has a high public profile, it has no formal competence to put in place new metropolitan policies. Rather, it acts primarily as a forum for deliberation and an agency for agenda setting. In a departure from the decades-long taboo among regional representatives of collaborating with Paris, today a new consensus is emerging regarding the need for some metropolitan institutions.

Partisan ideologies

The final dimension in the ongoing process of building metropolitan Paris is that of ideological cleavages. The city of Paris has traditionally been a stronghold of the right surrounded by the so-called "red belt" of the *petit couronne*, but the 2001 mayoral victory of Socialist Delanoë marked a watershed leftward shift in the city at both the local and national levels. The regional council also reflected this change, with a left alliance in power since 1998. The path to the making of the metropolis has been forged in this highly politicized atmosphere. Among local officials, divergent visions of the goals of metropolitan life and growth have prevented, for example, a clear agreement on entrepreneurial spatial policies, on the one hand, and fiscal redistribution to address social, environmental, and economic disparities on the other. This partisan impasse was amplified under Sarkozy, when conservative state-led urban policies conflicted with the center-left plans of Jean-Paul Huchon and the Regional Council, and the more progressive agenda of Paris Métropole.

Relative to American development regimes or British locational policies, the multi-scalar urban governance arrangements in Paris were able to prioritize social concerns, even in a neo-liberal climate. However, in general, these collaborative forms of governance still remained, even under socialist governments, more conducive to nationally steered collective growth strategies. Wealth distribution, environmental regulation, and socio-spatial marginality remained sticking points in the Île-de-France that could not be adequately addressed through voluntary collaboration. Debates about solidarity, territorial inequalities, and the power balance between the central state and local collectivities came to the fore in 1994 over the regional master plan for the Île-de-France, which pitted environmental concerns against competitive strategies for economic development. In general, since the 1990s, more right-wing factions have sought a viable metropolitan platform to organize economic development activities and enhance global competitiveness. Left networks, on the other hand, favor a more democratic municipalism and the institution of concerted social welfare measures, as well as measures for regional fiscal redistribution.

While ideological cleavages are an impediment to institutional reform, there is new momentum for compromise. Elements of the left and right seem to favor the construction of a metropolitan government and policies. At the same time, however, the national collapse of the Socialist Party and the rise of other parties, such as the Front Nationale, En Marche!, La France Insoumise, and Europe Écologie (the Green Party), complicate these partisan relations and have the potential to recast the traditional ideological groundwork for or against the construction of metropolitan institutions.

The metropolitics of Grand Paris

The *mille-feuille* of governance that defines metropolitan Paris today is a product of all four of these constitutive tensions. Together these dynamics combine in a process of "unregulated competitive decentralization" (Kantor et al., 2012, p. 171) the outcome of which is a fragmented political landscape where neither the state nor local actors have sufficient capacity or legitimacy to govern unitarily. A brief sketch of recent changes in this fragmented landscape will clarify the conditions of possibility for the emergence of the official Métropole du Grand Paris.

The Métropole du Grand Paris

In 2012, Paris Métropole released a green paper on metropolitan governance to establish a new direction for discussions and action, especially in the key issue areas of housing, mobility, revenues, and attractiveness. On the specific question of a new institutional authority, Paris Métropole proposed three main figures of metropolitan governance. The first aimed to unify governance and to simplify the institutional map through an "integrated metropolis." This proposition emphasized the need for streamlined institutions, a reduction in collectivities, clear decision-making procedures, and unitary and strong leadership. The second, a "concerted metropolis," suggested an evolved institutional system arranged through existing structures. This metropolitan scenario was founded on the constitutional principle of collectivities' free administration and valued the communes' capacity to create their own development. It aimed not at dismantling existing arrangements, but at the development of a sharing culture of negotiated metropolitanization through exchange, interconnection, and dynamic collaboration. The third scenario, that of a "confederated metropolis," sought the creation of a metropolitan institution through the coordination of existing collectivities. This middle ground proposal maintained the polycentrism of the concerted version but within a unified and integrated government capable of more binding decisions (Paris Métropole, 2013). These options thus represented different means of balancing institutional flexibility and local autonomy with coordinating capability.

The green paper set in motion a national legislative bill (Law on the Modernization of Territorial Public Action and the Affirmation of Metropolitan Areas, or MAPTAM) by Prime Minister Ayrault on the creation of new metropolitan government structures nationally, with special provisions for Paris. Through the opposition it identifies between the integrated and concerted metropolitan figures (the confederated version largely collapsed into the latter in subsequent discussions), the green paper also set the terms of the legislative debate and public discourse. Leading up to the law, the battles over the institutional form of the metropolis were oriented around these two main competing visions, each corresponding to authority's different principles and modalities.

On one side, proponents called for the dismantling of existing administrative organizations and the obligatory regrouping of all territorial arrangements into a united metropolitan institution with significant powers that would be located at a scale between departmental and regional governments. This streamlined and simplified version of the metropolis would, they said, be better situated to manage the "big" problems of housing, territorial inequality, unemployment, and environmental policy that have thus far eluded decision-makers. A coherent single-tier institutional arrangement would also be effective in bringing economic stakeholders (development agencies, transportation authorities, and city boosters) into metropolitan development initiatives. Here supporters, including national and local Socialist Party representatives, claimed that ad hoc collaborations that had thus far defined metropolitan governance were insufficient. They argued that a high degree of centralization and obligatory integration was necessary for ambitious policy, lasting decisions, and political legitimacy, and that parochialism had to be checked by formal interdependencies. Many Front de Gauche representatives opposed such a strong metropolis on ideological grounds for its dismissal of local autonomy. At the same time, many municipal UMP officials who gained in the previous election and would risk losing power back to the left if a metropolitan authority were created also opposed the integrated institution on strategic grounds.

On the other side, supporters of the federated version emphasized the importance of existing intercommunal structures and the culture of grassroots partnership. The federated metropolis, a "Metropolitan G20," would respect the powers of local collectivities, they said, while organizing them into a larger administration with capacities for planning at the metropolitan scale. The federated and multi-tiered arrangement sought to maintain local policy-making and deliberation within existing municipal structures while enabling cooperation and administrative functions at the metropolitan scale. Plaine Commune in particular mobilized in support of a radically federated institution that would maintain local power, claiming that intermunicipality was the pertinent scale at which to articulate metropolitan strategies and urban projects. These forces feared that a centralized metropolitan strategy – presumably with Paris at its core – would worsen territorial inequalities and prevent a truly polycentric emergence. More conservative politicians also maintained that communal autonomy and identity would be sacrificed in the face of a top-down mandate.

After prolonged debates, MAPTAM was adopted and ratified on January 27, 2015 (Act III of decentralization). The final act was a compromise in terms of both the structures it proposed for the new authority and the ends of metropolitanization at which it aimed. The MGP came in existence in January 2016.

The new MGP covers an administrative area that includes Paris, the three departments of the *petit couronne*, and several additional municipalities of the *grand couronne*. The law thus establishes for the first-time an institution capable of governing the roughly seven million inhabitants who make up the urban core of the Île-de-France. Administratively, the MGP takes over several functions "in the metropolitan interest" from member municipalities, particularly in metropolitan development, housing, urban and environmental policy, crime prevention, and economic and social development. The MGP is responsible, for example, for the development of a Metropolitan Master Plan, a metropolitan housing policy, economic development, participation in mega-event applications and planning, and devising a climate and energy plan. The MGP will also have tax-raising capacities to fulfill these activities, and MAPTAM permits tax harmonization and revenue sharing among members.

The MGP thus outlines significant new functions for metropolitan management but also raises new concerns. For Lacoste (2013), in privileging the urban but not the wider suburban and exurban region (the *grand couronne*), the MGP excludes vital parts of the territory and prevents truly metropolitan coordination. The MGP risks, in other words, widening the gap between the most urbanized areas of the Île-de-France, where jobs and enterprises are concentrated, and the periphery, thus extending existing spatial segregation (IAU-IDF, 2014). This is especially troubling, as the political space of the MGP does not sit neatly atop the much broader poly-centric economic area envisioned by the Grand Paris Express's new transit scheme. Many questions about urban-regional governance also remain. These questions include how competencies should be transferred from local to metropolitan levels, how to ensure a balance between the new responsibilities of the MGP and the resources at its disposal, how to ensure cooperation and equity between the demarcated territory of the MGP and the rest of the Île-de-France, and how to arbitrate divergent interests and territorial relations. Crucially, it is also unclear what buy-in exists among the restive citizens of Greater Paris, to see themselves as part of a shared political community with common goals and bonds.

The MAPTAM law is also significant for what it excludes. In terms of urban planning, the new metropolis must elaborate a Metropolitan Plan for Sustainable Development that is compatible with local and regional master plans; however, it does not alter the public development agencies' functions, which constitute the most powerful levers of urbanization within the region. In addition, the national legislature was clear that even though the new MGP has responsibilities for spatial planning and economic development, it will not have power over transportation, calling into question the extent to which transport and development would be integrated at the metropolitan scale (Enright, 2016).

Conclusion

Historically, Paris has been essential to French statecraft. Today Paris suggests a new order, that of the global city metropolis, as a novel form of political organization. In a globalized world the nature of the city and the nature of the nation are changing in concert, and the creation of the Parisian metropolis provides a unique window into this incipient reality. The political conflicts constitutive of the Métropole du Grand Paris certainly concern the uniquely French processes of contractualization, negotiation, and partnership that increasingly define decision-making at the metropolitan scale. They also concern the more general nature of territorial autonomy, the prospects of local democracy, the legitimacy of the state, and the pragmatic ability to address big issues that threaten the future of the planet. The contemporary questions surrounding the governance of the Paris metropolis are about Paris's relationship to France and the world at large (Veltz, 2012). They are also questions about the kind of polities that will organize collective life in the twenty-first century.

Guide to further reading

Cole, A. (2008) *Governing and Governance in France*. Cambridge, MA: Cambridge University Press.

Enright, T. (2016) *The Making of Grand Paris: Metropolitan Urbanism in the Twenty-First Century*. Cambridge, MA: MIT Press.

Fourcaut, A., Bellanger, E., and Flonneau, M. (eds.) (2007) *Paris/Banlieues: Conflits et solidarities*. Paris: Editions Créaphis.

Gilli, F. (2014) *Grand Paris: L'émergence d'une metropole*. Paris: Presses de Sciences Po.

Kantor, P., Lefèvre, C., Saito, A., and Savitch, H.V. (2012) *Struggling Giants: City-Region Governance in London, New York, Paris, and Tokyo*. Minneapolis, MN: University of Minnesota Press.

Subra, P. (2012) *Le Grand Paris: Géopolitique d'une ville mondiale*. Paris: Armand Colin.

References

Bernier, L.L. (1991) "Decentralizing the French State: Implications for Policy." *Journal of Urban Affairs*, 13(1), pp. 21–32.

Brenner, N. (2004) *New State Spaces: Urban Governance and the Rescaling of Statehood*. New York: Oxford University Press.

Cerny, P.G. (1989) "The 'Little Big Bang' in Paris: Financial Market Deregulation in a Dirigisme System." *European Journal of Political Research*, 17, pp. 169–192.

Enright, T.E. (2013) "Mass Transportation in the Neoliberal City: The Mobilizing Myths of the Grand Paris Express." *Environment and Planning A*, 45(4), pp. 797–813.

Enright, T.E. (2014) "Illuminating the Path to Grand Pari(s): Architecture and Urban Transformation in an Era of Neoliberalization." *Antipode*, 46(2), pp. 382–403.

Enright, T.E. (2016) *The Making of Grand Paris: Metropolitan Urbanism in the Twenty-First Century*. Cambridge, MA: MIT Press.

Fourcaut, A., Bellanger, E., and Flonneau, M. (eds.) (2007) *Paris/Banlieues: Conflits et solidarities*. Paris: Editions Créaphis.

Gilli, F. and Gonguet, J.-P. (2015) "Le Grand Paris, une métropole de techniciens sans vision politique." *La Tribune*, January 16.

Gravier, J.-F. (1947) *Paris et le desert Français*. Paris: Editions Flammarion.

IAU-IDF (Institut d'Aménagement et d'Urbanisme de la Région d'Île-de-France) (2014) "Métropole du Grand Paris et mobilité: Quels impacts? Quels enjeux?" *Note rapide*, 664. Available at: www.iau-idf.fr/savoir-faire/nos-travaux/edition/metropole-du-grand-paris-et-mobilite-quels-impacts-quels-enjeux.html.

Jouve, B. and Lefèvre, C. (2002) *Métropoles ingouvernables*. Montréal: Édition Elsévier.

Kantor, P., Lefèvre, C., Saito, A., and. Savitch, H.V. (2012) *Struggling Giants: City-Region Governance in London, New York, Paris, and Tokyo*. Minneapolis, MN: University of Minnesota Press.

Keating, M. (1983) "Decentralization in Mitterrand's France." *Public Administration*, 61, pp. 237–252.

Lacoste, G. (2013) "La Métropole du Grand Paris: Intégration ou confédération?" *Métropolitiques*, September 9. Available at: www.metropolitiques.eu/La-Metropole-du-Grand-Paris.html.

Lefèvre, C. (1998) "Metropolitan Government and Governance in Western Countries: A Critical Review." *International Journal of Urban and Regional Research*, 22(1), pp. 9–25.

Le Galès, P. and Vitale, T. (2013) "Governing the Large Metropolis: A Research Agenda." Working papers of the Cities Are Back in Town program, Presses de Sciences Po, Paris.

Mairie de Paris (2013) "Innovations and the Making of Metropolitan Identity International Conference." Conference Program, Paris, 26–27 November.

Mansat, P. (2012) "Opening Speech of the Seminar." *Mairie de Paris, Les Cahiers de la Métropole*, 2, pp. 12–13.

Offner, J.-M. and Gilli, F. (2010) *Paris métropole: Hors les murs*. Paris: Presses de Sciences Po.

Paris Métropole (2013) *Paris Métropole in Brief*. Paris: Paris Métropole.

Sarkozy, N. (2010) "Exclusive Interview with Nicolas Sarkozy President of the French Republic." *AA: L'Architecture d'Aujourd'Hui*, February, pp. 42–58.

Storper, M. (2014) "Governing the Large Metropolis." *Territory, Politics, Governance*, 2(2), pp. 115–134.

Subra, P. (2012) *Le Grand Paris: Géopolitique d'une ville mondiale*. Paris: Armand Colin.

Veltz, P. (2012) *Paris, France, monde: Repenser l'économie par le territoire*. Paris: Editions de l'Aube.

21

The French banlieue

Renovating the suburbs

Juliet Carpenter

Introduction

In a volume about the suburbs, inevitably many different interpretations of the word "suburbs" are brought to the fore, depending on the specific context and culture of the cases in question. But perhaps no greater difference can be found within the Global North, than between the Anglo-Saxon understanding of the word "suburbs" and its French equivalent: "les banlieues." Typically, the word "suburbs" in the United States and UK conjures up images of wealthier neighborhoods, detached houses set in well-kept gardens and occupied by middle- and upper-income families (although as this volume illustrates, there are also many other varieties of suburbs in the Anglo-Saxon context). Contrast this portrait of a "respectable" neighborhood with the traditional "received image" of the French suburbs or "banlieues," with their high-rise tower blocks, bleak open spaces, boarded-up shops, and groups of young people of color congregating in stairwells. Again, there are clearly peripheral areas around Paris and other French cities that don't match this description, but typically the word "banlieue" is associated with images of large-scale post-war concrete housing estates, characterized by a concentration of poverty and inequality, as well as anti-social behavior, crime, and social disintegration (Kokoreff and Lapeyronnie, 2013).

This chapter aims to provide insights into the French banlieue, in particular to trace how and why these characteristic banlieues grew up, mapping their history from initial construction in the post-war period, to their decline towards the end of the twentieth century. Recently, efforts have been made to regenerate some of the most deprived peripheral housing estates on the edge of French cities. This chapter will explore how this redevelopment has been characterized, and what it means in particular for local residents who live there and have been affected by the regeneration of their neighborhoods.

History of the banlieue

Starting in the mid-1950s, as a response to the post-war housing crisis, large-scale social housing estates were constructed on the edge of many French cities, through a house-building program of HLM (Habitat à Loyer Modéré, or Low-Cost Housing) under the Plan Courant of 1953. Often

Figure 21.1 Balzac, France

Source: Petit Louis. 2010. Creative Commons 2.0.

built in a highly functionalist style initially inspired by Le Courbusier, these "cités" (housing estates) were built rapidly, often with poor quality materials, on a scale that had never been witnessed before, with sometimes thousands of households living in one block. Figure 21.1 shows one such housing block, the "Balzac" Tower in La Courneuve on the outskirts of Paris, part of the "4000" housing estate.

Initially, the residents of these public housing estates were relatively diverse (Tissot, 2007). Following World War II, there were significant housing shortages, particularly in major cities, first as a result of the extensive damage from the war, and second because of the significant rural-to-urban migration prompted by employment opportunities in growing industrial sectors. Many middle-income families saw these new housing projects as attractive places to live. They represented modernity, offering light, space, and comforts such as a bathroom and central heating, often in stark contrast to some of the inner-city and rural housing to which newcomers had become accustomed. Furthermore, they symbolized the importance of the welfare state within society, both in facilitating access to housing, as well as in promoting economic growth through a government-subsidized mass housing construction program. During this period from 1953 to 1973, an average of 300,000 housing units were constructed per year (Bertagnini, 2013, p. 10).

However, during the 1970s the situation in these social housing estates began to shift, primarily due to three factors (Tissot, 2007). First, up until the 1970s foreign nationals had almost no access to public housing. There was significant discrimination against immigrants, particularly against those from the former French colonies in North and West Africa; until the 1970s, many non-French nationals lived in substandard slum housing and bidonvilles (informal shantytowns) in and around the city. However, in the early 1970s the government launched a major slum clearance program, and as a consequence, social housing landlords were subsequently obliged to house immigrants in their cités.

Second, there was a shift in France's house-building policy in the early 1970s, with a halt on the construction of large-scale public housing, coupled with incentive programs encouraging homeownership of individual houses through low-interest loans. So as middle-class households moved out of public housing estates into home ownership, migrant families were being rehoused from slum dwellings into the cités.

Third, the socioeconomic status of the banlieue residents was also shifting. Many cité residents were employed as low-skilled manual workers in factories, particularly around Paris. From the 1970s, with the global downturn following the 1973 oil crisis and subsequent industrial restructuring, many employees were made redundant, with foreign workers often among the first to lose their jobs. Thus, from the auspicious beginnings of the cités as places of modernity and optimism, the banlieue housing estates were increasingly characterized by deprivation, a high ethnic minority population, and economic and social exclusion, or what Wacquant (1996) has termed "advanced marginality," associated with the rise of a neoliberal economy.

However, it was not until the 1980s that the government recognized that a number of problems were concentrating in the banlieue as a result of the isolation, both physical and metaphorical, experienced by local residents. Growing resentment, particularly among young people who felt excluded from mainstream French society, was manifest in outbreaks of civil disturbances in the early 1980s, initially in Les Minguettes, a social housing estate in the suburbs of Lyon in 1981 and 1982, as well as in other housing estates around Paris and elsewhere. Institutionally, the government response was to restructure national agencies to have a more urban focus, with the creation in 1988 of three new governance bodies responsible for cities, the Comité Interministériel des Villes (CIV – Interministerial Committee for Cities), the Délégation Interministérielle à la Ville (DIV – Interministerial Delegation for Urban Affairs), and the Conseil National des Villes (CNV – National Council for Cities). These bodies were charged with the delivery of an urban policy specifically aimed at disadvantaged neighborhoods, the Politique de la Ville.

La Politique de la Ville

Historically, questions of poverty in France, including urban policy related to poverty in the city, have been addressed by programs aimed to reduce inequalities, taking a "color-blind" approach without reference to the role of ethnicity in reinforcing inequalities. This approach dates back to the French Republican ideals of "Liberty, Equality, and Fraternity," which focus on the universal citizen, rather than on citizens defined by their ethnicity or religion. This in theory guarantees equality for all and facilitates the integration of immigrants into French society.

However, in the context of debates related to the banlieue during the 1980s, for the first time ethnicity was introduced into discourses around tackling the "social problems" of the banlieue. As Tissot (2007) argues, narratives at the time from both politicians and the press drew a direct link between the emerging challenges of the banlieue and the issue of immigration, which in turn influenced the development of France's urban policy for disadvantaged neighborhoods, the Politique de la Ville of the late 1980s and 1990s.

Initially, the emphasis of the Politique de la Ville was on local social development, strengthening social ties, promoting community links, and enhancing civic participation. There was minimal physical intervention limited to minor refurbishment and occasional demolition. Rather than being coordinated by national agents, urban policy was to be administered in partnership with local stakeholders, including public, private, and civil society actors, in order to promote civic participation and social diversity as a means of addressing poverty and unemployment in the banlieue (Tissot, 2008).

Up until the 1990s, the Politique de la Ville was characterized by a diversity of approaches, addressing social, economic, environmental, and physical dimensions of disadvantaged neighborhoods through an integrated approach to tackling urban deprivation (Busquet et al., 2016). However, in the late 1990s there was a marked change in direction. It was felt that the previous approaches, particularly those related to social development measures designed to promote neighborhood cohesion, had failed to solve the "problems" of the banlieues (Lelévrier, 2004).

These problems were deemed to stem from the concentration of specific groups in certain areas, which could be addressed through the objective of so-called "social mixing" ("mixité sociale"). Rather than community development to support deprived neighborhoods, the Politique de la Ville shifted its focus towards a program of demolition and rebuilding, so-called "urban renewal" involving comprehensive housing diversification, which was seen as the appropriate response to achieve a "social mix" and thus to address the "problems" of the banlieue (Gilbert, 2009).

Social mixing is an ambiguous term. Officially, it refers to a mix of housing tenures, income groups, and classes that are found in any one area. The definition of social mixing from the "Critical Dictionary of Housing Conditions and Home" states:

> Social mixing (mixité sociale) is the objective of a social policy that aims to bring together different social classes to coexist within a given urban unit [e.g., a neighborhood] mainly through implementing housing programs.
>
> *(Bacqué, 2002, p. 297, author's translation)*

As noted, the French Republic does not permit distinctions to be made between social groups along lines of ethnicity or religion; this is the reason why there are no census data collected on ethnicity in France. However, the term "social mixing" in the context of urban redevelopment can also imply "ethnic mixing" and the implicit aim of encouraging a more diverse ethnic mix in areas characterized by a high proportion of non-French nationals (Kipfer, 2016). The rise in socioeconomic inequalities and a growing feeling of discontent in France has been exploited in political terms by the Far Right, led by the National Front but also echoed by other right-wing parties. The National Front's discourse, focusing on immigration as one of the key sources of French society's troubles, has concentrated its attention on immigrants and their spatial concentration (Gilbert, 2009). The French Republican tradition considers every citizen to be a member of the "national community" rather than any other religious or ethnic minority community, and a narrative has developed around the concentration of immigrants in the banlieue that represents a "tribalism" ('communautarisme') that cultivates difference and serves as a "threat" to national unity (Dikeç, 2007). Thus, within mainstream policy discourses related to the Politique de la Ville, the term "ghetto" was increasingly used to characterize the concentration of ethnic minority groups in the banlieue, although as Wacquant (1992) has shown, the comparison between the French situation in the banlieue, and the extreme social exclusion and racial tensions of the black American ghettos is far from justified. It was these so-called "ghettos" that the remodeled Politique de la Ville sought to break up by encouraging middle-income households to move into targeted neighborhoods through "social mixing." The next section outlines these more recent developments, including the impacts of the "social mixing" policy.

Recent policy approaches to the banlieue

Shortly after Chirac's right-wing government took power in 2002, a new approach to urban policy was introduced through the Borloo Act of 2003, named after Minister Jean-Louis Borloo, who at the time was responsible for City and Urban Renewal. This marked a shift in France's urban renewal policy, from a more holistic vision to one dominated by demolition and reconstruction, particularly in the industrial working-class housing estates of the banlieue (Dikeç, 2006). The first National Urban Renewal Program (PNRU – Program National de Rénovation Urbaine) launched in 2005 had a double remit, focusing both on creating mixed-income neighborhoods ("mixité sociale"), as well as on promoting sustainable development by targeting neighborhoods classified as "Zones Urbaines Sensibles" (ZUS – Deprived Urban Neighborhoods). These were often areas with concentrations of non-French nationals and their French-born descendants, where the government used a strategy that involved significant elements of housing demolition and rebuilding. Figure 21.2 shows the demolition of the Balzac Tower in 2011, as part of the urban renewal strategy for La Courneuve. In 2014, the PNRU was extended by the Socialist Government up until 2024 to include a wider remit, but also with fewer resources and far more focused territorial interventions (Gouvernement de France, 2014).

The "Agence Nationale de Rénovation Urbaine" (ANRU – the National Agency for Urban Renewal) coordinates the Program, which oversees the rehabilitation of social housing units. The PNRU initially targeted a population of over six million people, with other goals including the demolition of some 250,000 housing units over eight years, the renovation of around 400,000 housing units, the provision of community infrastructure, as well as a focus on cultural facilities and in particular, employment opportunities.

Through the PNRU, demolition was no longer seen as "taboo" for addressing the challenges of peripheral social housing estates (Baudin and Genestier, 2006; Driant, 2012, Kipfer, 2016). It was a legitimate method to follow through the "social mixing" policy, by redesigning the urban environment to diversify the housing supply, including the construction of different architectural designs (more low-rise buildings and individual housing), different sized properties, and different forms of tenure, including intermediate and market rent housing as well as homeownership.

The Borloo Act was partly introduced in reaction to the Socialist government's Law on Solidarity and Urban Renewal (Loi Relative à la Solidarité et au Renouvellement Urbain – SRU) dating from 2000, which made it compulsory for all municipalities in large metropolitan areas to provide at least 20 percent of their housing stock as social housing by 2020, for communes with at least 3,500 inhabitants (1,500 in the Greater Paris area) included in a metropolitan area of more than 50,000 residents (Desponds, 2010). The Borloo Act, coming from the right-wing government, took a different approach, prioritizing the demolition of social housing, particularly in communes where there was a concentration of public sector housing, in favor of a more varied housing offer, in order to diversify the residential population.

However, a number of critiques have been leveled at the approach the Borloo Act has taken. The National Urban Renewal Program (PNRU) could be seen as introducing a series of elements that work against the principles of social sustainability in a neighborhood context. By focusing on demolition rather than rehabilitation, communities have been broken up and social ties severed, weakening residents' social capital, connections, trust, and networks. Lelévrier (2008) has shown that the process of rehousing through the PNRU has been particularly unsettling for the most vulnerable households, for whom severing ties with social networks and familiar places has the most detrimental effects. In the reconstruction phase, there has also been a bias towards market and intermediate housing, which many

Figure 21.2   Demolition of the Balzac Tower, France

Source: Diego BIS. 2011. Creative Commons 2.0.

original residents are unable to afford, thus eroding the stock of affordable housing (Kipfer, 2016). Furthermore, in redesigning the urban environment, emphasis has been placed on issues of security, separating public and private space, and "privatizing" previously "open access" space (Epstein, 2013).

Gilbert (2009) identifies two key objectives of the PNRU; first, destigmatizing neighborhoods by reshaping their urban form, and second, modifying residents' behavior by transforming the urban environment and its social mix. The first objective aims to rid neighborhoods of their negative image and dispel the stigma associated with the area as a result of the "dysfunctionalities" of post-war physical planning. As Wacquant et al. (2014) identify, territorial stigmatization can have a significant impact on residents' everyday lives, such as on finding a job or looking for alternative accommodation, and it can also act as a deterrent to middle-income households moving into the area. Measures to address this stigmatization through the PNRU have included demolishing high-rise housing blocks and replacing them with smaller scale housing units; redesigning public spaces with streets, squares, and more "welcoming" public areas; and opening up neighborhoods through clearer and more accessible links to neighboring districts and the city center.

However, as Wacquant (2008) suggests, place-based policies often build on and reinforce territorial stigmatization, as they highlight the negative labels public authorities have assigned to an area, such as "problematic" and "worthy of destruction." Urban policies aimed at "changing the image" of an area often serve to underline the very issues that the demolition policy aims to address. A number of seminal sociological studies have demonstrated the negative effects of demolition programs on local communities, including the erosion of social ties based on neighborhood proximity (Coing, 1966; Young and Willmott, 1957). These have also been backed up by studies related to the PNRU (Lelévrier, 2008). Veschambre (2008) highlights the symbolic importance of recent neighborhood demolition policies, and the detrimental effect that this can have on residents' sense of home, community, and belonging.

The second objective of the PNRU identified by Gilbert (2009) relates first to modifying residents' behavior through the transformation of the urban environment, and second to diversifying an area's social mix. Policies aimed at transforming the urban environment have involved reconfiguring public spaces, in particular to address crime and anti-social behavior. However, critics have suggested that this involved the privatization of public space, where previously open spaces are now subject to restricted access, with increasing use of keypad codes, CCTV, and other tools to restrict and control residents' behavior.

The second aspect of this objective relates to social mixing, which has been implemented through a range of different measures. First, the PNRU aimed to reconfigure the urban form away from high-rise apartment blocks to a more "humane" scale of housing that aims to attract more middle-class households. Second, following demolition, the policy aims to reconstruct a more mixed tenure offer, including owner occupation, intermediate ownership, and private renting, in order to attract households with a wider range of incomes. Third, alongside the diversification of housing tenure, the PNRU aims to provide infrastructure, such as cultural and sporting facilities, to attract people from outside the area during the daytime. Figure 21.3 illustrates the type of housing that has replaced tower block estates, here showing new social housing in the Balzac neighborhood of La Courneuve, in 2014.

The aim is to increase the area's social mix, with the expectation of a "role-model" effect (Gilbert, 2009); that is, the idea that spatial proximity will promote the diffusion of "middle-class norms" within these areas. There are also expectations that school attainment levels will increase, given the higher proportion of children from middle-class households. However, such policies have generated considerable debate regarding evidence to support the "role-model" effect and the benefits for lower-income groups. As Kleinhans's review shows, merely living in close proximity in urban renewal areas does not necessarily result in social interaction between residents in different tenure types (Kleinhans, 2004).

Figure 21.3 Social housing at Balzac in La Courneuve, the 4000 Housing Estate, France
Source: darksabine. 2014. Creative Commons 2.0.

Impact of urban renewal on banlieue residents

In relation to social effects, Gilbert (2009) also identifies three impacts of the urban renewal process on housing and community. First is the impact on affordable housing availability. The demolition of low-cost social housing in urban renewal areas and its replacement with housing of different tenure types, often aimed at those with greater financial resources (intermediate housing and homeownership), has reduced the stock of affordable housing in targeted areas. This has inevitably led to the displacement of more precarious and low-income households outside the renewal area, hindering the right to decent housing for the very poor (Blanc, 2010). Second, the urban renewal process has impacted social capital, in particular kinship and support networks for precarious households (Bonvalet, 2003). Social networks have been weakened in renewal areas for households moving out of the area, and for those remaining whose family and neighbors have moved away. The third impact relates to the integration of new residents with the original community. Gilbert (2009) argues that, based on research in Les Minguettes housing estate in Lyon, newcomers adopt a distanced attitude toward their neighbors, as their stable income and employment situation often distinguish them from the majority of the local population. However, he points out that local residents rehoused in the new buildings hold the key to bridging connections between newcomers and "established" residents. This group is crucial in achieving the supposed positive effects of social mixing policy promoted through PNRU's urban renewal demolition program (Lelévrier, 2013).

Despite the significant funding and extensive renewal programs implemented under the PNRU, the Cour des Comptes (2012) found, in its review of 10 years of the Politique de la Ville (2003–2012), that there had been little impact on reducing urban inequalities. It found that in the targeted neighborhoods, where over eight million people live, unemployment is still twice the rate it is elsewhere, the proportion of people living below the poverty threshold has increased, and the average quality of life for households is less than half the national average. In relation to social sustainability, the report also found that the PNRU policy has not led to social integration as intended, but rather social fragmentation due to the increased supply of intermediate social and market housing in the target areas, the out-mobility of residents with higher incomes, and the difficulty of attracting new residents to some sites (Kipfer, 2016). This supports well-established studies based on research in the grands ensembles that show that "spatial proximity does not reduce social distance" (Chamboredon and Lemaire, 1970; see also Bernard, 2009), with no guarantee of the anticipated positive cooperation between local people and newcomers. Social mobility and the "social distinction" between groups depends more on broader processes of structural inequality, rather than on narrower contextual processes (Bourdieu, 1979). Indeed, the President of the Evaluation and Monitoring Committee of the ANRU in 2012 claimed that the objective of social mixing (that is, neighborhood diversity) hadn't been met by the PNRU: "We've remade ghettos, but only cleaner this time" (Le Monde, 2012).

One of ANRU's underlying principles for its urban renewal programs is public engagement; the objective is to involve residents in projects at the earliest possible opportunity in their development, as a key factor for success in both the short and long-term. However, certain commentators have raised questions about how far "involvement" actually goes. As Arnstein (1969) pointed out almost half a century ago, there are different levels on the "ladder of participation," with "involvement" often being little more than consultative, rather than true empowerment through engagement and participation in decision-making.

In terms of governance, projects under the PNRU are drawn up through a Contrat (Agreement) between the key institutional actors to set out the proposed program of work: between the State represented by the regional Prefect (Préfet), ANRU, the City represented by the Mayor, and the social housing landlords (bailleurs sociaux). From 2014, all priority neighborhoods under the Politique de la Ville have also been given the opportunity to set up Citizens' Councils (Conseils Citoyens) through which residents can engage with the urban renewal process. However, it is still too early to assess their impact, and how involved residents' associations, tenants' groups, and neighborhood committees are in decision-making. First analyses suggest that municipalities have a strong influence on Citizens' Councils (Talpin, 2014). Some have also suggested that in the lead up to urban renewal operations, site analysis is carried out by experts and technicians without reference to the daily experiences of local residents (Bertagnini, 2013). Thus, the discourse is mediated through the views of institutional actors and experts, who justify demolition through a narrative related to "opening up the neighborhood" (désenclavement), introducing more "mixité sociale" and addressing the area's "ghetto"-like aspects. It is only at this late stage that residents are invited to be involved, when a demolition program is already in place and a process of rehousing is already planned (Bertagnini, 2013, p. 13).

Examples of resistance to demolition are limited. Kipfer (2016) highlights cases of anti-demolition protests, where local resistance to redevelopment projects has had a direct impact, either stopping full demolition or reducing the number of demolished units, improving the rehousing process as well as bettering the quality of reconstructed housing units. However, examples of such resistance are limited.

More generally, however, it has been suggested that in some places urban renewal projects have had a positive effect in building capacity locally, in particularly strengthening the attachment that some residents feel towards their neighborhood, contributing to what Deboulet (2010) refers to as a new form of urban citizenship ("citoyenneté urbaine"), involving an individual and collective commitment to the city. This re-appropriation of place can be interpreted as residents' response to their threatened expropriation, who express a new willingness to participate in the construction of their everyday worlds (Bertagnini, 2013).

Conclusion

In many French cities, post-war urban economic growth was met with rapid expansion on the periphery of towns and cities. Low-cost solutions to house the growing population, coming both from rural hinterlands as well as from outside France, resulted in large-scale public housing projects on the edge of cities, characterized by generally poor quality high-rise housing developments, with few public services and often little access to public transportation.

Over time, these factors have been coupled with a concentration of low-income households, often of ethnic minority origin, and increasing socioeconomic exclusion, isolated by both visible and invisible borders. A feeling of disillusionment with and abandonment by the French political class, as well as alienation from mainstream society, led to civil disturbances in the early 1980s, early 1990s, and during the autumn of 2005 for around a month across 40 towns and cities (Tshimanga et al., 2009).

Various urban renewal programs have been instigated in these areas, in an attempt to address the escalating challenges associated with the banlieue. However, as this chapter has illustrated, a number of these programs' objectives have not been met. One of the key challenges is territorial stigmatization in the banlieue, but studies have shown that area-specific policies tend to reinforce the "outsider's" perception that the neighborhoods concerned are "problematic." Evidence shows that there is a need to improve housing conditions, but this objective appears to be secondary to the policy imperative of "social mixing." Research also highlights the problem of potentially increasing the precarity of the banlieue's most disadvantaged through a reduction in affordable housing stock, which can cause displacement and undermine social capital and local kinship ties through rehousing.

The impact of the first 10 years of PNRU's urban renewal program is beginning to be felt. Learning from the past decade, a number of recommendations have been put forward for the second program, 2014–2024 (Kipfer, 2016). Rather than emphasizing demolition, experience from the first program suggests that more reconstruction would produce better outcomes for residents. It is also important to focus on small-scale interventions, including social development actions, rather than purely physical mechanisms of demolition and rebuilding higher income housing. In order to build banlieues that are socially sustainable, there needs to be investment in community-led actions to promote social cohesion, connecting different social groups and building trust through engagement with residents. These bottom-up initiatives will construct more socially cohesive neighborhoods that will create the banlieues of the future, by and for residents.

Guide to further reading

Busquet, G., Hérouard, F., and Saint-Macary, E. (2016) *La Politique de la ville. Idéologies, acteurs et territoires.* Paris: l'Harmattan.

Desponds, D., Auclair, E., Bergel, P., and Bertucci, M.-M. (eds.) (2014) *Les Habitants: Acteurs de la Rénovation Urbaine?* Rennes: PUR.

Dikeç, M. (2007) *Badlands of the Republic: Space, Politics and Urban Policy.* Oxford: Blackwell Publishing.

Donzelot, J. (ed.) (2012) *A Quoi Sert la Rénovation Urbaine?* Paris: Presses Universitaires de France.

Kokoreff, M. and Lapeyronnie, D. (2013) *Refaire la Cité: L'avenir des banlieues*. Paris: Seuil.

Lelévrier, C. (2013) "Social Mix Neighborhood Policies and Social Interaction: The Experience of New-comers in Three New Renewal Developments in France." *Cities*, 35, pp. 409–416.

References

Arnstein, S. (1969) "A Ladder of Citizen Participation." *Journal of the American Planning Association*, 35, pp. 216–224.

Bacqué, M.-H. (2002) "Mixité Sociale." In: Segaud, M., Brun, J. and Driant, J.C. (eds.) *Dictionnaire critique de l'habitat et du logement*. Paris: Armand Colin, pp. 297–298.

Baudin, G. and Genestier, P. (2006) "Faut-il vraiment démolir les grands ensembles." *Espaces et Sociétés*, 2(3), pp. 207–222.

Bernard, M. (2009) *Sarcellopolis*. New ed. Bordeaux: Finitude.

Bertagnini, E. (2013) "The French Banlieues between Appropriation and Demolition." *Planum: The Journal of Urbanism*, 27(2), pp. 10–16.

Blanc, M. (2010) "The Impact of Social Mix Policies in France." *Housing Studies*, 25(2), pp. 257–272.

Bonvalet, C. (2003) "La famille-entourage locale." *Population*, 58, pp. 9–43.

Bourdieu, P. (1979) *La Distinction, Critique Sociale du Jugement*. Paris: Minuit.

Chamboredon, J.-C. and Lemaire, M. (1970) "Proximité spatiale et distance sociale. Les grands ensembles et leur peuplement." *Revue Française de Sociologie*, 11(1), pp. 3–33.

Coing, H. (1966) *Rénovation urbaine et changement social*. Paris: Les Editions Ouvrières.

Cour des Comptes (2012) *La Politique de la Ville: Une Décennie de Réformes*. Paris: Cour des Comptes.

Deboulet, A. (2010) "La Rénovation Urbaine Entre Enjeux Citadins Et Engagements Citoyens: Rapport de Recherche Puca." Paris.

Desponds, D. (2010) "Effets paradoxaux de la loi Solidarité et Renouvellement Urbains (SRU) et profil des acquéreurs de biens immobiliers en Île-de-France." *Espaces et Sociétés*, 1(140–141), pp. 37–58.

Dikeç, M. (2006) "Two Decades of French Urban Policy: From Social Development of Neighborhoods to the Republican Penal State." *Antipode*, 38, pp. 59–81.

Dikeç, M. (2007) *Badlands of the Republic: Space, Politics and Urban Policy*. Oxford: Blackwell Publishing.

Driant, J. (2012) "Défaire les grands ensembles." In: Donzelot, J. (ed.) *A Quoi Sert la Rénovation Urbaine*. Paris: Presses Universitaires de France, pp. 13–24.

Epstein, R. (2013) *La Rénovation urbaine: Démolition-reconstruction de l'État*. Paris: Presses Universitaires de France.

Gilbert, P. (2009) "Social Stakes of Urban Renewal: Recent French Housing Policy." *Building Research and Information*, 37(5–6), pp. 638–648.

Gouvernement de France (2014) "Loi n° 2014–173 du 21 février 2014 de programmation pour la ville et la cohésion urbaine." Available at: www.legifrance.gouv.fr/affichTexte.do?cidTexte=JORFTEXT000028636804 [Accessed 17 May 2017].

Kipfer, S. (2016) "Neocolonial urbanism? *La Rénovation Urbaine* in Paris." *Antipode*, 48(3), pp. 603–625.

Kleinhans, R. (2004) "Social Implications of Housing Diversification in Urban Renewal: A Review of Recent Literature." *Journal of Housing and the Built Environment*, 19(4), pp. 367–587.

Kokoreff, M. and Lapeyronnie, D. (2013) *Refaire la Cité: L'avenir des banlieues*. Paris: Seuil.

Lelévrier, C. (2004) "Que reste-t-il du projet social de la politique de la ville?" *Espirit*, 303, pp. 65–78.

Lelévrier, C. (2008) *Mobilités et trajectoires résidentielles des ménages relogés lors d'opération de renouvellement urbain. Synthèse de travaux menés entre 2004 et 2007*. Paris: PUCA.

Lelévrier, C. (2013) "Social Mix Neighborhood Policies and Social Interaction: The Experience of New-comers in Three New Renewal Developments in France." *Cities*, 35, pp. 409–416.

Le Monde (2012) "Yazid Sabeg: Avec la rénovation urbaine, on refait du ghetto, mais en plus propre." Available at: www.lemonde.fr/societe/article/2012/03/16/yazid-sabeg-avec-la-renovation-urbaine-on-refait-du-ghetto-mais-en-plus-propre_1670653_3224.html [Accessed 18 May 2017].

Talpin, J. (2014) "A quelles conditions participation peut-elle accroitre le pouvoir des habitants dans les quartiers populaires?" La participation: Laboratoire de la politique de la ville? Actes de la Rencontre régionale du, 25 novembre 2014, pp. 25–35. Available at: www.professionbanlieue.org/c__7_44_Publication_2654__1__La_participation_laboratoire_de_la_politique_de_la_ville_70_p_a_telecharger.html [Accessed 24 July 2017].

Tissot, S. (2007) *L'Etat et les quartiers. Genèse d'une catégorie d'action publique.* Paris: Le Seuil.

Tissot, S. (2008) *"French Suburbs": A New Problem or a New Approach to Social Exclusion?* CES Working Paper Series 160. Available at: http://aei.pitt.edu/11792/ [Accessed 20 July 2017].

Tshimanga, C., Gondola, D., and Bloom, P.J. (eds.) (2009) *Frenchness and the African Diaspora: Identity and Uprising in Contemporary France.* Bloomington: Indiana University Press.

Veschambre, V. (2008) *Traces et mémoires urbaines. Enjeux sociaux de la patrimonialization et de la démolition.* Rennes: Presses Universitaires de Rennes.

Wacquant, L. (1992) "Pour en finir avec le mythe des 'cités-ghettos': Les différences entre la France et les États-Unis." *Annales de la Recherche Urbaine*, 54, pp. 21–30.

Wacquant, L. (1996) "The Rise of Advanced Marginality: Notes on Its Nature and Implications." *Acta Sociologica*, 39, pp. 121–139.

Wacquant, L. (2008) *Urban Outcasts: A Comparative Sociology of Advanced Marginality.* Cambridge, MA: Polity Press.

Wacquant, L., Slater, T., and Pereira, V. (2014) "Territorial Stigmatization in Action." *Environment and Planning A*, 46, pp. 1270–1280.

Young, M. and Willmott, P. (1957) *Family and Kinship in East London.* London: Routledge.

22

Shrinking suburbs in a time of crisis

*Justin B. Hollander, Colin Polsky,
Dan Zinder, and Dan Runfola*

Introduction

In recent years, increased scholarly attention has been paid to the fall-out from the 2008 sub-prime lending debacle, a national collapse of the housing market that resulted in massive fore-closures and widespread housing vacancy throughout the United States (Immergluck, 2011; Hollander, 2009). Its effect on perennially growing areas such as Sunbelt cities and suburban living was unprecedented (Goodman, 2007; Leland, 2007; Dash, 2011). From Atlanta to Fort Meyers to Phoenix, massive new housing developments sat largely unoccupied while older housing sat abandoned due to foreclosure (Runfola and Hankins, 2010). With the housing market in a tailspin, cities in the Sunbelt faced depopulation and housing vacancy akin to that observed in the early stages of industrial Rustbelt cities' decline when their major industries began to falter.

It appeared that the Great Recession (the interval between 2006 and 2009) brought on a new era of shrinkage in formerly growing cities. In this chapter, we define shrinkage as net decline in occupied housing units. Lucy and Phillips (2006) note that decline in housing occupancy is a more meaningful indicator of decline than population loss, since the latter may indicate that units are being used differently rather than, for example, a community having more single or dual occupancy units where previously most units housed an entire family. In that instance, population decreases but the neighborhood has not necessarily experienced any destabilizing effects. The circumstances surrounding these new declines differ from Rustbelt population declines that date from the 1950s. In the Rustbelt, shrinkage has been driven by global economic conditions that pushed many manufacturing industries overseas – an economic trend unlikely to reverse in the near future. In contrast to this, the sun will always shine on the Sunbelt, likely driving an eventual return of the demand for retiree housing, even if tempered lending practices do not permit the same rate of consumption without capital.

Even if the circumstances around decline and shrinkage differ in these regions, we can still learn lessons from the Rustbelt about the process of shrinkage and the planning methods already undertaken in that region to encourage recovery. It has been suggested that once individual neighborhoods reach a certain threshold of vacancy, the likelihood of reversing that trend dimin-ishes greatly (Hoyt, 1933; Wilson and Kelling, 1982; Wallace, 1989; Temkin and Rohe, 1996;

Keenan et al., 1999). Even as the United States as a whole recovers, individual pockets, neighbor-hoods, or regions that are still declining will likely persist in doing so, posing increasingly daunt-ing challenges such as crime, poverty, depopulation, and ultimately revenue loss to local planners and policymakers.

As planners in this new group of shrinking places develop strategies to address decline, there is much they can learn from the emerging body of literature on smart shrinkage (Pop-per and Popper, 2002; Schilling and Logan, 2008; Hollander et al., 2009). Smart shrinkage is a set of policies that help areas with declining populations manage the associated land use changes. Instead of fighting population loss, smart shrinkage begins with the idea that maintaining a high quality of life for the remaining residents can be achieved without growth.

Central to smart shrinkage is recognizing the "fallibility of the myth of endless growth" (Popper and Popper, 2002, p. 23). Smart shrinkage was mentioned as early as 1989, when Clark encouraged preserving declining areas for "parkland and recreational spaces" (p. 143) – a sug-gestion echoed recently by Schilling and Logan (2008). Armborst, D'Oca, and Theodore (2005) introduced the idea of widespread acquisitions of vacant lots as a means of expanding average lot sizes and better managing shrinking populations, a process they described as "blotting."

A number of policymakers and lobbying groups have already implemented or considered smart shrinkage policies. Community leaders in Youngstown, Ohio, a city that has lost half of its population since 1950, adopted this approach with a new Master Plan to address its remain-ing population of 74,000. In the Plan, the city came to terms with its ongoing population loss and called for a "better, smaller Youngstown," focusing on improving the quality of life for existing residents rather than attempting to grow the city (City of Youngstown, 2005; Hol-lander, 2009; Schatz, 2010). In Philadelphia, the nonprofit Public/Private Ventures office issued a report calling for "the consolidation of abandoned areas and, in some cases, the relocation of those households that remain in blocks that too often look like Dresden after the Second World War" (Hughes and Cook-Mack, 1999, p. 15). City leaders in New Bedford, Massachu-setts, and Rochester, New York have also explored the potential of smart decline (Goodnough, 2009; Fairbanks, 2010).

As shrinkage spreads beyond a handful of Rustbelt capitals, the smart shrinkage approach may gain increasing prominence in planning practice. Before studying smart shrinkage as a solution, however, planners need to better understand the process of shrinkage. When places lose a sizable number of people, the first thing that changes is housing demand. Glaeser and Gyourko (2005) have demonstrated the durability of housing using economic analysis – that is, as people leave a place, homes do not leave in concomitant levels synchronously. In time, many factors will result in fewer occupied housing units, including abandonment, demolition, arson, or an unwillingness of landlords to lease their property.

In this chapter, we seek to understand the physical impacts of economic contraction on hous-ing occupancy patterns before and after the Great Recession. Additionally, we ask whether or not different census-defined density-determined regions – urbanized areas, metropolitan statistical areas, and rural areas – were affected uniformly during economic contraction.

We answer these questions by exploring household residential delivery data acquired from the U.S. Postal Service for February of 2000, 2006, and 2011; these years roughly mark the beginning of the real estate boom, peak of the real estate market, and years of decline and instability in the housing market following that peak. These data contain household delivery counts for every zip code in the country. In analyzing this dataset, net changes in occupied housing by zip code were tabulated and mapped, and patterns of spatial clustering were explored using Global and Local Moran's I statistics and Local Indicators of Spatial Association (LISA).

The planning context

Planners and developers have, for decades, been caught up in a false dichotomy: when a community grows in population it prospers, and when its population declines it suffers. E. F. Schumacher challenged that false dichotomy in 1973 with his volume *Small is Beautiful*. Popper et al. followed with *Urban Nongrowth* in 1976. But the on-the-ground world of practice never really responded to those critiques, and the growth/decline dichotomy prevailed until just recently.

Now, some local officials are asking if their communities can thrive and improve while staying small or even declining in population. In many ways, it is out of desperation driven by the profound failure of economic development strategies to arrest decline in many urban areas over the last several decades. The infusion of public monies into new stadiums, job training centers, infrastructure, and new housing in cities big and small has had a positive effect on some cities, but not all. Called the "forgotten cities" by Hoyt and Leroux (2007), the power and success of economic turn-around by building additional infrastructure has simply not worked everywhere. In fact, there is evidence that economic development has failed to reverse structural economic conditions contributing to decline more often than it has succeeded (Schumpeter, 1934; Boyer, 1983; Logan and Molotch, 1987).

Urban population decline has a bad reputation. Beauregard's 2003 book *Voices of Decline* documented in fastidious detail the ways that the discourses of decline were developed and positioned in American culture. Beauregard concludes that modern society's drive for bigger, faster, more of everything required that population and employment loss in mid-twentieth century cities was to be viewed in antipode to the growth and vitality of the suburbs. Lucy and Phillips's 2000 work further illustrated the plight of suburbs facing disinvestment and decline. They argued that booming, growing suburbs were at the greatest risk for decline, because they generally lacked the sense of place that can continue to attract new residents. Lang and Lefurgy's 2007 "boomburbs" analysis found that these growing suburbs were a significant phenomenon; they found 53 such boomtowns – places at high risk for decline in the event of economic contraction.

Little is known about how the Great Recession affected growing places like "boomburbs": whether they lost housing, how pervasive that loss was, and whether it was geographically clustered or dispersed. As such, the Great Recession provides a useful natural experiment for studying how an economic crisis influences shrinkage.

Alternative methods of calculating neighborhood decline

Bowman and Pagano (2004) conducted an exhaustive study on this topic, seeking to understand the vacancy problem's extent. They administered written surveys to local officials and assembled a database of abandoned buildings and vacant lot counts across more than a hundred U.S. cities. Unfortunately, this survey-based method was proven unreliable when cross-checked against housing unit counts from the Decennial Census (Hollander, 2009). Local officials across the country use very different strategies to account for vacancy and abandonment, making the use of locally distinct administrative data sources a suboptimal approach for producing national generalizations (Bowman and Pagano, 2000). Hillier et al. (2003) examined Philadelphia's housing databases to track vacancy and abandonment data, but their systems are not interoperable, making comparative analysis practically impossible. Wilson and Margulis (1994) developed a similarly localized analysis in Cleveland. Runfola and Hankins (2010) conducted on-the-ground fieldwork in Atlanta to tally abandoned and derelict housing, but their method was time-intensive and thus only conducted in a limited number of census block groups.

Many remote-sensing and GIS-based studies have had some success measuring urban population change, examining areas experiencing growth throughout the duration of the studies (Weber and Puissant, 2003; Xiao et al., 2006; Yang and Lo, 2002). Examining urban shrinkage, Ryznar and Wagner (2001) attempted to study the effects of population decline but could only measure net change in forested and agricultural land, extrapolating their findings to housing and commercial land use changes. Banzhaf et al. (2007) explored shrinkage in Leipzig, Germany, but found that the necessary data to validate their findings was lacking.

While remote-sensing-based approaches suggest that urban change may one day be measured accurately by aerial and satellite imagery, the data and technical requirements can be out of reach for many planning departments, and widely replicable methodologies have not yet been developed. One possible solution is to reconsider some of the available data from the U.S. Decennial Census. Data from the census provides total counts of occupied housing units for neighborhood-level census tracts every 10 years. Each housing unit in the U.S. is classified as either occupied or vacant. If vacant, the Census Bureau devised several possible classifications to reflect different reasons including for sale, seasonal home, or a catchall category, "other vacant," which researchers use to indicate abandoned homes (Hollander, 2010).

The challenge of using this dataset to produce insights into events such as the economic downturn is the process of population movement unfolds on a continuous basis whereas the census is only collected every 10 years. Other census data sources, such as the American Community Survey, provide annual updates; however, the finest scale on which these housing occupancy data are released is the municipal level, making this option inadequate for a neighborhood-level analysis.

Methodological approach

Occupied housing units offer a clear and useful path for studying abandonment at a national scale. This study explores a relatively unstudied data source, the U.S. Postal Service Delivery Statistics, as an alternative to the more spatially and time constrictive datasets described in the previous section. Six days a week, every week of the year, USPS sends a postal worker to walk up and down nearly every street in America and collects data on total deliveries for each zip code. The USPS regularly releases datasets that provide information on occupied housing units for each U.S. zip code. When a postal worker determines that a housing unit has not been occupied for more than 90 days, the USPS removes the address from its active inventory, data which is aggregated to zip codes and made available on a monthly basis.

Basic tabulation

For this study, three USPS datasets were analyzed: February 2000, February 2006, and February 2011. Nearly all zip codes in the lower 48 states were included in the analysis. Two time intervals were selected for analysis: February 2000 to February 2006 and February 2006 to February 2011. The first interval corresponds with the housing boom in the decade's first half and the second interval corresponds with the approximate tipping point of the boom into the foreclosure crisis and recession.

Change for each time interval was calculated by subtracting total occupied housing at the beginning point of each interval from occupied housing at the interval's end. The later time intervals total households from the earlier time intervals (e.g., total households in Feb 2000 for zip code X subtracted from total households in Feb 2006 from that same zip code, produces a measurement of occupied housing change for the interval between 2000–2006, where a negative value implies shrinkage and a positive value implies population growth). Change in housing

Justin B. Hollander, et al.

occupancy was measured nationally (see Table 22.1) and for each of the four major census regions: Northeast (see Table 22.2), South (see Table 22.3), Midwest (see Table 22.4), and West (see Table 22.5). Additionally, the national and regional datasets were divided into three sub-regions based on "urban-ness" to test the severity in which more densely populated urban cores were impacted relative to less densely populated suburban and rural areas. Because the census categorizes regions as "Urbanized Areas" and "Metropolitan Statistical Areas" by population density, we used these boundaries as a rough approximation of different components of the metropolitan region for our study. Areas outside of Metropolitan Statistical Areas were identified as "rural and small towns."

Data mapping

Zip codes were also mapped nationally to show zip codes with net gains and declines for each interval (see Figures 22.1 and 22.2). Data from Figures 22.1 and 22.2 were combined into a third

Table 22.1 Summary statistics for national zip code district housing occupancy

		Urbanized Areas*	Metropolitan Statistical Areas**	Rural and Small Towns	Total
Interval 1:	**Total Count of Zip Codes**	6686	12071	9855	28612
2000–2006	**Count of ZDHOs***	2000	1361	2250	5611
	Percentage ZDHOs	29.9%	11.3%	22.8%	19.6%
Interval 2:	**Total Count of Zip Codes**	6243	11447	9186	26906
2006–2011	**Count of ZDHOs**	2239	1945	2425	6609
	Percentage ZDHOs	35.9%	17.0%	26.4%	24.6%
Change in ZDHOs		239	584	175	998
Adjusted Change in ZDHOs		256	616	188	1061
% Change in ZDHOs		12.8%	45.2%	8.3%	18.9%

* Within Metropolitan Statistical Areas
** Excluding zip codes located in Urbanized Areas
*** Zip codes with a net decline in housing occupancy

Table 22.2 Summary statistics for Northeastern zip code district housing occupancy

		Urbanized Areas	Metropolitan Statistical Areas	Rural and Small Towns	Total
Interval 1:	**Total Count of Zip Codes**	1918	1966	1298	5182
2000–2006	**Count of ZDHOs**	434	185	170	789
	Percentage ZDHOs	22.6%	9.4%	13.1	15.2%
Interval 2:	**Total Count of Zip Codes**	1859	1874	1221	4954
2006–2011	**Count of ZDHOs**	560	274	212	1046
	Percentage ZDHOs	30.1%	14.6%	17.4%	21.1%
Change in ZDHOs		126	89	42	257
Adjusted Change in ZDHOs		130	93	45	269
% Change in ZDHOs		30.0%	50.5%	26.3%	34.1%

Table 22.3 Summary statistics for Southern zip code district housing occupancy

		Urbanized Areas	Metropolitan Statistical Areas	Rural and Small Towns	Total
Interval 1: 2000–2006	Total Count of Zip Codes	2048	4788	3044	9880
	Count of ZDHOs	561	439	468	1468
	Percentage ZDHOs	27.4%	9.2%	15.4%	14.9%
Interval 2: 2006–2011	Total Count of Zip Codes	1889	4611	2863	9363
	Count of ZDHOs	621	579	553	1753
	Percentage ZDHOs	32.9%	12.6%	19.3%	18.7%
Change in ZDHOs		60	140	85	285
Adjusted Change in ZDHOs		65	145	90	301
% Change in ZDHOs		11.6%	33.1%	19.3%	20.5%

Table 22.4 Summary statistics for Midwestern zip code district housing occupancy

		Urbanized Areas	Metropolitan Statistical Areas	Rural and Small Towns	Total
Interval 1: 2000–2006	Total Count of Zip Codes	1504	3379	4169	9052
	Count of ZDHOs	660	480	1378	2518
	Percentage ZDHOs	43.9%	14.2%	33.1%	27.8%
Interval 2: 2006–2011	Total Count of Zip Codes	1395	3286	3941	8635
	Count of ZDHOs	679	766	1362	2807
	Percentage ZDHOs	48.7%	23.3%	34.6%	32.6%
Change in ZDHOs		19	286	-16	289
Adjusted Change in ZDHOs		20	294	-16	303
% Change in ZDHOs		31.0%	61.3%	-1.1%	12.0%

Table 22.5 Summary statistics for Western zip code district housing occupancy

		Urbanized Areas	Metropolitan Statistical Areas	Rural and Small Towns	Total
Interval 1: 2000–2006	Total Count of Zip Codes	1203	1925	1320	4448
	Count of ZDHOs	344	255	352	951
	Percentage ZDHOs	28.6%	13.2%	26.7%	21.4%
Interval 2: 2006–2011	Total Count of Zip Codes	1083	1681	1150	3914
	Count of ZDHOs	373	324	299	996
	Percentage ZDHOs	34.4%	19.3%	26.0%	25.4%
Change in ZDHOs		29	69	-53	45
Adjusted Change in ZDHOs		32	79	-46	51
% Change in ZDHOs		9.4%	31.0%	-13.1%	5.4%

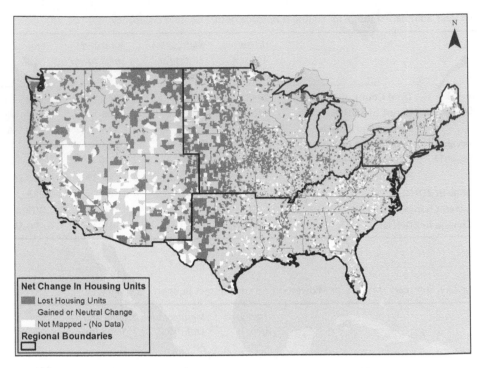

Figure 22.1   Change in occupied households 2000–2006

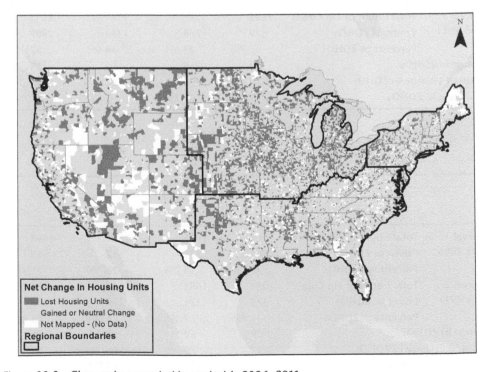

Figure 22.2   Change in occupied households 2006–2011

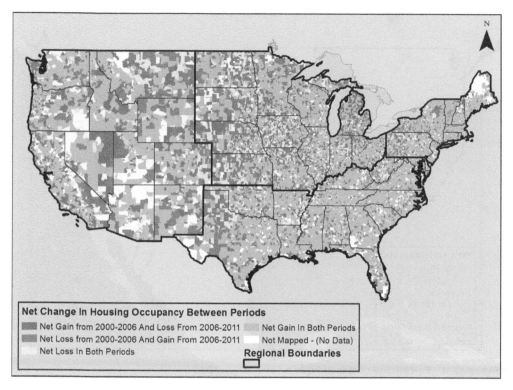

Figure 22.3 Occupied housing trends: comparing 2000–2006 with 2006–2011

national summary map (see Figure 22.3) showing trends over the full 11-year span by highlighting zip codes that declined in both intervals, Interval 1 only, and Interval 2 only.

To quantify severity of declines, maps were also created for all declining zip codes in each interval, showing what housing percentage was lost in these zip codes (see Figures 22.4 and 22.5). Declining zip codes from Interval 1 were divided into quartiles based on the percentage of the total housing stock lost during that interval. The quartile cutoff points for the first interval were applied to Interval 2 as well, to contrast how severe declines were for each interval.

Global and local indicators of spatial autocorrelation

Prior research has shown contagion effects for decline at the sub-neighborhood level (Harding et al., 2009). To test this, two univariate measures of spatial autocorrelation, Global Moran's I and a Local Indicator of Spatial Association (LISA), are used to explore spatial clustering of USPS Housing Unit Occupancy Change (cf. Anselin, 1995). These tests identify geographic clustering of zip codes showing statistically similar occupancy trends. Identifying adjacent declining clusters not only shows where "contagion effects" have been prevalent, but also elucidates "hot spots" of shrinkage (and growth) across the United States for each time interval.

The Global Moran's I analysis is employed to examine if household mail recipient losses and/ or gains occurred in neighboring zip codes. Where the global statistic is statistically significant, the LISA analysis maps the clusters by partitioning them into groups of zip codes exhibiting anomalously high or low values, relative to the national mean.

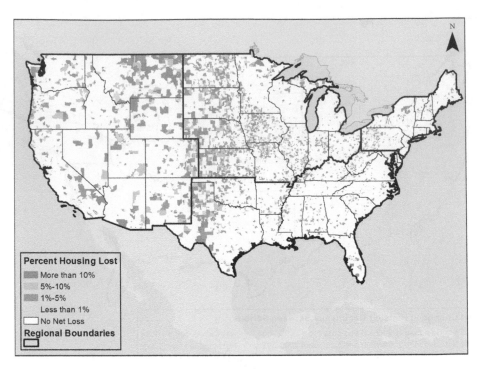

Figure 22.4 Percentage occupied housing lost 2000–2006

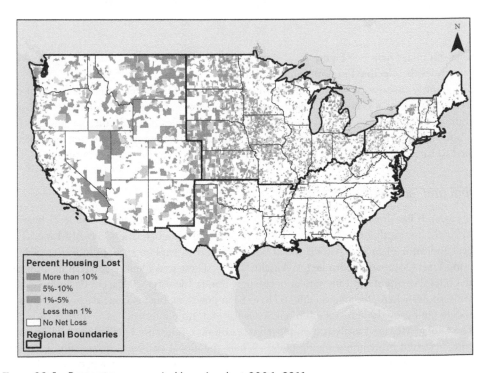

Figure 22.5 Percentage occupied housing lost 2006–2011

Each zip code in a cluster is labeled according to four possible categories in the LISA test:

1) High-High clustering – zip codes that experienced anomalously high growth surrounded by zip codes that also experienced anomalously high growth;
2) Low-Low clustering – zip codes that experienced anomalously high loss surrounded by zip codes that also experienced anomalously high loss;
3) Low-High clustering – declining zip codes surrounded by gaining zip codes;
4) High-Low clustering – gaining zip codes surrounded by declining zip codes.

Of these results, only low-low results, indicating homogeneous local clustering of zip codes anomalously below the national mean were mapped. Only zip codes that shared a border (the "Queen's" first-order contiguity matrix) were defined as being "neighbors."

Methodological considerations

Similar to remotely sensed and census data sources, using zip codes to measure urban shrinkage comes with a number of methodological concerns. Zip code boundaries occasionally change. In some cases, zip codes are eliminated; in other cases, two or more neighboring zip codes may be merged. In both cases, our calculations for such zip codes might indicate spurious losses or gains. Luckily, the Postal Service makes these changes available through their biweekly publication, *The Postal Bulletin*, enabling an analysis of how much these changes might impact analysis. Between 2000–2011, of the roughly 30,000 zip codes in the United States, roughly 130 zip codes on average changed boundaries per year. Any zip code that had a boundary change was omitted from the analysis.

Additional zip codes were omitted from the 2006–2011 interval in instances where the housing total went from a positive number in 2006 to zero occupied housing units in 2011. While it is conceptually possible that a zip code could lose its entire housing stock, our preliminary analysis identified 356 egregious cases where a zip code containing many housing units in 2006 – thousands in some cases – would be reduced to 0 in 2011.

In addition to these challenges, zip codes face many issues common to census, town, or otherwise zonal datasets. The number of occupied housing units within a zip code is a one-dimensional metric, and negative changes may be unrelated to neighborhood decline. For instance, a zip code could lose households if a new civic center is built and residences demolished or if multi-family homes are converted into single-family.

Findings

Absolute change in housing occupancy

This analysis of the USPS dataset revealed a number of trends that provide new and expanded insight into the spatial context in which we understand the Great Recession and subprime lending crisis' impact. Additionally, it provides a previously unavailable, imitable tool for periodic analysis of national and regional housing trends. As expected, more zip codes had a decline in occupied housing in the 2006–2011 interval than the 2000–2006 interval (see Table 22.1). Nationally, 998 more zip codes lost housing in the latter interval, resulting in a 19.6 percent increase in declining zip codes over the previous interval.

Each of the national sub-regions examined in this study experienced an increase in Zip Code District Housing Occupancy (ZDHOs) in the second interval. However, the magnitude of that change was not uniform across urban areas. In both intervals, Urbanized Areas had the highest percentage of ZDHOs, rural areas and small towns the second highest, and Metropolitan Statistical Areas the lowest. However, Metropolitan Statistical Areas had far and away the largest increase in ZDHOs in the second interval. The finding that less densely populated sections of metropolitan areas outside of the urban core showed the greatest percentage change in ZDHOs was consistent with findings limited to the subdivision of Urbanized Areas (Zinder, 2009).

All four of the regions examined experienced an increase in ZDHOs. The Midwest had the greatest overall percentage of ZDHOs in both intervals, followed by the West, Northeast, and South, respectively. However, the West and Midwest had the smallest percentage increase of ZDHOs between the two intervals. The Northeast had the greatest percentage increase in ZDHOs. The two largest subregion percentage changes in ZDHOs occurred within Metropolitan Statistical Areas in the Midwest and Northeast, respectively. The only two sub-regions that did not show an increase in ZDHOs were Midwestern and Western Rural and Small Towns. The second interval was not only notable for an overall increase in ZDHOs but also for an increased migration in the distribution of ZDHOs from urban centers into suburban and exurban areas.

Comparing spatial distribution of ZDHOs between intervals

Only 9.3 percent of zip codes nationally declined in both intervals (see Figure 22.3). The majority of the ZDHOs only declined in one of the intervals, with 11.1 percent of zip codes nationally only declining in the first interval and 15.2 percent of the zip codes only declining in the second interval. The Midwest had the most continuity between the two intervals, but each region showed volatility in new areas for the second interval. Thus, the second interval was not only marked by a general increase in ZDHOs, but by a widespread number of zip codes and regions whose housing occupancy reversed from growth to decline. Conversely, many previously declining regions, notably the High Plains and Pacific Northwest, began to see reversal patterns towards growth in the second interval.

Magnitude

While the second interval has been thus far characterized as having more widespread and new instances of decline, the first interval was better characterized as having the sharpest declines. Housing occupancy of 50 percent of all ZDHOs declined by more than 3.5 percent in the first interval whereas only 29.9 percent declined at that rate in the second interval. These trends were fairly consistent regionally for the first interval with the Midwest and South, as well as rural and small towns experiencing slightly disproportionate numbers of ZDHOs that lost over 3.5 percent of their housing stock. Regions with a high number of zip codes that declined only in the second interval tended to have a high percentage of ZDHOs with housing occupancy losses below 1.4 percent.

Global Moran's I and LISA

The results of the Moran's I global autocorrelation test indicated gaining and losing zip codes tended to cluster on the landscape. This was true in both time intervals, with Moran's I results 0.16 ($p < 0.05$) for the 2000–2006 interval and 0.24 ($p < 0.05$) for the 2006–2011 interval. This prompted us to perform a Local Indicator of Spatial Autocorrelation (LISA) to determine where this spatial clustering occurred. These results are seen in Figures 22.3 and 22.4, focusing only on statistically significant ($p < 0.05$) zip codes that had a net loss of occupied households.

Dark red areas show zip codes that experienced high losses of people receiving mail (relative to all zip codes) for a time interval, surrounded by other zip codes that also experienced high loss.

Centralized (clustered) housing loss occurred in both intervals; however, similar to ZDHO distribution, clustering patterns differed greatly between the two intervals. Clusters in the Great Plains, Mississippi and Missouri valleys, and along the Great Lakes stand out in the first interval. Clustering along the Great Lakes and in the Phoenix Metro area expanded in the second interval and popped up in Sun Belt metropolitan areas such as Las Vegas, Tampa, Miami, throughout California, and (for different reasons), New Orleans.

Conclusion

This research shows that the distribution of housing occupancy declines shifted between the 2000–2006 interval and the 2006–2011 interval. Many areas that previously exhibited perennial growth, particularly low-density suburban and exurban areas, and real estate-fueled markets in the Sun Belt, began to experience the initial stages of unoccupied housing challenges previously associated solely with centralized, densely populated urban areas. While housing occupancies have continued in many urban cores, particularly those within the Rust Belt and Great Plains, the rate and centralization of housing loss in these regions has tempered. Conversely, declines in lower-density areas across the United States and Sun Belt cities have not spiraled to the same extent that urban cores did in the late twentieth century. However, declines in these areas have become widespread.

Through the lens of housing occupancy, this chapter serves to initiate a first step in developing a methodological approach to identifying vulnerable neighborhoods and regions in a timely and accurate manner, and it further addresses a critical indicator of neighborhood health.

Understanding the scale and scope of housing unit declines throughout the United States contributes to the growing literature on smart shrinkage. The findings here lay a foundation for planners to begin to measure, and in turn manage, the physical changes that are occurring in their communities due to shrinkage. Additionally, the potential exists for USPS data to assist communities and RPOs in monitoring and analyzing shrinking areas within their boundaries.

Determining where these shrinkage hot spots are can aid planners and policymakers in developing tailor-made smart shrinkage strategies. For example, zip codes with high levels of shrinkage could be targeted for land banking, housing demolition, community-based agriculture, or expanded park systems. Additionally, such places might be receptive to a new form of zoning: relaxed zoning. Relaxed zoning addresses the fundamental economic problem in shrinking cities: excess supply of structures (housing, stores, schools) relative to demand (the number of people) (Hollander, 2011). Falling demand for housing will result in falling rents and house values (Hoyt, 1933; Temkin and Rohe, 1996). As prices fall, the ability to sell is impaired (especially if the mortgage exceeds the value of the property). As prices fall, the ability of landlords to recoup the costs of protecting and maintaining their properties in rents is likewise compromised (Keenan et al., 1999). A spiral of declining values, disinvestment, and deteriorating housing stock typically destroys stable neighborhoods and spreads dereliction (Bradbury et al., 1982).

This chapter examined, at the national scale, the patterns of urban growth and shrinkage from 2000 to 2011. Findings suggest that declining zip codes in the 2006 to 2011 interval were more dispersed and often in new territory, in comparison with 2000 to 2006. Further, more zip codes declined from 2006 to 2011 than the preceding interval. Future research could employ multivariate statistical techniques to explore the key attributes that cause places to decline and in such a clustered fashion. Focused studies on specific cases of intra-metropolitan shrinkage could provide policymakers with guidance on how shrinkage happens on a more local scale.

Justin B. Hollander, et al.

Guide to further reading

Hollander, J.B. (2011) *Sunburnt Cities: The Great Recession, Depopulation and Urban Planning in the American Sunbelt*. New York: Routledge.

References

Anselin, L. (1995) "Local Indicators of Spatial Autocorrelation: LISA." *Regional Research Institute Research Chapter*, 9331, pp. 1–26.

Armborst, T., D'Oca, D., and Theodore, G. (2005) "However Unspectacular." In: Oswalt, P. (ed.) *Shrinking Cities, Vol. 2: Interventions*. Ostfildern, Germany: Hatje Cantz Verlag.

Banzhaf, E., Kindler, A., and Haase, D. (2007) "Monitoring, Mapping and Modeling Urban Decline: A Multi-Scale Approach for Leipzig, Germany." *EARSeL eProceedings*, 2, pp. 101–114.

Beauregard, R.A. (2003) *Voices of Decline: The Postwar Fate of U.S. Cities*. 2nd ed. New York: Routledge.

Bowman, A.O'M. and Pagano, M.A. (2000) *Vacant Land in Cities: An Urban Re Source*. Washington, DC: Brookings Institution Press.

Bowman, A.O'M. and Pagano, M.A. (2004) *Terra Incognita: Vacant Land and Urban Strategies*. Washington, DC: Georgetown University Press.

Boyer, M.C. (1983) *Dreaming the Rational City*. Cambridge, MA: MIT Press.

Bradbury, K.L., Downs, A., and Small, K.A. (1982) *Urban Decline and the Future of American Cities*. Washington, DC: Brookings Institution Press.

City of Youngstown (2005) *Youngstown 2010 Citywide Plan*. Youngstown, OH.

Dash, E. (2011) "Banks Amass Glut of Homes, Chilling Sales." *The New York Times*, May 23.

Fairbanks, P. (2010) "Shrinking of Cities Catches Traction." *Buffalo News*, August 3.

Glaeser, E.L. and Gyourko, J. (2005) "Urban Decline and Durable Housing." *Journal of Political Economy*, 113(2), pp. 345–375.

Goodman, P.S. (2007) "This Is the Sound of a Bubble Bursting." *The New York Times*, December 23.

Goodnough, A. (2009) "Hard Times for New England's 3-Deckers." *The New York Times*, June 20.

Hall, P. (1997) "Modeling the Post-Industrial City." *Futures*, 29(4/5), pp. 311–322.

Harding, J.P., Rosenblatt, E., and Yao, V.W. (2009) "The Contagion Effect of Foreclosed Properties." *Journal of Urban Economics*, 66, pp. 164–178.

Hillier, A.E., Culhane, D.P., Smith, T.E., and Tomlin, D.C. (2003) "Predicting Housing Abandonment with the Philadelphia Neighborhood Information System." *J Urban Affairs*, 25(1), pp. 91–105.

Hollander, J.B. (2009) *Polluted and Dangerous: America's Worst Abandoned Properties and What Can Be Done about Them*. Burlington, VT: University of Vermont Press.

Hollander, J.B. (2010) "Moving towards a Shrinking Cities Metric: Analyzing Land Use Changes Associated with Depopulation in Flint, Michigan." *Cityscape: A Journal of Policy Development and Research*, 12(1), pp. 133–151.

Hollander, J.B. (2011) *Sunburnt Cities: The Great Recession, Depopulation and Urban Planning in the American Sunbelt*. New York: Routledge.

Hollander, J.B., Pallagst, K., Schwarz, T., and Popper, F. (2009) "Planning Shrinking Cities." *Progress in Planning*, 72, p. 3 (special issue: Emerging Research Areas).

Hoyt, H. (1933) *One Hundred Years of Land Values in Chicago*. Chicago, IL: University of Chicago Press.

Hoyt, L. and Leroux, A. (2007) *Voices from Forgotten Cities: Innovative Revitalization Coalitions in America's Older Small Cities*. Cambridge, MA: MIT Press.

Hughes, A.H. and Cook-Mack, R. (1999) "The City Needs a Fresh Approach to Dealing with Vacant Property." *The Philadelphia Inquirer*, May 23.

Immergluck, D. (2011) *Foreclosed: High-Risk Lending, Deregulation, and the Undermining of America's Mortgage Market*. Ithaca, NY: Cornell University Press.

Keenan, P., Lowe, S., et al. (1999) "Housing Abandonment in Inner Cities: The Politics of Low Demand for Housing." *Housing Studies*, 14(5), pp. 703–716.

Lang, R.E. and LeFurgy, J. (2007) *Boomburbs: The Rise of America's Accidental Cities*. Washington, DC: Brookings Institution Press.

Leland, J. (2007) "Officials Say They Are Falling Behind on Mortgage Fraud Cases." *The New York Times*, December 25.

Logan, J.R. and Molotch, H.L. (1987) *Urban Fortunes: The Political Economy of Place*. Berkeley, CA: University of California Press.

Lucy, W.H. and Phillips, D.L. (2000) *Confronting Suburban Decline: Strategic Planning for Metropolitan Renewal.* Washington, DC: Island Press.

Lucy, W.H. and Phillips, D.L. (2006) *Today's Cities, Tomorrow's Suburbs.* Chicago, IL: Planners Press.

Oswalt, P. (ed.) (2006) *Shrinking Cities, vol. 2: Interventions.* Ostfildern: Hatje Cantz Verlag.

Popper, D.E. and Popper, F.J. (2002) "Small Can Be Beautiful." *Planning,* July, pp. 20–21.

Runfola, D.M. and Hankins, K. (2010) "Urban Dereliction as Environmental Injustice." *ACME: An International E-Journal for Critical Geographies,* 9(3), p. 345.

Ryznar, R. and Wagner, T. (2001) "Using Remotely Sensed Imagery to Detect Urban Change." *Journal of the American Planning Association,* 67, pp. 327–336.

Schatz, L. (2010) *What Helps or Hinders the Adoption of "Good Planning" Principles in Shrinking Cities? A Comparison of Recent Planning Exercises in Sudbury, Ontario and Youngstown, Ohio,* PhD diss. Waterloo: University of Waterloo.

Schilling, J. and Logan, J. (2008) "Greening the Rust Belt." *Journal of the American Planning Association,* 74(4), pp. 451–466.

Schumpeter, J. (1934) *The Theory of Economic Development.* Cambridge, MA: Harvard University Press.

Temkin, K. and Rohe, W.M. (1996) "Neighborhood Change and Urban Policy." *Journal of Planning Education and Research,* 15(3), pp. 159–170.

Wallace, R. (1989) "Homelessness, Contagious Destruction of Housing and Municipal Service Cuts in New York City: 1. Demographics of a Housing Deficit." *Environment and Planning A,* 21, pp. 1285–1603.

Weber, C. and Puissant, A. (2003) "Urbanization Pressure and Modeling of Urban Growth: Example of the Tunis Metropolitan Area." *Remote Sensing of Environment,* 86(3), pp. 341–352. Available at: http://linkinghub.elsevier.com/retrieve/pii/S0034425703000774.

Wilson, D. and Margulis, H. (1994) "Spatial Aspects of Housing Abandonment in the 1990s: The Cleveland Experience." *Housing Studies,* 9(4), pp. 493–511.

Wilson, J.Q. and Kelling, G.L. (1982) "Broken Windows: The Police and Neighborhood Safety." *The Atlantic,* March.

Xiao, J., et al. (2006) "Evaluating Urban Expansion and Land Use Change in Shijiazhuang, China, by Using GIS and Remote Sensing." *Landscape and Urban Planning,* 75(1–2), pp. 69–80. Available at: http://linkinghub.elsevier.com/retrieve/pii/S0169204605000058.

Yang, X. and Lo, C. (2002) "Using Satellite Imagery to Detect Land Use Land Cover Changes in Atlanta, Georgia Metropolitan Area." *International Journal of Remote Sensing,* 23, pp. 1775–1798.

Zinder, D.H. (2009) *Through the Rings: A Study of Housing Occupancy Declines across Major Urbanized Areas in the United States,* Unpublished Master's Thesis. Department of Urban and Environmental Planning and Policy. Medford, MA: Tufts University.

23

Redesigning the suburbs

New town and master-planned suburbs of the 1960s and 1970s

Lisa Benton-Short

The planning context for master-planned communities

Urban historians have written about the historical development of suburbs in the United States. Kenneth Jackson's (1985) *Crabgrass Frontier* chronicles U.S. suburbanization from the early enclaves of the nineteenth century up to the more contemporary 1980s. In their early history, suburbs were "bourgeois utopias" available only to society's elite (Fishman, 1987). Examples of well-to-do suburbs include Riverside in Illinois, Llewellyn Park in New Jersey, Roland Park in Baltimore, and Beacon Hill in Boston. Later early suburbs were designed for an emerging middle class; others were working-class, self-built suburbs. Even relatively early in the history of suburbs there was diversity of form and function. As Jackson (1985, p. 2) notes, "American suburbs come in every type, shape, and size: rich and poor, industrial and residential, new and old."

Suburbanization in the United States grew dramatically after World War II. Housing and housing development underwent a transformation; the demand for housing and the expansion of the nation's interstate and highway systems facilitated the suburbs' growth. By the 1950s, suburbs became more widespread and accessible for working- and middle-class families. Cheap land, cheap gasoline, and reforms in mortgage practice spurred suburban development.

Post-war suburbs had several distinct characteristics. First, while some were cookie-cutter subdivisions, many suburbs were a mix of housing styles and neighborhood types, built piece-meal and added to over time. Second, many were "one-off" developments that had little relation to other nearby subdivisions. Third, many were built with no controls, no provisions for open space, and no mandate to develop or preserve community character. Fourth, most suburbs were primarily single-family residences and had little commercial or industrial activities. This was because land developers and home builders built housing, while retail developers built shopping areas; neither group typically had enough land to even consider planning an entire town (Ward, 2006). This changed in the 1960s, however, with the emergence of "planned communities," or the "new town" movement.

Although "planned communities" became a new form of suburban development in the 1960s, they have their roots in the historical "Garden City" movement of Ebenezer Howard and Patrick Geddes. Howard's Garden City model envisioned an alternative to London's rapid, uncontrolled growth. His vision was of a central city inside concentric rings of housing and factories, surrounded by greenbelts or open spaces designed for community gardens and other services. The design of Radburn, New Jersey, embodied this model, and included green space, a complete separation of pedestrians and automobiles, and curving streets, in contrast to the traditional urban grid pattern. The Garden City movement was realized in the 1920s and 1930s with the building of towns such as Greenbelt, Maryland, Radburn, New Jersey, and Park Forest, Illinois, but they never became the dominant form of suburban design.

Levittowns in Long Island and Pennsylvania were also precedents for master-planned communities. Built starting in 1947, Levittowns are regarded as the archetype for American post-war suburbs. The houses were not prefabricated, but they were among the first built with mass production techniques (Schmitz, 1998). The company purchased many building items in bulk directly from manufacturers: lumber, windows, appliances, and other materials. Building hundreds of houses at the same time allowed them to scale up housing construction at a time when demand was booming. Initially, Levittown houses sold for $8,000 each, and came with fully-stocked kitchens and a television in the living room. The houses were designed in a network of sidewalk-lined curving streets linked to arterial "parkways." The communities included some common public amenities such as retail stores, schools, swimming pools, and playgrounds, but these were generally scattered around the developments. Once sold, Levitt & Sons released these developments to various towns and municipal boundaries, resulting in administrative challenges regarding infrastructure such as sewage and water systems.

Most of the 1950s building-boom suburbs did not appear to be orderly, thoughtful new towns but a sprawl of cookie-cutter houses surrounded by highways, leapfrogged by other suburbs (Bookout, 1992). They were critiqued for ugliness, cultural conformity, social isolation, and environmental problems (Forsyth, 2005).

In response to the chaotic suburban development of the 1950s, in the 1960s the "new town movement" was an effort to "reform suburbia" (Forsyth, 2005, pp. 2–5). It incorporated the Garden City ideas of green space and the Levittown model of mass production of housing. New towns of the 1960s and 1970s designed communities on an even larger scale and introduced comprehensive master-planning. For some, new town living represented an escape from the dying central cities and the unplanned agglomerations and suburbs; for others, it represented a bold and innovative way of living. These new towns would redefine the nature of suburbanization, place-making, and real estate from 1960 to the 1990s.

New towns and master-planned communities of the 1960s

Unlike previous suburban developments, new towns or master-planned communities occur on very large tracts of land beyond the urban fringe, often more than several thousand acres. Developers created whole communities rather than simple subdivisions, avoiding many of the problems of uncoordinated, incremental growth (Forsyth, 2005). Communities built by private large-scale builders and developers used production techniques that standardized single-family housing. As a result, many master-planned new towns are characterized by widespread tract-style housing.

Master-planned communities are typically a product of long-term, multi-phase development programs that combine a mix of land uses. Multi-phase development means that plans are implemented in logical stages over time, in some cases over a 20-year period. Unlike previous suburban developments, their large size and scale, and the fact that these are long-term projects, mean developers take a significant financial risk in building an entire community according to the set of aesthetic, social, and economic designs they initially adopt. This can be particularly risky

Lisa Benton-Short

Table 23.1 Planning concepts of new towns and master-planned communities

- Large-scale
- Comprehensively planned
- Planned Unit Development
- Phased development and building
- Built in greenfields
- Housing grouped into clusters, allowing for more open space
- Located on the edges of metropolitan areas
- Designed as largely self-sustaining
- Some mixing of income levels
- Communitywide zoning, homeowner's associations and Covenants, Conditions ,& Restrictions (CC&Rs)
- Attention to the street front
- Curved streets/loops/cul-de-sac (curvilinear)
- Wide streets
- Open spaces (greenbelts, golf courses, hiking trails, etc.)
- Deep setbacks for houses
- Single-use neighborhoods that included non-residential land uses such as shopping and employment centers
- Attempt to balance residential, commercial, and industrial activities
- Enclosed malls
- Gathering places/public spaces that invite community socialization, and can host a diverse range of festivals and events
- Deliberate efforts to create a strong sense of place and community

because many master-planned communities feature a unifying character or design element that may endure or become dated quickly. In theory, a master-planned development has flexibility to respond to market conditions through a phased development strategy. Some master-planned communities followed incremental planning; others stuck to the master plan.

New towns were based on comprehensive plans that attempted to consider every element in community design. They were based on the concept of self-contained villages in which residents could, ideally, live, play, and work. While the physical form is suburban in character, what was distinctive about them is the emphasis on green space and social planning.

Many new towns included a broad range of social, economic, and physical activities within a defined area of land (Campos, 1976). Social activities included schools, health services, recreational facilities, civic organizations, and religious institutions. Economic activities included industrial parks and retail and commercial centers. Physical activities included the provision of roads, utilities, and housing. Conceptually, these three activities are cohesive and coherent, and developers of master-planned communities believed they created a strong sense of community. Table 23.1 highlights the planning concepts for the master-planned communities of the 1960s and 1970s.

Planned Unit Development

Although the exact origin is not known, the Planned Unit Development (PUD) concept was first used by the new town movement in response to criticism of the 1950s suburban boom. Up until this time, the dominant land use approach was based on individually owned lots set apart from other uses (such as apartments, commercial, etc.); this is how many post-war suburbs developed. The rise of the master-planned community movement rejected the more rigid and standard zoning practices, and sought a more creative and flexible alternative to land use control that allowed the developer to mix land uses within a larger area. Rather than subdividing a parcel into individual lots and building on each separately, the development is planned and approved as one contiguous parcel (Purcell, 1986).

A PUD is a type of zoning classification that contains both residential and non-residential buildings and provides more flexibility in the design process. Generally, a PUD is used for a new subdivision on vacant land and allows a large area to be planned all at once so land uses complement each other. A PUD allows a developer to design and group a variety of land uses such as housing, recreation, commercial centers, and industrial parks all within one contained development or subdivision. This allows hospitals, shopping centers, convenience stores, restaurants, professional offices, and even industrial uses to be permitted within the neighborhoods and villages. Developers argued that this type of zoning fosters a more comprehensive community environment.

An advantage of a PUD is that it can protect open spaces by allowing slightly higher residential density than standard "grid" zoning. For example, a developer can propose the PUD to be clustered so that the individual lots are smaller but more open space is preserved. Open space, recreational areas, and other amenities can be bargained for during the negotiation, which is a key part of the review and approval process (Purcell, 1986).

The PUD concept can also be applied to encourage creative mixes of land uses, by permitting certain non-residential uses (or a mix of different kinds of residences, such as single and multi-family) in the development (McMaster, 1994). For master-planned communities with a long-range development plan, a PUD project might take longer to generate profits, but the cost of development can be lower because the developer can cluster building units.

A PUD developer or company will create a PUD that is operated by a homeowner's association (HOA), to which every homeowner contributes monthly or quarterly fees based on their percentage ownership in the overall project, which operates the PUD on a day-to-day basis after completion. Like a condo project, those HOA fees cover amenities (such as parks and recreational areas) and maintaining services. While the new town and master-planned communities of the 1960s and 1970s pioneered PUD's use for large-scale developments, today PUDs have evolved into all shapes and sizes, and are tailored to a variety of land use developments.

Early prototypes of master-planned suburbs

Although there are numerous new towns across the U.S., the Washington DC metro area, New York, Texas, and California were among the earliest to develop them. Early leaders in the new town movement include Reston in Virginia, Columbia in Maryland, and Mission Viejo and Irvine in California. Many developers won design awards for these communities. Table 23.2 highlights these early master-planned suburbs.

Reston, Virginia is located about 20 miles west of Washington, DC near Dulles International Airport. Reston was first conceived as a planned community in 1962 by architect, economist, sociologist, and planner Robert E. Simon, who articulated seven goals that served as the blueprint for the community's development (see Table 23.3).

Reston is a site of over 7,000 acres (compare that to the 640 acres of Radburn, New Jersey's Garden City). The Garden City model certainly influenced Reston's planning (Robert Simon's father had been involved in the development of Greenbelt, Maryland). Although Reston's overall density was comparable to conventional suburban zoning, the concentration of residential

Table 23.2 New towns and master-planned communities in the 1960s and 1970s

Westlake, California (1963)	Reston, Virginia (1962)
Valencia, California (1965)	Columbia, Maryland (1966)
Irvine, California (1970)	St. Charles, Maryland (1965)
Mission Viejo, California (1965)	Audubon New Community, New York (1973)
Laguna Niguel, California (1971)	Riverton, New York (1973)
Rancho Bernardo, California (1962)	Woodlands, Texas (1971)

Table 23.3 Robert Simon's seven goals for Reston

1	Provide a wide range of recreational and cultural facilities
2	Make it possible for anyone to remain in a single neighborhood throughout his/her life by providing a full range of housing styles and prices
3	The focal point of planning is the importance and dignity of each individual
4	That people be able to live and work in the same community
5	That commercial, cultural, and recreational facilities be available at the outset of the development – not years later
6	That beauty is a necessity of the good life
7	That Reston be a financial success

development in higher densities in selected areas preserved significant areas of open landscape (Ward, 2006). The Reston plan, considered a pioneering approach, inspired widespread enthusiasm and has won more than two-dozen awards in real estate development, planning, urban design, and architecture. Today, Reston is home to over 60,000 residents, has five village centers, a Metro station, and a wide array of commercial, cultural, and recreational amenities. It is the largest planned community in Virginia.

Columbia, Maryland was initiated in 1966 by mortgage banker and shopping center developer Jim Rouse, who said his planning goals were "to create a social and physical environment which works for people and nourishes human growth and to allow private venture capital to make a profit in land development and sale" (Bailey, 1971, p. 16). Columbia spanned a total of 17,000 acres, with 2,600 acres set aside for industrial use. The plan called for the phased development of seven villages, each of which would have a population between 10,000 and 15,000; each village would consist of three to four neighborhoods. At the center of each village was a shopping area featuring a grocery store, dry cleaners, and other retail. The social focal point of Columbia is the downtown, which is anchored by the Columbia Mall, an indoor mall.

Planned communities in southern California benefited from the Mediterranean climate, a coastal location, and the California sun. In particular, Orange County evolved from citrus groves into suburbs and became home to the largest concentration of planned communities, although the majority of these developments were middle- and upper-class residential. Two new towns in Orange County, Irvine and Mission Viejo, are noteworthy.

One of the oft-cited master-planned new towns is Irvine, California, a former ranch of 110,000 acres. In 1878, James Irvine established the Irvine Ranch; in the early twentieth century his son created the Irvine Company. By the late 1950s, the Irvine Company and the ranch began to diversify away from agriculture; the Irvine Company sold parcels of land to the U.S. government for two Marine Corps Air Stations. In 1960, the Irvine Company sold the University of

California 1,000 acres for a new "Irvine" campus for a dollar. And consistent with new town principles, the Irvine family bequeathed the new town some 60,000 acres of wild lands and parks designated as permanent open space throughout Orange County. Planning for a new town started in 1960 and would develop with the university as its nucleus. By 1970, the plan called for the development of 50,000 people, and houses and neighborhoods replaced citrus trees, as Irvine became California's fastest-growing city.

Unlike some of the other nearby new towns such as Mission Viejo, the master plan for Irvine balanced residential areas with industrial parks and proposed the development of major industrial areas, including the Irvine Industrial Complex near the airport. Newport Center was envisioned as "a downtown for the south coast of Orange County" (Liebeck, 1990). The 700-acre site was to contain high-rise office buildings, medical centers, service businesses, and a regional shopping center. Over the next several decades, numerous aerospace companies, electronics firms, and research and development industries were located in the Irvine Industrial Complex. Many experts say that Irvine has been so successful because of this attention to jobs and industrial parks.

Like Reston and Columbia, Irvine was planned as a "city of villages." The first phases of the villages of Turtle Rock, University Park, Westpark (then called Culverdale), El Camino Real, and Walnut were completed by 1970. Each was initially planned to have a distinct architectural theme that would create a distinct sense of place (designs included Spanish, Tuscan, California Modern design, etc.). Unlike many nearby master-planned communities, Irvine had a checkerboard pattern of residential and commercial/industrial park areas.

The planned communities of Reston, Columbia, and Irvine shared, along with others, an emerging articulation of design and community-building principles. However, it is one thing to design a new town; it is another to sell it. I now turn to the master-planned community of Mission Viejo, California, as an exemplar of a strategically marketed and sold community, and the place where I grew up from 1969 to 1986.

Mission Viejo, California: marketing and selling the new town

The making, marketing, and selling of Mission Viejo is implicitly – if not explicitly – the intersection of geography and community. In this part of the essay, I link the development of Mission Viejo's master plan with the new town principles; I then critically analyze the marketing and selling of Mission Viejo by looking at the deliberate efforts to do so, including slogans and advertisements, and wider recognition through Rose Parade float entries. At the time, these methods were innovative; later they became industry standards.

The Mission Viejo master plan: embodying new town principles

In 1963, the area that would become Mission Viejo was part of the Rancho Mission Viejo, a sprawling area of walnut and citrus groves, eucalyptus windbreaks, olive trees, California oaks, tomato farms, sheep, and cattle pastureland. The ranch dates back to the King of Spain and was at that time owned by the O'Neill family.

In the 1950s and 1960s, as many as 50,000 newcomers moved into Orange County each year. The completion of Interstate 5 linked northern Orange County to San Diego, and the completion of Interstate 405 linked Orange County's coastline to San Diego. The I-5 freeway was the main catalyst that would facilitate the development of southern Orange County and the O'Neill Ranch. Anthony Moiso, great-grandson of the original rancho owner, Richard O'Neill, noted,

"All of a sudden we found ourselves in a world that we didn't even know existed – a world with land values going up as developers came in to buy the properties" (Walker, 2005). Ranchos in Orange County, including Rancho Mission Viejo, Irvine Ranch, and Moulton Ranch, recognized that residential development might be inevitable. The O'Neill family commissioned an engineering study that concluded urbanization would continue to spread south throughout the county. At the time, the O'Neill Ranch comprised 52,000 acres. The family decided to sell 11,000 acres for the first foray into nonagricultural development.

In 1963, a group of unknown, would-be developers approached the O'Neill family and offered $7,000 for each of the 11,000 acres adjacent to the I-5 freeway. They also offered the family 20 percent of the corporation and day-to-day involvement. Wanting to remain involved with development, the family also felt that the buyers would preserve the ranch's heritage. The group of three, along with the O'Neills, formed the Mission Viejo Company.

The first rule of land development is to sell land for more than you bought it. The second rule is to develop raw land intelligently and to merchandise what it has to offer in terms of market needs. The founders of the Mission Viejo Company did both. They conducted a meticulous, two-year, long-range planning process. They conducted feasibility studies, market analyses, planning research, and engineering surveys that informed their priorities and emerging new town principles: to provide well-designed homes in a variety of price ranges; to develop a balanced community of homes, schools, churches, shopping, and recreation; and to foster a strong sense of community identity as quickly as possible (Breton, n.d.a.). They drew up a master plan with goals and objectives, creating one of the earliest master-planned communities in the country.

The master plan divided the 11,000 acres into: 5,400 acres for residential use, 3,000 acres for parks, recreation, and open space, 700 acres for school and churches, and 900 acres for business properties. The plan involved controlled growth by implementing development in phases. The Mission Viejo Company (MVC) got the ranch owners to agree to split the land into parcels; the MVC would then purchase on an as-needed basis. This shrewdly allowed them to generate the capital they needed from the previous development to fund the next one.

What is distinctive about Mission Viejo, compared to many master-planned communities, is that the Mission Viejo Company never deviated from the 1965 plan. As Mission Viejo Company spokeswoman Wendy Wetzel Harder noted, "To our knowledge, Mission Viejo is the only planned community in the nation that was developed according to the original master plan, under the original company, and according to the original guidelines" (Walker, 2005, p. 50). Within a year, where only Indians and cowboys, cattle, and horses had trod for centuries, the first 90 homes sold within three hours. For the next several years, prospective buyers would stand in line for days.

The Mission Viejo Company emphasized residential development while other Orange County developers at the time, such as Irvine, as noted previously, focused more on attracting businesses. The MVC attempted to create a completely self-contained community; however, by focusing predominately on residential housing and ignoring commercial and industrial interests, Mission Viejo became a primarily residential town. This followed the MVC's mission statement to create

> pleasant neighborhoods with adjacent parks for each and schools and churches nearby where family could recreation, worship and work toward mutual values in their lives in a beautiful rural setting where the price of housing was amenable to those to whom it appealed.
>
> *(Walker, 2005, p. 48)*

From the beginning, Mission Viejo's new town emphasized creating neighborhoods of reasonably priced single-family homes. In the 1960s, many of the single-family detached homes sold in

Mission Viejo ranged in price from \$34,000 to \$43,000, which at the time was well below the Orange County average price of \$60,000. Breaking with development convention, the Mission Viejo Company built more affordable housing first, then moved to expensive housing.

Like many new towns, Mission Viejo was conceived as a network of neighborhoods in which each would have amenities close to home. They planned for schools to be within walking distance of the residences and for each neighborhood to have a small center with groceries, dry cleaners, restaurants, and other amenities. Over time, the Mission Viejo Company would build two golf courses, four recreation centers, 40 improved parks, an equestrian center, a bowling alley, off-road bicycle paths, hiking trails and the 124-acre Lake Mission Viejo. Philip Reilly, then President of the MVC, noted, "Our plans included from the very beginning, all the ingredients that people need for a healthy, thriving town. Our projects, even in the early stages, were more than just houses. They included schools, churches, shops, parks, and business" (Walker, 2005, p. 47).

One measure of success was how rapidly new phases of development sold; in the early years, people would camp for several days outside the sales trailers to purchase their lot of choice. At one point in time, housing tracts sold out before construction even began on them. Another indication of success was that a major corporation, the Philip Morris Company, bought the Mission Viejo Company in 1970; this gave the MVC the financial expertise, security, and necessary capital to expand.

Creating community and sense of place

The development of Mission Viejo included several innovative suburban design elements (see Table 24.4).

A priority of the Mission Viejo Company was to develop community identity and distinctiveness while also paying homage to the ranch's legacy. A distinctive feature became a unifying Spanish theme by linking Early California motif and design elements throughout the community. Spanish Heritage Design included:

- Custom street lights replicating the El Camino Real bells along the CA mission trail – their first use in the United States as street lights
- The flavor of old Spanish mission and rancho days retained in architecture and street names (romantic tones of the Spanish)
- "Barcelona walls" or monument markers at neighborhood entrances
- Split-rail fencing borrowed from the cattle ranch

Table 23.4 Innovative design elements in new towns

- Divided parkways on major arterial streets
- Commercial signage strictly controlled by size, height, and structure
- All utility lines underground
- Tracts that were self-contained neighborhoods with associations, greenbelts, and CC&Rs
- Recreation centers for each cluster of neighborhoods
- Special emphasis on sports and recreation
- Encouragement of community participation
- Creation of hometown atmosphere with community celebrations and events
- Roads that were contoured in curves over the rolling hills
- Landscaping of the median strips and hillsides

Source: Walker, D. (2005)

Figure 23.1 Official seal of Mission Viejo, California

Source: City of Mission Viejo, Courtesy of Mission Viejo Library

Two Spanish elements became the community emblems. The first is the "Rafter M" brand, used for decades by Rancho Mission Viejo as the working cattle brand. The second was the Mission Bell street light, inspired by the old El Camino Real bells along the mission trail. These became a community trademark and today adorn the city's seal (Figure 23.1).

The Mission Viejo Company even hired a person to carefully choose street names, because "new homeowners should acquire positive address images along with their residence. The streets should have positive meaning as well as sound nice" (Walker, 2005, p. 61). Early on most of the streets had Spanish names such as Cervantes, El Crego, Segovia, Coronado, and Cortez; Acapulco, Veracruz, and Mazatlán were an homage to Latin America. Major arterial parkways were named for members of the O'Neill family, including Marguerite (the wife of Richard O'Neill, Jr.) and their two children (Alicia and Jerome). Ironically, in proper Spanish, the town name should have been Mission Vieja, but like many of the California rancho names, it came down to us by way of the U.S. Land Commission of the 1850s, which garbled some of the old Spanish names and created a number of new ones (Brigandi, 2006).

In the first phases, Spanish architectural design features included Spanish-tiled roofs, tiled courtyards, fountains, balconies, thick stucco walls, textured exterior finishes, an earth-tone slump stone custom-designed for the Mission Viejo Company, wrought-iron gates and railings, and Spanish archways and verandas (Figure 23.2). Landscaping included olive, palm, and jacaranda trees, and lots of bougainvillea and lantana.

Mission Viejo's "front door" is the La Paz entry from the San Diego Freeway; La Paz refers to one of the old ranchos, which had an original name meaning "peace." Oso Parkway reminds residents of the grizzly bears along nearby Trabuco Creek, named for a lost Spanish musket. Beyond street names, the Mission Viejo Company devised names for neighborhoods. These included: Alicante, Andalucía, Barcelona, Catalonia, Cadiz, Castile, Cordova, Finisterra, Madrid, Seville, and Valencia. Within the neighborhoods, the entry streets were named for Spanish regions and

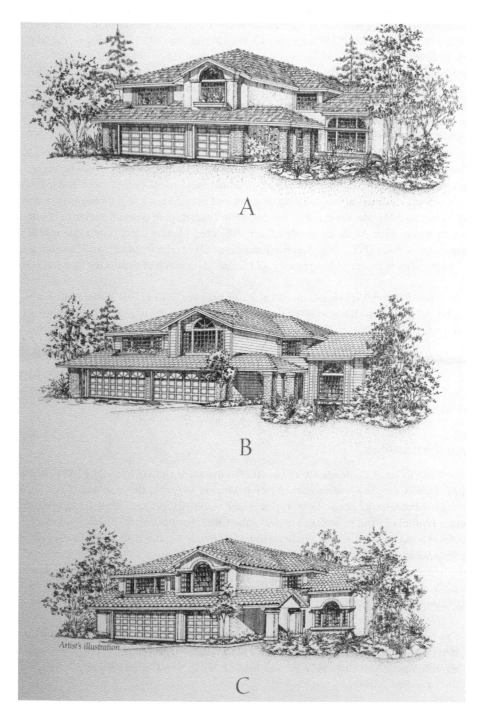

Figure 23.2 Architectural drawing of homes for sale in Mission Viejo, California
Source: Lisa Benton-Short

radiating streets were given names of the cities within that region (Walker, 20065). This proved to be an important way for residents to identify themselves. It was not uncommon to hear neighbors introduce themselves with, "We live in the La Paz Homes," or "We're moving to the Madrids" (Benton, 2016). The MVC also conceptualized what they called "Barcelona walls," a Spanish architectural touch marking the entrances into the neighborhoods. This particular innovation quickly became an industry standard around Southern California.

Establishing a sense of community was an important aspect of place-building and marketing. In the early days, the MVC established the Mission Viejo Activities Committee to organize special events that would provide a sense of hometown spirit and community pride. These included an annual Easter Egg Hunt, Five Nights of Christmas (with a live nativity scene), the Cinco De Mayo Celebration, and in September a five-day celebration called the Mission Viejo Days, a week-long event for families to participate in athletic and social events, including log-sawing and tug of war, watermelon seed-spitting contests, and greased pig and jumping frog contests.

MVC invented Mission Viejo and its "traditions" in an attempt to establish a sense of community, continuity, and stability. As the geographer Karen Till noted, identity is not only a question of time, but a question of space (Till, 1993). And while there is much to criticize in terms of "inventing traditions," these efforts at design and community sold houses. Thousands of them. One new resident said,

> We think Mission Viejo is the nicest rural area available. It's not just a bunch of houses, like other developments. It not only has the recreation center and golf course, but also many parks and schools. It's just so well planned. You can really see someone is trying to do a good job.
>
> *(City of Mission Viejo, 2016)*

Local area writer Doris Walker said, "Finally, the great ranchos, those last outposts of the past, have felt the subdivider's knife. . . . If there must be subdividing and developing – and we all plead guilty to moving here, then I say: this is the kind it should be!" (Walker, 2005).

Selling the new town

It is one thing to build a community and another to market it and sell it, and the MVC was adept at all of them. Company leaders wanted to put Mission Viejo on the map and they conceived early on of strategies to create local, regional, and nationwide name-recognition. As suburbs became a symbol of "the American Dream" – the belief that no matter their origins, Americans can achieve social eminence and a richer standard of living – more than anyplace else, Southern California became that symbol of post-war suburban culture.

In 1972, billboards along I-5 and I-405 advertised, "The California Promise: Live it in Mission Viejo" (see Figure 23.3). Harvey Stern, then executive vice-president of the Mission Viejo Company, explained,

> You expect, when you come to California, that you're going to have a certain kind of lifestyle and a certain physical beauty. It's not available in every place, but it is available in Mission Viejo. And that's why we call Mission Viejo 'the California Promise.' It's kind of like a promise fulfilled.
>
> *(Breton, n.d.b.)*

Beyond the sheer audacity of the claim, the campaign became so successful that several decades later when the City Council was searching for a city motto, it chose the slogan originated by the Mission Viejo Company: "The California Promise."

Figure 23.3   A billboard marketing Mission Viejo as 'the California Promise'
Source: Lisa Benton-Short

In addition to "The California Promise," there were other slogans used by the Mission Viejo Company (see Table 23.5; Figure 23.4). The slogan "You have arrived: Mission Ridge" was designed to appeal to the family ready to move up in status. The Mission Ridge neighborhood was a collection of quasi-custom homes set in a circle high on a hill. For each of the advertising slogans, the MVC sent residents dozens of blank postcards featuring views of Mission Viejo to mail to friends living elsewhere, essentially creating thousands of "community boosters."

MVC also sought to elevate Mission Viejo's reputation and invested in a range of novel efforts that included hosting the Virginia Slims tennis tournament and the original "Battle of

Table 23.5 Advertising Slogans in New Towns

- *Mission Viejo: Where Home is a Home Town Again*
- *Live the California Promise*
- *It's So Nice to Have Mission Viejo Around the House*
- *Mission Viejo: Cradle of the Good Life in South Orange County*
- *It Happens in Mission Viejo*
- *A World of Difference, Because It's Not a World Apart*
- *You have arrived: Mission Ridge*

Source: Breton, D. (n.d.)

Figure 23.4 Marketing for Mission Viejo, California

Source: Lisa Benton-Short

the Network Stars" competition. However, as much as the MVC effectively used slogans and advertisements, nothing did more to broaden the community's image than entering floats in the annual Pasadena Tournament of Roses Parade. The MVC spent as much as $500,000 dollars to design, build, and promote these floats, which featured a theme connected to the community, a not-so-subtle form of advertising (Benton, 2017). Each year from 1977 to 1986 hundreds of community volunteers assembled the floral pieces for the floats. The first float's theme, "A Dream Come True" (1977), depicted the columns and circular walls typical of Mission Viejo's street intersections. At the edge of the float were an elderly man and woman, and a young child. At their feet was, "Mission Viejo California," with the city's logo left of the words. Subsequent float themes included "Day of Fiesta" (1978), and "Water babies" (1979), which saluted the then-famous Olympic swimmers who trained with the Mission Viejo Nadadores Swim Club. The marketing efforts paid off. By 1998, Mission Viejo had been completely developed according to the master plan and a group of residents led the movement for cityhood.

Critiquing the new town

Many scholars have written extensively critiquing the suburbs for being socially alienating, geographically isolated, and environmentally harmful. They have criticized design for nostalgic romanticism; they have even noted that many new towns and master-planned communities became places of conformity, exclusivity, materialism, and homogeneity.

Geographer Karen Till explored many of these criticisms in her research on Rancho Santa Margarita, a development south of Mission Viejo. Till notes that using early Californian Spanish colonial architecture to provide a sense of identity for the planned community is problematic. These images are more manufactured than authentic, the so-called "traditional design guidelines" recent inventions, and the "histories" that accompany them fabricated (Till, 1993, pp. 719–720).

Despite initial planning objectives, many new towns became exclusive and unaffordable. During the early years of development, developers favored the middle- and upper-income sectors of the market to improve the revenue flow and offset some of the front-end costs of providing services and infrastructure. The challenge can be, however, that negative inertia will build up; for the residents who purchased during the early phase realized their homes would not appreciation rapidly if lower-cost homes were built nearby. In some of these new communities, residents resist low-and middle-incoming housing after the first wave of high-middle and middle-middle income families have made their purchases.

Although Mission Viejo touted affordable housing in 1969, today the California statewide median-priced home is $496,620; Orange County set a new record of $675,000 in 2016; and the median-priced home in Mission Viejo is $690,000. My family's old four-bedroom home sold recently for more than a million dollars; my parents originally bought it for $270,000 in 1978. A recent study by ApartmentList.com noted it could take an average millennial 20 years to save up for a 20 percent down payment on a house or a condominium in Orange County, leading one commentator to state that California is not experiencing a housing crisis, but a housing catastrophe (Smith, 2017).

Conclusion

In the 1960s and 1970s new towns and master-planned communities redefined the nature of suburbanization, place-making, and real estate. However, these new towns have a mixed record. They succeeded in creating community identity and preserving open spaces, but they often failed at housing affordability and transportation choices.

The legacy of master-planned communities continues to be a significant feature of the suburban landscape particularly in the U.S. For example, many of the innovations in the new town movement have become industry standard practices today – slogans, advertising, community building and deliberate efforts to design community character – are found in many recent developments.

The principles and characteristics of the new town movement also influenced the neo-traditional town planning movement, new urbanism, and smart growth movements that emerged more forcefully in the 1990s. In many ways, the master-planned communities were among the first to implement smart growth practices on a large-scale, and many addressed issues of sustainability far before this concept entered the wider lexicon.

Today's popular planning trends run counter to some of the design elements of the new towns. New urbanism has reintroduced high-density and walkability, a counter to the meandering curving streets, cul-de-sacs, and automobile dependency of the new towns.

The master-planned communities of the 1960s and 1970s left a legacy that remains an important part of the story of suburbs. Indeed, recently there has been an explosion of scholarly research on the new towns and master-planned communities in Australia, which should remind us that it is best to see "suburbs" as a dynamic urban form, designed, redesigned, and redesigned again.

Guide to further reading

Forsyth, A. (2005) *Reforming Suburbia: The Planned Communities of Irvine, Columbia and the Woodlands*. Berkeley, CA: University of California Press.

Jackson, K. (1985) *Crabgrass Frontier: The Suburbanization of the United States*. New York: Oxford University Press.

Schmitz, A. and Bookout, L. (1998) *Trends and Innovations in Master-Planned Communities*. Washington, DC: Urban Land Institute.

Till, K. (1993) "Neotraditional Towns and Urban Villages: The Cultural Production of a Geography of "Otherness." *Environment and Planning D: Society ad Space*, 11, pp. 709–723.

Walker, D. (2005) *Mission Viejo: The Ageless Land from Prehistory to Present*. Mission Viejo, CA: The City of Mission Viejo.

References

Bailey, J. (ed.) (1971) *New Towns in America: The Design and Development Process*. New York: John Wiley and Sons, Inc.

Benton, B. (2016) personal interview. Bonnie Benton is my mother, and a real estate agent and sales and marketing strategist; she worked for the Mission Viejo Company, the Irvine Company and many other developers in Orange County throughout her career.

Benton, B. (2017) personal interview. Bonnie Benton noted: $500,000 sounds like a lot, but advertising is expensive; Raoul Rodrigues (the famous Rose float designer) is paid $200,000 or thereabouts just for his award-winning design work. And then the cost of the flowers, transporting them, fabricating the infrastructure, etc.

Bookout, L. (1992) "Neotraditional Town Planning: A New Vision for the Suburbs?" *Urban Land*, 51(1), pp. 20–26. Washington, DC: Urban Land Institute.

Breton, B. (n.d.a) "Community Spirit: Creating That Home-Town Feeling." *Mission Viejo Library*. Available at: http://cityofmissionviejo.org/community-spirit-creating-home-town-feeling [Accessed 22 May 2017].

Breton, B. (n.d.b) "Keepers of the Promise: Mission Viejo Company." *Mission Viejo Library*. Available at: http://cityofmissionviejo.org/keepers-promise-mission-viejo-company [Accessed 22 May 2017].

Breton, D. (n.d.) "Community Distinction: Marketing the Mission Viejo Miracle." *Mission Viejo Library*. Available at: http://cityofmissionviejo.org/community-distinction-marketing-mission-viejo-miracle [Accessed 1 March 2016].

Brigandi, P. (2006) *Orange County Place Names A to Z*. San Diego: Sunbelt Publications.

Campos, C. (1976) *New Towns: Another Way to Live*. Reston, VA: Prentice Hall.

City of Mission Viejo (2016). Personal interview.

Fishman, R. (1987) *Bourgeois Utopias: The Rise and Fall of Suburbia*. New York: Basic Books.

Forsyth, A. (2005) *Reforming Suburbia: The Planned Communities of Irvine, Columbia and the Woodlands*. Berkeley, CA: University of California Press.

Jackson, K. (1985) *Crabgrass Frontier: The Suburbanization of the United States*. New York: Oxford University Press.

Liebeck, J. (1990) *Irvine: A History of Innovation and Growth*. Houston, TX: Pioneer Publications. Available at: http://irvinehistory.org/irvine-ranch-gives-way-to-urban-development/ [Accessed 1 August 2017].

McMaster, M. (1994) "Planned Unit Developments." *Planning Commissioners Journal*, 15, p. 14. Available at: http://plannersweb.com/1994/07/planned-unit-developments/ [Accessed 25 June 2017].

Purcell, A. (1986) *An Analysis of the Planned Unit Development Concept and Its Application in Rhode Island*, Master's thesis. University Rhode Island. Available at: http://digitalcommons.uri.edu/cgi/viewcontent.cgi?article=1692&context=theses [Accessed 30 June 2017].

Schmitz, A. (1998) "New Visions in Community Design." In: Schmitz, A. and Bookout, L. *Trends and Innovations in Master-Planned Communities*. Washington, DC: Urban Land Institute, pp. 83–98.

Smith, E. (2017) "Housing Crisis? What California Has Is a Housing Catastrophe!" *Sacramento Bee*, June 2. Available at: www.sacbee.com/opinion/opn-columns-blogs/erika-d-smith/article153899264.html#emlnl=Todays_Top_Stories [Accessed 20 July 2017].

Till, K. (1993) "Neotraditional Towns and Urban Villages: The Cultural Production of a Geography of 'Otherness'." *Environment and Planning D: Society and Space*, 11, pp. 709–723.

Walker, D. (2005) *Mission Viejo: The Ageless Land from Prehistory to Present*. Mission Viejo, CA: The City of Mission Viejo.

Ward, A. (2006) "Certainty to Flexibility: Planning and Design History, 1963–2005." In: Ward, A. (ed.) *Reston Town Center: A Downtown for the 21st Century*. Washington, DC: Academy Press, p. 33.

24

Keeping up with the Joneses

Residential reinvestment and mansionization in American inner-ring suburbs

Suzanne Lanyi Charles

Introduction

"A house may be large or small, but as long as the surrounding houses are equally small, it satisfies all social requirements of a dwelling place. But let a palace arise by the side of this small house, and it shrinks from a house into a hut."

(Marx, 1891, p. 35)

American suburban single-family houses have grown substantially larger over time. The median floor area of new single-family homes completed in the United States reached an all-time high of 2,500 square feet in 2015, nearly 1,000 square feet larger than in the 1970s. New large single-family houses, pejoratively termed "McMansions," "monster homes," or "starter castles," form what Knox (2005, p. 42) terms "vulgaria," a suburban landscape characterized by "inert and pretentious neighborhoods that are irradiated by bigness and spectacle." In a process known as "mansionization," tracts of new, exceedingly large single-family houses are built on previously undeveloped land at the edges of metropolitan areas. Mansionization also takes place within older inner-ring suburbs as infill development upon previously developed single-family residential parcels. This conspicuous form of incremental suburban housing reinvestment occurs through the complete demolition of the original houses and the rebuilding of new and much larger houses in their place. These houses are known by regionally varied terms such as "teardowns" or "knock-down-rebuilds" (KDR). The process is common in many suburban neighborhoods across North America and around the world (Pinnegar et al., 2010, 2015; Wiesel et al., 2013; Randolph and Freestone, 2012; Charles, 2013, 2014, 2017).

The basic mechanics of teardown mansionization are seemingly straightforward. When the difference between the economic return from a property in its original state and the potential return if a property were developed to its most profitable use becomes large enough for one to purchase the property, pay to redevelop it, and then sell it for an acceptable profit, redevelopment will occur (Smith, 1979). While Smith defined this process in response to neighborhood change in cities, a "rent gap" and its implications for reinvestment can now be seen in inner-ring suburban areas as well. Houses built in the post-war era on subdivided suburban greenfield land reflected the tastes and expectations of suburban homebuyers of that time, but over time as household preferences change, they may no longer represent the most profitable use of that property. Reinvestment in the original house through remodeling may increase the house's value, but property owners and developers may take the more comprehensive tactic of demolishing the original house and building an entirely new one, often one that is the maximum size legally allowed.

Reactions to teardown mansionization by neighborhood residents range from indifference to organized resistance. Residents react most strenuously to the physical changes, arguing that the new houses loom over their neighborhoods, diminishing light and privacy. They complain of noise and inconveniences during construction, and they lament the loss of older houses and those houses' role in the neighborhood's collective character. And homeowners harbor fears that the physical changes will decrease the value of their own properties. Residents' vocal concerns have led some municipalities to refine their zoning codes to control the size of new houses. Suburban residents and municipal leaders do not generally cite gentrification as a primary concern; however, researchers question the social sustainability of teardown mansionization and warn of the potential exclusionary displacement associated with it (Charles, 2017; Hanlon, 2015). Despite these concerns, in many places teardown mansionization occurs unabated. It replaces older, sometimes deteriorating housing with new houses that help suburbs modernize their housing stock and adapt to changing household preferences, potentially attracting new, younger families. Municipal officials, who appreciate the increased competitiveness and property tax revenue that larger new single-family housing may bring, often welcome this change.

This chapter examines suburban teardown mansionization through a study of the process in three inner-ring suburbs of Chicago. It demonstrates how the process occurs in suburbs that vary socio-economically and physically. I focus on Winnetka, La Grange, and Norridge, Illinois, three similarly-sized suburbs with high rates of teardown mansionization but significant differences with respect to socio-economic characteristics. The chapter reveals how the physical manifestations of teardown redevelopment vary, and draws upon theories of conspicuous consumption and positional goods as a framework for understanding the teardown mansionization process.

Conspicuous consumption

Cohen (2004, p. 6) describes post-war American suburbs as a "landscape of mass consumption." They provide the physical and social setting for the consumption of houses and the objects to fill them. Veblen (1899) identifies a consumer-oriented society's need to outwardly express success through what he termed "conspicuous consumption" – expenditures on goods and services that are motivated by the public display of wealth and status. Conspicuously consumed goods signal one's position in a consumer society (Schor, 1998). Hirsch (1976, p. 2) identifies these as "positional goods," which he defined as those for which "the satisfaction that individuals derive from goods and services depends in increasing measure not only on their own consumption but on consumption by others as well."

Visibly consumed goods are more positional than less visible goods; housing is thus considered to be a highly positional good. The evaluation of a positional good is based largely on its context and how it compares relative to that consumed by others (Frank, 1985b). As Frank (2007b) writes, given the choice, most people would prefer a 3,000-square-foot house in a neighborhood of 2,000-square-foot houses to a 4,000-square-foot house in a neighborhood of 6,000-square-foot houses. The latter option results in greater absolute consumption; the former results in greater relative consumption. Research suggests that happiness is affected by relative, rather than absolute, income levels (Easterlin, 1995), but consumption of positional goods is a zero sum game. The increase in status that comes with possession of a positional good occurs at the expense of others, whose status declines in relative terms (Frank, 1985a; Easterlin, 1995; Marx, 1891).

Greater spending on positional goods by affluent households affects the less affluent group slightly below them, which affects the less affluent group below them, and so on, setting off an "expenditure cascade" (Frank, 2007b). Consumers compare themselves to people above them on the socioeconomic ladder, and aspire to emulate and join them through the consumption of positional goods (Frank, 2007b). Expenditure cascades occur not because middle-class households want to emulate the highly affluent but because households imitate those nearest to them. Frank (2007b) attributes the steady increase in single-family housing's size over time to this expenditure cascade. Rising income inequality has led to less affluent households spending beyond their means to keep up with the consumption patterns of those with greater wealth; as top income levels rise, less affluent households allocate a higher percentage of their household budget to more visible goods (Bertrand and Morse, 2016).

In general, households overspend on positional goods, such as housing, and underspend on non-positional goods, such as leisure activities or savings (Frank, 1999). This leads to negative positional externalities at the individual and societal level. Increases in the consumption of positional goods are associated with more bankruptcies, a higher divorce rate, and a lower savings rate (Bertrand and Morse, 2016; Frank et al., 2014). Household overconsumption of positional goods collectively results in a "positional arms race" (Frank, 2007a). A prisoner's dilemma arises. One person working longer hours to afford a larger, more expensive house creates a situation in which others must also work longer hours just to maintain their positions (Frank, 2007a).

Teardown mansionization

Teardown mansionization is common but not ubiquitous. It is highly spatially clustered and acts like a contagion – where one teardown occurs, others are likely to follow (Charles, 2014). Spatial clusters of teardowns occur in a variety of neighborhoods; high rates of redevelopment occur in very affluent places as well as in places with more modest household incomes and house values (Charles, 2014). In their study of Australian suburbs, Pinnegar et al. (2010) find that although the teardown process appears to be more prevalent in affluent neighborhoods, teardowns also occur in less wealthy places. Particularly in suburbs with lower property values, owner-occupiers who reinvest their capital in the neighborhoods where they already live undertake teardown redevelopment (Pinnegar et al., 2010, 2015). While this type of incumbent upgrading occurs in the United States as well, developer-driven redevelopment – which occurs when private entities buy properties, redevelop them, and then sell them for a profit – is also common (Charles, 2013). Hanlon (2015) and Charles (2017) contend that redevelopment may threaten the social sustainability of inner-ring suburban neighborhoods through exclusionary displacement that can lead to gentrification.

Residential reinvestment in the inner-ring suburbs of Chicago

This chapter focuses on residential reinvestment through teardown mansionization in the Villages of Winnetka, Norridge, and La Grange, three inner-ring suburbs of Chicago, Illinois, using quantitative exploratory data analysis methods and qualitative semi-structured interviews with local municipal officials. I draw upon a dataset of all single-family teardown redevelopments in suburban Cook County from January 1, 2000 through December 31, 2015. In all, 7,405 single-family houses were demolished and replaced in the inner-ring suburbs of Chicago from 2000 through 2015. They were located in 84 different municipalities and unincorporated areas of Cook County. The three suburbs were selected because they are among the suburbs with the overall highest rates of teardown mansionization and they are similar in size, but they differ considerably in the socioeconomic characteristics of the households that live there and in the physical characteristics of the original housing stock.

Winnetka, La Grange, and Norridge each have a teardown mansionization rate – the percentage of single-family residential parcels upon which a house was entirely demolished and then rebuilt – in the highest quintile of all suburban Cook County municipalities. Rates of teardown mansionization vary from 4.7 percent to 5.6 percent in Norridge and La Grange, respectively, to 12.5 percent in Winnetka, which is the highest rate of redevelopment in the county. From 2000 through 2015, 206 houses were demolished and replaced in Norridge, 230 in La Grange, and 463 in Winnetka. Each municipality is located within a statistically significant spatial cluster of teardown redevelopment (Charles, 2014).

The three suburbs are differentiated by geographic and socioeconomic factors. Winnetka is located on the North Shore, north of Chicago on Lake Michigan; Norridge is located northwest of the Chicago Loop, directly bordering the city boundary; and La Grange is located immediately east of a group of very affluent western suburbs straddling Cook and DuPage Counties, as illustrated in Figure 24.1. Although their locations within the metropolitan area vary, each of the three suburbs is roughly equidistant from the Chicago Loop. Driving distances are within 15 to 20 miles, and each is served by commuter rail with direct service to the Chicago central business district (CBD). Winnetka – one of the most affluent municipalities in the state – is representative of other affluent North Shore suburbs, which have incomes greater than U.S.$100,000, according to the 2000 U.S. Census. The most publicized teardown redevelopments take place in such suburbs, yet they account for only 24 percent of the suburbs in the highest quintile of teardown rates. Norridge is representative of less wealthy and occupationally diverse suburbs in which households earn a median income of less than U.S.$60,000. This type of suburb makes up 33 percent of the highest quintile. La Grange, with a median household income of U.S.$80,000, is representative of the remaining 43 percent of the highest quintile, with household incomes in the upper-middle-class range.

Two distinguishing characteristics between the three suburbs are household incomes and house values. Winnetka has the highest household incomes and house values of all three suburbs. The 1999 median household income, measured immediately prior to the period studied in this chapter, was U.S.$167,000, and the median house value in 2000 was U.S.$737,000; both are amongst the highest in the state. Norridge has the lowest household incomes and house values of the three, with a median household income of U.S.$48,000 and a median house value of U.S.$198,000, both of which are similar to the median for suburban Cook County but less than 30 percent of the median for Winnetka. La Grange falls in the middle, with a median household income of U.S.$80,000 and a median house value of U.S.$256,000. Thus, suburban municipalities with the highest rates of teardown redevelopment include those that are the wealthiest and have the most expensive housing, as well as more modest middle-class places.

Other characteristics that set the three suburbs apart from each other are the occupations, races, and ethnicities of their residents. Ninety-three percent of Winnetka residents are employed in white-collar occupations, in contrast to 58 percent of Norridge's residents; the remaining 42 percent work in service, construction, production, and transportation occupations. La Grange is in the middle, with 84 percent of residents employed in white-collar jobs. In terms of race and ethnicity, none of these suburbs are particularly diverse. All three are at least 91 percent white. La Grange and Norridge have relatively high percentages of Hispanic residents compared to Winnetka, 3.7 percent and 3.8 percent, respectively, versus 1.3 percent in Winnetka. La Grange has the highest percentage of black residents, 6 percent compared to less than 1 percent in Winnetka and Norridge; still, this percentage is very low compared to the overall Chicago metropolitan area. Norridge, unlike the other two suburbs, has a high percentage of foreign-born residents and residents who speak a language other than English at home.

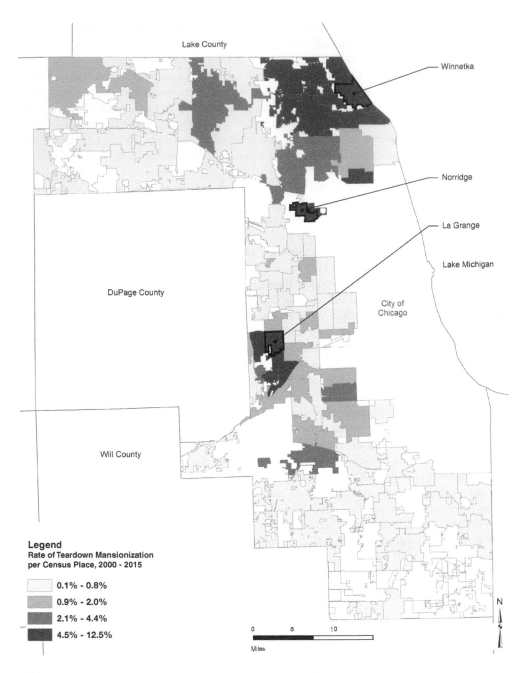

Figure 24.1   Teardown mansionization in metropolitan Chicago

Source: Suzanne Lanyi Charles

During the first two years of the 2000 through 2015 study period, relatively few teardowns took place in suburban Cook County. The number increased rapidly to a high point in 2005, then rapidly fell to a low point in 2010 before experiencing a modest resurgence in the years that followed. Teardown mansionization in Norridge parallels that in suburban Cook County overall,

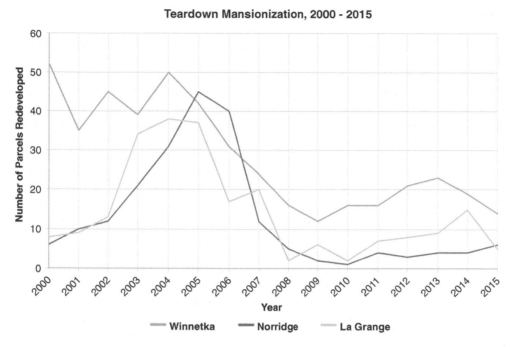

Figure 24.2 Teardown mansionization, 2000–2015

Source: Suzanne Lanyi Charles

as shown in Figure 24.2. After two initial years of few teardowns, the number per year increased rapidly, reaching a high of 45 properties redeveloped in 2005. From that peak, the number of teardown redevelopments dropped rapidly to a low of only one property redeveloped in 2010. It remained low, i.e., under ten annually, through the end of 2015. The temporal pattern in La Grange is similar to that of Norridge, although the apogee is lower and occurs one year earlier. The pattern of redevelopment in Winnetka is markedly different. Teardown redevelopment occurred in large numbers at the beginning of the decade and had been occurring prior to the study period, in the late 1990s (M. Donofrio, personal communication, January 8, 2010). The number of redevelopments per year in Winnetka varied between 35 and 52 from 2000 through 2004. From that point, the number of redevelopments fell to a low point in 2009. The number of redevelopments per year in Winnetka has not been fewer than 12 annually from 2000 through 2015.

Although the overall number and the rates of redevelopment in highly affluent Winnetka are higher than those in Norridge, at the peak of the housing boom in 2005 there were more teardown redevelopments in Norridge. Moreover, although La Grange is more affluent than Norridge, the number of houses redeveloped there annually remained fewer than in Norridge throughout the housing boom's peak years. From the low point in 2009 onward, redevelopment has made a modest resurgence in Winnetka and La Grange, but it remains low in Norridge.

The physical manifestations of teardown mansionization

The teardown process varies considerably in the three suburbs studied in terms of the sizes of the houses demolished and the sizes of the new houses. Overall, the houses demolished were significantly smaller in Norridge and La Grange. In fact, although La Grange is an upper-middle-class suburb much like Winnetka, the physicality of teardown redevelopment there is similar to that in working-class Norridge. The median area of the demolished houses in Norridge and La Grange is 1,110 square feet and 1,250 square feet, respectively, while the median area of the demolished houses in Winnetka is considerably larger at 2,160 square feet. The median floor area of the new houses in all of suburban Cook County is 3,540 square feet, ranging from 3,340 square feet in Norridge to 3,600 square feet in La Grange and 3,970 square feet in Winnetka. Although the median size of the new houses is roughly similar for all three suburbs, one difference is the range of the new houses' sizes. In Winnetka the largest new house is an extraordinarily large 17,900 square feet, and six redeveloped properties are larger than 10,000 square feet. The largest rebuilt properties in Norridge and La Grange are similar in size, at 6,630 and 6,620 square feet, respectively.

The ratio of the new house's square footage to the original house's is the smallest in Winnetka, where, on average, new houses are 2.1 times larger than demolished houses. In Norridge and La Grange, redeveloped houses are 3.1 and 3 times larger than the original houses, respectively. Although the largest teardown redevelopments, measured by the floor area, are in Winnetka, the highest teardown ratios – the ratio of the floor area of the new houses to that of the original houses – are in La Grange and Norridge. While new houses in the highest teardown ratio quintile in Winnetka are from 2.5 to 5.2 times larger than the original houses, new houses in the highest teardown ratio quintile in Norridge are from 3.9 to 6.2 times greater than the original houses. The variation is in part due to the original house's square footage, the parcel's size and configuration, and the local zoning regulations. In Winnetka, the median size of demolished houses is larger than in Norridge and La Grange, and parcel size, parcel configuration, and zoning restrictions limit the size of the new houses. Conversely, in Norridge, smaller original houses combined with relatively large parcel sizes – and in some cases more permissive zoning regulations – permit relatively larger new houses.

Each of the three suburbs has high teardown rates, and each is within one of the ten statistically significant spatial clusters of teardown mansionization (Charles, 2014); however, rates of redevelopment are not uniform within the municipalities. When we examine the rates of redevelopment on a smaller scale, a more finely grained pattern appears. The overall teardown redevelopment rate within the Village of Winnetka is 12.5 percent, but at the census block group level, rates of redevelopment range from 6.9 percent to 19.1 percent. In La Grange rates range from less than 1 percent to 9.2 percent per block group. And in Norridge teardown mansionization is even more uneven. Rates of teardowns per block group range from less than 1 percent to 15.0 percent. Thus, there are areas of Norridge and La Grange with very little teardown redevelopment, and other areas with rates greater than in areas of Winnetka. In the following sections, I describe the three very different physical manifestations of teardown mansionization in Winnetka, La Grange, and Norridge.

Winnetka

The new houses in highly affluent Winnetka are larger than the houses they replace; however, due to lot size and configuration, and local zoning regulations, in many cases the difference is not particularly apparent when the houses are viewed from the street. Figure 24.3 provides an example of two new houses, center and right, juxtaposed against an original, 77-year-old house on the left. All three houses, as well as the others on the block, are two-story structures. The new houses have considerably more complex roof forms and front façades. However, the overall massing – the shape, form, and size – of the new houses is roughly similar to other houses on the street. In this example, details such as the front window mimic historical details on the original house next-door. The redeveloped houses are built to the maximum floor area and height

Figure 24.3   Winnetka neighborhood in suburban Chicago

Source: Suzanne Lanyi Charles

allowed under the local zoning code, but their side and front yard setbacks are somewhat similar to the original houses. The floor area of the house in the center of Figure 24.3 is 3,660 square feet. The floor-area-ratio (FAR) of the house is 0.33; the FAR of the original house on the left is 0.31. Thus, in this example, the new house encompasses only slightly more of the parcel than its original neighbor does; however, the house pictured is significantly larger – 1.6 times larger – than the house it replaced. The 2017 estimated market value of the new house in the center of Figure 24.3, according to the Cook County Assessor's Office (CCAO), is 1.2 times greater than that of the original house on the left. The neighborhood is served by a network of alleys that run parallel to the residential blocks, allowing vehicular access to the rear of the lot from which the detached garages are accessed. Thus, the garage is not a prominent feature of the streetscape; garages are not visible, and there are few curb cuts to interrupt the street's pedestrian experience.

La Grange

In the Village of La Grange, the original houses are generally a mix of one-, one-and-a-half-, and two-story structures. New houses, such as the one pictured on the right in Figure 24.4, are taller and have a substantially larger footprint than the houses they replaced. The new house in this example is built closer to the property lines than the original house, maximizing the footprint allowable under the zoning code. The new house has a roof form that is roughly similar to its neighbor's, i.e., it has a similar pitch and dormers, but the house is considerably larger in scale.

Figure 24.4   Village of La Grange neighborhood in suburban Chicago

Source: Suzanne Lanyi Charles

The floor area of the house pictured on the right of Figure 24.4 is 3,080 square feet. The FAR of the house is 0.47; that of the original house on its left is 0.22. The new house, which is 1.8 times larger than its predecessor, is more conspicuous than the example in Winnetka, as observed from the street. The 2017 estimated market value of the new house on the right of Figure 24.4, according to the CCAO, is 2.3 times greater than that of the original house on the left. As in the neighborhood in Winnetka, the lots on this street are 50 feet wide, but unlike Winnetka's original houses, these were originally designed with detached garages at the rear of the lot accessed by a driveway from the street. The new house has a two-car garage integral to the house, with a wider driveway and a more prominent garage than the original houses.

Norridge

The physical manifestations of the teardown process in middle-class Norridge stand in stark contrast to those in more affluent Winnetka and La Grange. In some areas within the village, large and wide parcels, as well as the applicable zoning regulations, allow for new houses that are considerably larger in bulk and floor area than those they replace. The floor area of the house pictured in the center of Figure 24.5 is 5,120 square feet, which is 4.1 times larger than the house it replaced. The FAR of the new house pictured is 0.50; the FAR of its neighbor on the left, representative of the original houses in the neighborhood, is 0.17. Thus, not only is the new house significantly larger than the original houses, but it also encompasses a significantly larger share of

Figure 24.5   Norridge neighborhood in suburban Chicago
Source: Suzanne Lanyi Charles

its lot. Moreover, the 2017 estimated market value of the new house in the center of Figure 24.5, according to the CCAO, is 3.5 times greater than its next-door neighbor. The house is built closer to the side property lines, extends deeper into the backyard, and is substantially taller than the original houses. In this neighborhood most of the original houses are one-story structures, and the new houses typically have two stories, with a raised first floor and a full basement.

The original houses in Norridge were designed with a one-car detached garage at the rear of the lot, accessed via a driveway from the street; in contrast, the new house in Figure 24.5 has a three-car garage, which is part of the overall house volume. The garage is accessed directly from the street and features a circular driveway in front, which makes the garage and driveway much more prominent elements of the streetscape than those of the original houses. The new houses are designed with multiple complex roof forms, highly irregular asymmetrical street façades, multiple window shapes, and oversized front entries. The front of the house is where the greatest expense is focused; the less visible sides are constructed with cheaper materials and have fewer windows and embellishments. Teardown mansionization of this general style is not specific to Norridge; it appears in middle- and upper-income neighborhoods throughout the inner-ring suburbs.

Conclusion

Teardown mansionization is a particularly conspicuous form of housing consumption. Some of the characteristic architectural elements of teardown reinvestment, particularly those most visible to passersby, accentuate the house's bigness and spectacle, revealing its important role in the owners' outward expression of wealth and status. Teardown mansionization takes place in a wide variety of neighborhoods and occurs at equally high rates in highly affluent and middle-income suburban neighborhoods. Regardless of neighborhood socioeconomic status, new infill houses are often built to the maximum square footage that the parcel size and configuration and local zoning codes will allow. In some places, teardown mansionization results in a particularly acute contrast between the new house and its neighbors, leading to great absolute consumption but also great relative consumption; the new houses are objectively large as well as large in comparison to their neighbors. The positional nature of these conspicuously consumed houses explains, in part, their proliferation across the suburban landscape.

Demand for teardown mansions may be attributed to the relative benefits of owning one, rather than the absolute. Being able to afford a new larger house in a neighborhood of smaller original houses signals greater relative wealth. This may be a particularly important driver of teardown mansionization in middle-class neighborhoods. While households consider myriad reasons for choosing a neighborhood – social ties, proximity to work, public school quality, etc. – positionality may also enter the decision process, perhaps implicitly. A large new house in a middle-class suburb of small post-war single-family houses offers greater relative consumption than a similarly priced house in a more affluent suburb. The increase in relative wealth is made at the expense of the other residents of the neighborhood, which may contribute to the long-term residents' expressed disdain for them. Additionally, the relative benefits may be short-lived. As the teardown mansionization process continues and more houses are incrementally demolished and rebuilt as larger, more expensive houses, the relative wealth signaled by owning such a house decreases.

Teardown mansionization is highly spatially clustered across the suburban landscape. The contagion aspect of teardown mansionization may be explained in part by its positional nature. The consumption of positional goods, such as teardown mansionization, is highly dependent on neighborhood context. When a house is demolished and replaced with a much larger house, the neighborhood context changes incrementally, gradually creating a new normal. Subsequent housing consumption decisions are made based upon this new context, and the cycle continues. A house that once appeared large relative to its neighbors then appears less so. And this, in turn, may lead to the further proliferation of teardown mansionization.

The relative physical change in the housing stock – measured by the ratio between the new house and the one it replaced – varies considerably across the suburban landscape, yet the sizes of newly redeveloped houses are roughly similar. The median size of new houses in the suburbs studied varies from 3,340 square feet in middle-class Norridge to 3,970 square feet in very wealthy Winnetka. This can, in part, be explained by rising income inequality and expenditure cascades. When highly affluent households spend on positional goods, it affects the less affluent group slightly below them, which in turn affects the less affluent group below them, leading less affluent households to allocate a larger proportion of their household budgets to housing.

Teardown mansionization appeared first in the most affluent places well before 2000. During the housing boom, permissive mortgage lending practices made larger, more expensive housing financially feasible to a larger group of middle-class households. Thus, it became possible for these households to emulate the housing consumption of more affluent households, perhaps through increased debt, and teardown mansionization began to occur with great frequency in less affluent neighborhoods. The lending restrictions put in place after the housing crash corresponded with a reduction of teardown mansionization in all suburbs, but while teardown mansionization has since restarted in the most affluent places, it remains less frequent in many middle-class areas.

Despite being often maligned on aesthetic grounds, teardown mansionization is nevertheless common in some American inner-ring suburbs. Further research is necessary to understand the individual households' motivations and housing preferences; however, the theory of conspicuous consumption partially explains teardown mansionization as a public display of wealth and status consistent with that of positional goods. The neighborhood context of housing consumption decisions is important, as it leads to clusters of teardown mansionization and affects future housing consumption choices as residents attempt to "keep up with the Joneses."

Guide to further reading

Charles, S.L. (2017) "Teardowns and Reinvestment in the Inner-Ring Suburbs of Chicago." In: Squires, G., Heurkens, E. and Peiser, R.B. (eds.) *Routledge Companion to Real Estate Development*. New York: Routledge.

Frank, R.H. (1985) *Choosing the Right Pond: Human Behavior and the Quest for Status*. New York: Oxford University Press.

Frank, R.H. (1999) *Luxury Fever: Weighing the Cost of Excess*. Princeton, NJ: Princeton University Press.

References

Bertrand, M. and Morse, A. (2016) "Trickle-Down Consumption." *Review of Economics and Statistics*, 98(5), pp. 863–879.

Charles, S.L. (2013) "Understanding the Determinants of Single-Family Residential Redevelopment in the Inner-Ring Suburbs of Chicago." *Urban Studies*, 50(8), pp. 1505–1522.

Charles, S.L. (2014) "The Spatio-Temporal Pattern of Housing Redevelopment in Suburban Chicago, 2000–2010." *Urban Studies*, 51(12), pp. 2646–2664.

Charles, S.L. (2017) "Teardowns and Reinvestment in the Inner-Ring Suburbs of Chicago." In: Squires, G., Heurkens, E. and Peiser, R.B. (eds.) *Routledge Companion to Real Estate Development*. New York: Routledge.

Cohen, L. (2004) *A Consumers' Republic: The Politics of Mass Consumption in Postwar America*. New York: Vintage Books.

Easterlin, R.A. (1995) "Will Raising the Incomes of All Increase the Happiness of All?" *Journal of Economic Behaviour & Organization*, 27, pp. 35–47.

Frank, R.H. (1985a) *Choosing the Right Pond: Human Behavior and the Quest for Status*. New York: Oxford University Press.

Frank, R.H. (1985b) "The Demand for Unbservable and Other Nonpositional Goods." *The American Economic Review*, 75(1), pp. 101–116.

Frank, R.H. (1999) *Luxury Fever: Weighing the Cost of Excess*. Princeton, NJ: Princeton University Press.

Frank, R.H. (2007a) "Does Context Matter More for Some Goods Than Others?" In: Bianchi, M. (ed.) *Evolution of Consumption: Theories and Practices*, West Yorkshire, UK: Emerald Group Publishing Limited, pp. 231–248.

Frank, R.H. (2007b) *Falling Behind: How Rising Inequality Harms the Middle Class*. Berkeley, CA: University of California Press.

Frank, R.H., Levine, A.S., and Dijk, O. (2014) "Expenditure Cascades." *Review of Behavioral Economics*, 1(1–2), pp. 55–73.

Hanlon, B. (2015) "Beyond Sprawl: Social Sustainability and Reinvestment in the Baltimore Suburbs." In: Anacker, K.B. (ed.) *The New American Suburb: Poverty, Race and the Economic Crisis*. Burlington, VT: Ashgate, pp. 133–152.

Hirsch, F. (1976) *Social Limits to Growth*. Cambridge, MA: Harvard University Press.

Knox, P. (2005) "Vulgaria: The Re-Enchantment of Suburbia." *Opolis*, 1(2), pp. 33–46.

Marx, K. (1891) *Wage-Labor and Capital, Translated by J. L. Joynes*. Chicago, IL: Charles H. Kerr & Company.

Pinnegar, S., Freestone, R., and Randolph, B. (2010) "Suburban Reinvestment through 'Knockdown Rebuild' in Sydney." *Suburbanization in Global Society*, 10, pp. 205–229.

Pinnegar, S., Randolph, B., and Freestone, R. (2015) "Incremental Urbanism: Characteristics and Implications of Residential Renewal through Owner-Driven Demolition and Rebuilding." *Town Planning Review*, 86(3), pp. 279–301.

Randolph, B. and Freestone, R. (2012) "Housing Differentiation and Renewal in Middle-Ring Suburbs: The Experience of Sydney, Australia." *Urban Studies*, 49(12), pp. 2557–2575.

Schor, J. (1998) *The Overspent American: Upscaling, Downshifting, and the New Consumer*. New York: Basic Books.

Smith, N. (1979) "Toward a Theory of Gentrification: A Back to the City Movement by Capital, Not People." *Journal of the American Planning Association*, 45(4), pp. 538–548.

Veblen, T. (1899) *The Theory of the Leisure Class*. New York: Macmillan & Co.

Wiesel, I., Pinnegar, S., and Freestone, R. (2013) "Supersized Australian Dream: Investment, Lifestyle and Neighbourhood Perceptions among 'Knockdown-Rebuild' Owners in Sydney." *Housing Theory & Society*, 30(3), pp. 312–329.

25

Planning and the cultural landscapes of suburban Turkey

*Bahar Durmaz-Drinkwater,
Jaap Vos, and Asli Ceylan Öner*

Overview

Urla, once a rural town, has been undergoing several changes since the mid-1990s due to the migration from İzmir, other large cities in Turkey, and also from abroad. During the past 20 years, it attracted economically wealthy groups based on its environmental assets and proximity to İzmir. The construction of the İzmir-Çeşme expressway in the 1990s accelerated the change in Urla, and as Datta and Yucel Young (2007) argued, Urla has become a place of gated communities due to the influx of people mainly migrating from İzmir. In the last decade, Urla has attracted different groups, including daily, seasonal visitors, permanent or temporary residents, and newcomers to Turkey. Focusing on the mid-2000s to present, this chapter discusses how and why Urla became an important destination for the upper- and middle-income groups either living and working in İzmir, or living in Urla, and commuting to İzmir for work. It further aims to reveal the local authorities' role in planning and design decisions in terms of this migration to Urla.

Our research approach is a qualitative method of case study influenced by ethnographic observation and media analysis. Through the case study of Urla, the amenity migration concept sheds light on how different income groups transform rural landscapes, how they alter land use dynamics, and how strategies of local authorities (district- and city-level municipalities) affect this process. In the next section, we provide the concept of amenity migration as a theoretical framework for our analysis. After introducing the case of Urla, we discuss its planning and urban development process, the projects developed both by the community, and local initiatives that support amenity migration. Our aim in this section is to document the assets that made Urla an important amenity migrant destination, and to reveal the changing social and cultural dynamics in the cultural landscape. In the last section, we discuss Urla's case within the framework of amenity migration.

Amenity migration

Amenity migration is defined as the movement of people based on certain pull factors related to natural and cultural amenities, which might result in significant changes and possible conflicts in the ownership, use, and governance of rural lands and communities (Gosnell and Abrams, 2011;

Walker and Fortmann, 2003). As Gosnell and Abrams (2011) argue, amenity migration has been discussed in literature in a diffused manner, with academic research published in urban geography, natural studies, rural studies, aging, mountain research, and tourism-related journals (Gosnell and Abrams, 2011). The existing studies mostly focus on upper- or middle-class migration to a rural community for a different lifestyle as "newcomers," either staying seasonally in second homes (Kondo et al., 2012), or permanently (Smith and Krannich, 2000; Loeffler and Stenicke, 2007; Walker and Fortmann, 2003; Bryson and Wyckoff, 2010).

The groups of people that "migrate" to an amenity-rich area are those that aspire to experience the lifestyles offered, as well as investors and entrepreneurs that see the place's potential (Woods, 2011). Most of the amenity migration areas are attractive because their rural rustic charm offers a refuge from the chaos of large cities. In this sense, rural landscapes have transformed into a new social space with both long-term residents and recent amenity migrants (Öner and Vos, 2015). Most often, once these places become popular destinations for new, permanent, and seasonal residents, they also become tourist attractions. Thus, these once-quiet towns are slowly transformed by different interests – the permanent newcomers, seasonal newcomers, tourists, and the local community – pulling in many directions.

In the end, places of amenity migration and their land use patterns may become contested sites based on the conflicting demands of locals, lifestylers, and entrepreneurs (Öner and Vos, 2015). Planners and policymakers in local governments usually end up dealing with problems such as land use planning eroding the existing sense of place identity, gentrification, environmental degradation distribution, and exploitation of resources (Crot, 2006; Woods, 2011). Kondo et al. (2012) demonstrate through their case study how even though second homeowners in two counties of Washington State desire privacy and isolation, certain conflicts over land use can arise between long-time residents and newcomers. Locals might push for stricter land use decisions, and they can be more reluctant regarding further change, preferring to keep the area as it was when they first decided to purchase property there. Affluent new residents might perceive that "they are better able to protect natural resources than the long-term residents who have participated in traditional resource extraction economies" (Kondo et al., 2012, p. 175). This might also be connected to the notion that affluent groups typically have more legal power compared to the long-term local residents (Hay and Muller, 2012).

In addition to the established trends of amenity migration, Woods (2011) points to its changing dynamics as a result of globalization. He states, "recent observations have pointed to a new feature within amenity migration: not only is amenity migration booming, but it is becoming increasingly global in character" (Woods, 2011, p. 365). As the flow of capital and people increase in the era of globalization, multiple residency-based amenity migrations by the global elite are also shaping local land uses and property dynamics (Cadieux and Hurley, 2011), engaging these local areas in global networks and producing cultural hybridization (Woods, 2011). A few existing studies on the global elite's amenity migration examine why they are attracted to rural areas under the concept of "globalizing the rural countryside, hybrid countryside" (McCarthy, 2008; Woods, 2011; Woods, 2007). Thus, elite groups' amenity migration is also associated with aspirational ruralism (Gosnell and Abrams, 2011; Cadieux and Hurley, 2011).

In this context, the chapter analyzes Urla's evolution, which turned a once-small resort and rural town into a major amenity migration destination. The discussion aims to reveal which of Urla's characteristics attracted people from outside of the community and the landscapes created to cater to the affluents' expectations.

Suburban Turkey: the case of Urla

Urla is a coastal town, one of İzmir's 30 administrative districts, located outside the boundaries of İzmir Greater Metropolitan Municipality, on İzmir's western corridor, approximately 40 kilometers away from the city center. According to the 2016 census, Urla's population is 62,439 and has increased 20 percent since 2008; those residents are spread over 704 square kilometers (Nufusmobi, 2017; İzmir Kültür Turizm, 2017a).

Urla consists of administrative neighborhoods called *"mahalle,"* the smallest administrative boundary in the Turkish system. Some of these neighborhoods are located in the city center and some by the coastline. The coastal neighborhoods, such as Çeşmealtı, İskele, and Zeytinalanı, are where seasonal second homes are mostly located. The local population predominantly lives inland, around the town center. Small-scale industrial establishments, including agriculture, green housing, florists, trade, and tourism have been the traditional primary sources of income for Urla's residents.

Urla is surrounded by islands that also attract tourists. In the summer there are ferries from İzmir to some of these islands such as Yassıca Island. The other one, Quarentina Island, the most well-known, is connected to the mainland via a bridge. It was built in 1865 to prevent the spread of fatal diseases like cholera, plague, typhoid, typhus, and yellow fever, and lasted for 150 years, later accommodating the Government Hospital of Orthopedics (Milliyet, 2015).

The planning and urban development process

Urla was connected to İzmir as one of its districts in 1867 and became a district municipality in 1890 (İzmir Kültür Turizm, 2017a). Velibeyoğlu (2004) argued that the metropolitan planning process for İzmir started in 1968, which partially influenced Urla. The İzmir Metropolitan Planning Bureau prepared a plan for all the coastal areas of İzmir, including Urla. Furthermore, in 1976, Urla was included in the İzmir Tourism and Recreation Master Plan, which was approved in 1981 and accelerated Urla's urban development. With rapid development and increasing demand for secondary dwellings, a new development plan was approved in 1984, which widened the municipal borders and the coastal area (Velibeyoğlu, 2004).

Urla's urbanization runs parallel to developments in İzmir's other resort districts on the western axis, on the way to Urla. As these districts developed between the 1950s and 1980s, the population expanded to Urla as well, pushing the fringe areas towards Urla (Sönmez, 2009). By 1980s, Çeşme, the western administrative district of İzmir, became an internationally acclaimed tourist destination, with luxury hotels for the rich and foreign tourists, and a place for summer homes, especially for people living in İzmir and other large Turkish cities. The other coastal districts and towns of the İzmir peninsula followed this trend and became destinations for secondary summer dwellings (Datta and Yucel Young, 2007). In Urla, secondary dwellings increased in the 1980s as well (Velibeyoğlu, 2004). The location of these places on the İzmir Peninsula connects the towns and districts of Urla and İzmir.

Construction of the İzmir-Çeşme expressway in the 1990s accelerated the migration to Urla. In 1992, the Urla section of the expressway project became operational and in 1994 construction up to Çeşme was finished; the huge effect this had on land prices caused land speculations. Datta and Yucel Young (2007) argue that the construction was a strategic decision to develop tourism along İzmir's western axis in towns on the peninsula connected with this expressway. Starting off from Güzelbahçe, another western district of İzmir, the expressway extends to the affluent Çeşme district on the west. The expressway is surrounded by agricultural land, different types of housings, forests, and a variety of rich landscapes.

Mass Housing Legislation, passed in 1984, gave a new direction to housing developments in İzmir, which also affected the coastal fringe areas. While this legislation aimed to develop affordable housing, private groups and developers also benefited, and they built new residences for middle- and upper-middle class residents (i.e., gated apartment blocks and detached villas) located away from the city, which is also associated with the expressway's construction (Datta and Yucel Young, 2007).

After the passing of the Mass Housing Legislation, the number and different types of residential settlements increased with various design typologies and construction organizations. Large cooperative housing projects surrounding central Urla began to develop like "544 Houses" and "1000 Houses" (Velibeyoğlu, 2004). They were not enclaved housings but were located along traditional street patterns with public access. On the other hand, low-density single-family housing estates, defined by Datta and Yucel Young (2007) as "gated, luxury, single-family detached houses," began developing in the middle of agricultural land and forests. Some of these single-family estates were constructed as cooperative houses, some by individual developers, or by large construction firms (i.e., collaboration between the landowner and the construction firm). Almost all of the estates were private enclaves in the form of gated communities. Velibeyoğlu (2004) argued that an uncontrolled and poorly planned urban sprawl process caused these estates, and that people preferred living in them for several reasons: the houses had gardens and provided access to nature; were close to summer homes in Çeşme; friends and relatives were also living in these estates; and the houses came with security, prestige, maintenance, and other services. Velibeyoğlu (2004) reported that there were around 38 single housing estates in Urla by 2003 (gated enclaves with ten units or more).

As discussed by Datta and Yucel Young (2007) and Velibeyoğlu (2004), housing demand in Urla increased after the 1980s and 1990s; hence, local government suggested revising previous planning decisions to respond to growth pressures. A new *lejand* called "special yield zone" was denominated in the revised plans, which made construction on the agricultural lands and forests around the expressway possible. The aim of these plans was also to preserve agricultural land while providing space for low-density housing such as farmhouses. Velibeyoğlu (2004) argued that the "special yield zone" accelerated development. Unfortunately, as Datta and Yucel Young (2007) argue, the plans did not reach fruition due to problems in practice and implementation. Instead of farmhouses, gated luxury enclaves were built for the "urban elite."

In addition to these planning decisions and the construction of the expressway, Urla's geographic conditions increased its appeal to upper-income groups. Urla has a mild Mediterranean climate, slightly colder than İzmir due to the exposure to the northern winds, which influences people's choice of Urla as their second home during the summer days. Furthermore, people from İzmir visit Urla year-round due to its amenities. Urla has also become a permanent home for those who work in İzmir and commute every day.

Diversity of population

There are different groups living in and visiting Urla. The first group comprises the affluent suburban residents who live in the gated communities. Some of them commute to work in İzmir. The second group is made of wealthy seasonal residents with secondary summer dwellings in Urla, spending time there during the summer and sometimes visiting on weekends during the winter. The third group is the local population, which has been living in Urla's individual houses or apartment buildings, and both lives and works there. The fourth group comprises the daily or weekend visitors attracted to Urla for its amenities, as discussed next.

Not only do people from İzmir migrate to Urla, but the area also attracts people from İstanbul (Bakiler, 2017; Çakırcan, 2016). This context overlaps with the framework of the aforementioned groups contributing to amenity migration, including the permanent newcomers, seasonal new-comers, and tourists. Recently, residential developments, gated communities, the İzmir-Çeşme expressway, sociocultural, historic, and touristic amenities, as explained in detail as follows, and the existence of universities that attract the student population (İzmir Institute of Technology, Aegean University Faculty of Fishing and Eylül University Vocational School of Marine Management) have all contributed to population growth. In the next section, Urla's historic, geographical, sociocultural, and spatial characteristics, and the local government's initiatives promoting Urla are explained in detail.

Amenities in Urla

We have divided this section into two types of amenities. First are the amenities that have been embedded into the Urla community for years or developed by the locals and newcomers, form-ing local attraction points. We call this part "organic amenities." The other amenities are the projects initiated by the municipality for strategic development.

Organic amenities of Urla

İskele Mahallesi

İskele Mahallesi, one of the mahalles of Urla, has important historic potential. In the nineteenth century, export and import ran with the port of İskele's help. The İskele area included shops and stores for the export of products. In this area, economic activity was quite strong. İskele Mahallesi accommodates important historic sites and archeological excavation areas. Liman Tepe, Antique Olive Press, 360 History Research Association, and an ancient shipyard are all located in İskele Mahallesi. They attract a certain type of daily visitor, and the fish restaurants and cafes serve the needs of both locals and visitors (see Figure 25.1). People from İzmir, visit Iskele for leisure and coastal activities.

Liman Tepe archeological site

Urla Limantepe Mound, located in İskele Mahallesi, dates back to 3000 B.C. as one of the Ancient Klazomenai City's harbors and one of the Aegean's most important harbor cities (Ankusam, 2017). Painted vessels, rounded one-room or narrow-cornered homes with ovens, cook and mine molds, ceramic tools and animal sculptures, and a 6-meter city wall were excavated in Liman Tepe. Liman Tepe was first discovered in 1950, and the excavations started in the1980s and still continue (İzmir Kültür Turizm, 2017b). The other historic site located in İskele Mahallesi is the olive oil press, which is known as the oldest olive oil press located in Anatolia, dating back to sixth century B.C. It served for 2,600 years. This oil factory was found during the excavation between 1992 and 1998 (Klazomeniaka, 2017), and it is now open to visitors as a touristic/archeological landmark.

360 History Research Association and ancient shipyard

The association was established in Urla-İskele in 2004, near Liman Tepe, by historians, arche-ologists, engineers, and sea lovers based in a shipyard, near the Liman Tepe excavation site. The association conducts research about maritime history, organizes exhibitions, symposiums,

Figure 25.1   Liman Tepe, Antique Olive Press, Ancient Shipyard

Source: Gaye Bezircioğlu

reconstructions, and re-animations of the historic sunken ships such as Kybele and Uluburun, and historic İzmir little boats called kayık. There are other projects such as the Phokaia-Marsillia Historic Sea-Journey Project, the Kaş Underwater Archeology Center, and the Mordoğan Artificial Reef Arkeopark Project (360 Derece, 2017; Urla Egemen Haber, 2016a). The Association collaborates with the Ankara University Mustafa Vehbi Koç Marine Archeology Research Center, which was established in Urla, Çeşmealtı in 2006 (Ankusam, 2017).

Urla town center

There have been several mosques, Turkish baths, mausoleums, and countless fountains located in Urla town center. The oldest mosques that have endured are Fatih Bey Mosque, dating back to the fourteenth century, and Kamanlı Külliyesi, dating back to the fifteenth century, both located in the Urla Town Center. Urla was one of the first towns in the Ottoman times, and the current municipality organization dates back to 1866. *The Old Urla Municipality Building* is located in the town center. *The gaswork building*, which was used as a kerosene storehouse around 1903, has been restored by the Urla Municipality and currently hosts art-related events, exhibitions, and courses for the community (Hürriyet, 2007). The building hosts different photographers every month to create a living and cultural activity space. In addition to this rich heritage, there are also examples of contemporary architectural design, such as the town center square. The project was initiated in 2007 and construction finished in 2011. The town center is a place of celebrations and social gatherings (see Figure 25.2).

Figure 25.2 Urla Town Center and the Square

Source: Lale Başarır

Figure 25.3 Arasta and Malgaca Bazaar

Source: Filiz Keyder Özkan

Malgaca Bazaar and Arasta, located in the Urla Town Center, is a little bazaar with narrow streets connected to a small square surrounded by various shops such as butchers, grocers, tailors, shoemakers, barbers, coffee houses, and small kiosks and markets. It is a good place for shopping and hanging out. There are small gourmet restaurants where people from İzmir also come for lunch (Urla Egemen Haber, 2016b) (see Figure 25.3).

These historic sites make Urla an important place to visit for those interested in history and archeology. Besides these historic assets, there are other amenities that attract people to Urla.

Famous people who lived in Urla and associated places

Urla has been home to many famous residents, including poets, singers, novelists, and important characters. Those famous people have left significant traces. They contributed to Urla's cultural landscape, i.e., one of the factors that made Urla an amenity-rich destination. There is a public park named after Turkish singer Tanju Okan, with a sculpture of him in the middle; the house where Greek poet Seferis lived has been converted to a hotel named after him that also accommodates an art gallery. Turkish writer Necati Cumalı's house was restored by the Urla Municipality in 2001 and is now the Necati Cumalı Culture and Memory House (DHA, 2014). Turkish sculptor, Mehmet Şadi Çalık, the pioneer of modern sculpture, has a street named after him by the Urla Municipality.

Breakfast places and fish restaurants

Urla and İskele Mahallesi are locations for daily activities, more over the weekends, when they are visited by middle-income residents from İzmir. People visit Urla for long Sunday brunches or to go to the fish restaurants located along İskele port. This has become a ritual or way of living for people in İzmir. Even in the winter days, people visit Urla on the weekends for the fish restaurants and various breakfast places. Due to these activities, the population of Urla increases during summers and weekends.

Art spaces and creative entrepreneur-led developments

In recent years, there have been several initiatives related to art and culture. These are mainly focused on young entrepreneurs who opened art spaces, galleries, different sorts of themed restaurants and cafes, and ateliers of individual artists, as listed in Table 25.1. In addition, there is a street devoted to art, called Urla Art Street, which accommodates different types of restaurants, cafes, and design stores mainly managed or established by people who migrated from İstanbul (Çakırcan, 2016). As Urla provides an alternative to urban living, this newly emerged art scene attracts other like-minded people, creating a snowball effect that forms another community, of which newcomers enjoy being part (Figure 25.4). In addition to these spaces, the event scene has grown richer and more diverse, not only due to the festivals organized by the local authorities, but also due to the individual initiatives that color the lives of both Urla's residents and those who just visit for a specific event.

These art spaces are not only located in the town center or the best-known seaside place, İskele, but also in nearby villages such as Barboros, Kuşçular, and Bademler Villages. Bademler Village is known as the art village, accommodating the very first village theater and library. Recently a Montessori School opened, offering an alternative method of primary and secondary

Table 25.1 Art spaces located in Urla

Ayerya Organic Product shop and café
Fırın Vourla
Onart Urla Art House
Çivit Mavi Atelier
Yıldızca Ceramic and Design Atelier
Karma Atelier
Nüans Art Gallery
Atelier Kırmızı
Urla Sanatsal Çevreler Art Gallery
Toprak Stage Art Performance Center
Necati Cumalı Culture House
Yorgo Seferis Art Gallery

Figure 25.4   A ceramic gallery used by local artists
Source: Filiz Keyder Özkan

school education (İzmir Dergisi, 2017). Young couples living in İzmir do consider these alternative lifestyles; Urla will be their way out from the standardized and hugely controlled education system in Turkey.

Urla design library

Another project initiated by a creative-entrepreneur and academic is the Urla Design Library. Restored from a sixteenth-century Ottoman School, the building was purchased by Professor of Architecture Tevfik Balcıoğlu in 2005 and converted into a library with a private collection of books. The building has been open for use since October 2017. As the first in Urla, this design library has the potential to attract researchers, students, and academics that might contribute to the town's changing cultural landscape.

Projects initiated by the municipality: strategic developments

Urla was the first district in Turkey that developed a *Local Agenda 21* in compliance with the 1992 Rio Summit's agreements to promote sustainable development. Agenda 21 is a program supported by the United Nations Development Program (UNDP) and Global Environment Facility-Small Scale Projects Program (GEF-SGF) (Datta and Yucel Young, 2007). The current local municipality and İzmir Greater Municipality are keen on preserving Urla's local assets and aim to develop projects creating sustainable local development. One of them is The Peninsula Project.

The Peninsula Project

The İzmir Peninsula includes administrative suburban districts such as Urla, Karaburun, Çeşme, Seferihisar, Urla, and Güzelbahçe; with an area of approximately 171,000 hectares, this is Turkey's largest peninsula, which also has easy access to the Greek Islands, such as Chios, and Lesvos (see Figure 25.1).

The Peninsula Project was initiated by the Greater Municipality of İzmir in 2008 with the Idea Competition "Urla-Çeşme-Karaburun Peninsula Idea Competition." The winner's visions and strategies directed the planning of the area for the Peninsula's *asset-based sustainable local economic* development (Velibeyoğlu, 2004). "İzmir Peninsula Sustainable Development Strategy" aims to develop the region and its natural, cultural, economic, and historic assets and amenities. The Project contributes to the local economy by helping people living in the region with their businesses, agriculture, and economic activities in order to sustain the peninsula's natural, cultural, and historic potential and contribute to İzmir's sociocultural and economic development. The strategies support local production, introduce vocational schools teaching basket weaving, wood burning, pension, and hotel management, fishnet weaving, ceramics, stone wall lining, cookery, baking, jam, and pickle making. Furthermore, the Project introduces newly designated thematic tour routes invested in the historic, gastronomic, and local assets of the region, such as the Ephesus-Mimas Route (which includes paths for walking and cycling), the Olive Grove Route, the Vineyard Route, and the Blue Route. As well as these routes, enhancing the peninsula's existing thematic bazaars, festivals, and carnivals is another goal of the project. All of these routes pass through Urla, making the town an important anchor point. Many of the events and some of the bazaars are located in Urla as well (Yarımadaİzmir, 2015). These routes are explained in turn.

Ephesus-Mimas Route

An ancient Iconic Cities Routes project was initiated as part of İzmir Greater Municipality's Peninsula Project (Yarımadaİzmir, 2015). The thematic route passes through six iconic ancient cities, such as Ephesos, Kolophon, Lenedos, Teos, Klazomenai, and Ertyhrai. The route includes informative road signs and signposts to direct travelers. Five of the 12 Iconic cities – Erythrai (Ildırı-Çeşme), Klazomenai (İskele-Urla), Teos (Sığacık-Seferihisar), Lebedos (Gümüldür-Seferihisar), Kolophon (Değirmendere-Menderes) – are located on the peninsula (Yarımadaİzmir, 2015).

Urla Wine/Vineyard Route

Beginning in the nineteenth century, Urla has been famous for its grapes and olive oil production, which have become important economic assets. The old port, İskele, used to welcome lots of ships, mainly from overseas, to import and export olives, grapes, and wine. In conjunction with the Peninsula Project, Urla Vineyard Tour Route reveals and sustains the wine production embedded in Urla's history. With the motto "There is history in every step/Her adımda tarih var," the vineyard route creates an alternative tourism movement. The outcome of an initiative led by the head of the Urla Wine Producers and Vinegrowers association, Can Ortabaş, the project takes inspiration from the gastro-tourist centers of Toscana, Bordeaux, Napa Valley and aims to brand Urla as an agricultural, gastronomy-based livable and sustainable town, "Toscana of İzmir" (Uras, 2016).

The wine route includes seven different wineries and vineyards such as Urlice Vineyards, USCA Winery (established in 2003, wine production in 2012), MMG Winery, Mozaik Winery (2006), Liman Tepe Vineyard, Urla Bağevi, and Urla Şarapçılık (Urla Bağ Yolu, 2017). Some of these have restaurants, cafes, and small boutique hotels as well as wine cellars open to visitors. Their contribution to Urla is not only economic and agricultural, but the design of some of these wineries comes from their contemporary architectural style. It is expected that the wine routes and the ancient cities routes project will increase the number of visitors. For example, last year, the wineries hosted 80,000 people, and this trend is expected to double in future years (Uras, 2016).

Olive Grove Route

As well as wine, olive production is one of Urla's strongest economic activities. The first modern olive oil production atelier is located in ancient Klazomenai city, in İskele. Historic sites and landscape elements located along this route include monumental olive trees, olive oil workshops, water springs, olive oil factories, observation terraces, olive press locations, rocks used for olive oil production, camping sites, wells, windmills, and watermills. The route's map is marked with the works of the Delice Olive Association, another example of the collaboration between the municipality and NGOs. This route is also connected to the "Delice Network of World Gourmet Cities" (Yarımadaİzmir, 2015; İzmirbel, 2015).

Blue Route

Blue Route involves the beaches, camping sites, fisherman shelters (balıkçı barınağı), and the waterfront areas where people fish. It aims to create an artificial reef project and support fishing, biological diversity, and diving. It plans to map these locations and prepare an inventory of the beaches with blue flags. Furthermore, the project aims to integrate Blue Route with the municipality's urban sea-transportation network (Yarımadaİzmir, 2015).

Thematic markets and festivals

Most of the thematic markets and festivals have grown organically; however, recently they have also been promoted by the municipality. These markets are located in different districts/towns and set up at different times of year, either periodically, seasonally, weekly, monthly or for specific occasions. There are open food markets, food bazaars, art markets, night markets, a women producers' market, farmers' market, fish market and auctions, or organic food markets. Local government, producers, and consumers come together at these markets (Yarımadaİzmir, 2015) and the various festivals as well. Most of the festivals take place in Urla's public spaces and venues. Table 25.2 displays a list of the festivals.

Table 25.2 Festivals in Urla

Traditional Herb Fest	March
International Urla Artichoke Fest	May
Hıdırellez Fest	May
International Village Theater Fest	August
Vintage Fest	August
Literature Week	September
Nohutalan Melon Fest	September
Sardine Fest	September
Urla Independence Day	September
Barbaros Village Scarecrow Festival	May
Urla Grape Harvest Festival	August
Open-air movie screenings	August

Source: Urla Etkinlikleri (2017)

Amenity migration in the (sub)urban context

Amenity migration is the movement of people based on certain pull factors related to natural and cultural amenities (Gosnell and Abrams, 2011; Walker and Fortmann, 2003). The people attracted to the area can be permanent residents, seasonal newcomers, and tourists. This can create land-related conflicts or synergy with the local community. In the end, gentrification is a very likely trend (Crot, 2006; Woods, 2011). On the other hand, this amenity migration-based gentrification has a rather interesting dynamic; the newcomers might resist any change of local character because they want to preserve the place to which they moved (Kondo et al., 2012). Conflicts between newcomers and locals are possible, although these are not the only groups that might fall into conflict. The permanent and seasonal newcomers might also have some conflicts of interest.

In this case study, we examined Urla's transformation. Once a small rural town of İzmir, today it is a more affluent locality. Although Urla is an administrative district with its own municipality, after the construction of the expressway in the 1990s, it has embraced the character of the new suburbia, based on multiple-housing communities (Datta and Yucel Young, 2007). This new character is perceived as a physical manifestation of modernity identified by Taylor (1999) as "consumer modernity" (Datta and Yucel Young, 2007). With the İzmir-Cesme expressway, the commuting patterns diversified, and people became attracted to Urla. There is ample land, opportunity to live in single-family homes, pleasant weather, and peace from chaotic city life. In this regard, Urla has become a residential suburb of İzmir since the 1990s.

Datta and Yucel Young (2007) surveyed Urla between the mid-1990s and mid-2000s, focusing on the developing residential patterns in what was once an agricultural community. A suburban perception of Urla is quite evident as the town transformed from a residential suburb to a multiple amenity-based destination. Urla does not rely on a single amenity but offers multiple amenities including design, art, gastro-tourism, organic agriculture, history, and the coastline's natural amenities.

Since the mid-2000s, the community-led organic amenities flourished as did government-led initiatives. The organic developments and strategic interventions have been parallel movements in the case of Urla, which could be a positive factor in preserving the place's locality. Both of them sprouted around the same time period, mid-2000s. Neither one led the other. Thus, by the mid-2000s, Urla as a suburban location experienced another transformation, from a healthy suburb to a multiple amenity-based destination, with the contribution of both organic and government-led strategic initiatives.

By June 2017, the data revealed the average square meter price increase in the İzmir metropolitan municipality was double the national average at 13 percent. Urla and its vicinity played an important role in this (Ticaret Gazetesi, 2017). In parallel to Woods's (2011) comment about amenity migration becoming more global in character, international capital has also made its way to Urla through international construction companies. Recent projects like Village Urla and Q205 Villa Projects were undertaken by a large-scale construction firm with a multinational board. There are other gated communities constructed by large multinational Turkish firms. Recently, it was speculated that Hollywood stars Angelina Jolie and Brad Pitt had acquired a 10,000 square meter property in Urla for 2.7 million dollars. This was the ultimate confirmation that Urla has become an important amenity migration area not only known nationally, but internationally as well (Dailysabah, 2015).

A transformation of this scale might result in conflicts between different interests; however, for now Urla seems to be managing well. As one artist demonstrated, many people came to Urla from the United States, Europe, and Istanbul with the hope of experiencing a more peaceful lifestyle. According to this artist, the local community did not oppose these incoming migrants; she said this might be because the locals are also educated and well cultured. After 2009, the arts really flourished in Urla, with Art Street housing many independent artists' studios. These changes are still very recent, and while for now Urla seems to be a place where locals and newcomers have created a cultural hybrid, it is important to see what the future holds. In this regard, the local and metropolitan municipalities need to balance the clear strategic plans and visions for Urla's future with organic developments while carefully monitoring future change.

Based on the numbers below, it is possible to argue that Urla is not yet as touristic as the nearby affluent town of Çeşme. The number of the hotels located in each town could be taken as an indicator for comparison. According to a simple search on the travel search engine Trivago, there are 397 hotels in Çeşme compared to 49 in Urla. Besides, according to the 2016 census, Urla's population is 62,439 whereas Çeşme's population is 40,312, which could mean there are more permanent residences living in Urla and fewer overnight visitors. Urla could consider being less "touristic" than Çeşme as an asset to protect.

The sprawl toward Urla, coupled with tourism and amenity migration, could be a future threat on its locality and environmental assets. Hence, the decisions toward any development or investment should be taken into consideration seriously in balancing different interests and responding to conflicts. It is important to maintain the balance between preserving the locality and assets of the place while also responding to the demands of tourism.

Guide to further reading

Freitag, U., Fuhrmann, M., Lafi, N., and Riedler, F. (eds.) (2011) *The City in the Ottoman Empire: Migration and the Making of Urban Modernity.* New York: Routledge.

References

360 Derece (2017) "Association of 360 Derece Research Group." Available at: www.360derece.info/grubu-muz.htm [Accessed 10 August 2017].

Ankusam (2017) "Ankara Universities Sualtı Arkeolojik Araştırma ve Uygulama Merkezi." Available at: http://ankusam.ankara.edu.tr/ [Accessed 10 August 2017].

Bakiler (2017) "İstanbullular Urla'yı keşfetti." Available at: www.gazeteduvar.com.tr/hayat/2017/05/27/istanbullular-urlayi-kesfetti/ [Accessed 28 August 2017].

Bryson, J. and Wyckoff, W. (2010) "Rural Gentrification and Nature in the Old and New Wests." *Journal of Cultural Geography*, 27(1), pp. 53–75.

Cadieux, K.V. and Hurley, P.T. (2011) "Amenity Migration, Exurbia, and Emerging Rural Landscapes: Global Natural Amenity as Place and Process." *GeoJournal*, 76(4), pp. 297–301.

Çakırcan (2016) "Ege köyüne yerleşen İstanbullu'nun Urla Günlüğü." Available at: www.arkitera.com/gorus/977/ege-koyune-yerlesen-istanbullunun-urla-gunlugu1 [Accessed 28 August 2017].

Crot, L. (2006) " 'Scenographic' and 'Cosmetic' Planning: Globalization and Territorial Restructuring in Buenos Aires." *Journal of Urban Affairs*, 28, pp. 227–251.

Dailysabah (2015) "Latest on Pitt-Jolie: Hollywood Power Couple Buys Villa in Turkey." *Dailysabah.* Available at: www.dailysabah.com/turkey/2015/11/02/latest-on-pitt-jolie-hollywood-power-couple-buys-villa-in-turkey [Accessed 26 November 2017].

Datta, A. and Yucel Young, S. (2007) "Suburban Development and Networks of Mobility: Sites in İzmir, Turkey." *Global Built Environment Review*, 6(1), pp. 42–53.

DHA (2014) "Necati Cumalı Anı ve Kültür Evi Urla'nın Gözbebeği." Available at: http://arsiv.dha.com.tr/necati-cumali-ani-ve-kultur-evi-urlanin-gozbebegi_806630.html [Accessed 24 August 2017].

Gosnell, H. and Abrams, J. (2011) "Amenity Migration: Diverse Conceptualizations of Drivers, Socioeconomic Dimensions, and Emerging Challenges." *GeoJournal*, 76(4), pp. 303–322.

Hay, I. and Muller, S. (2012) "'That Tiny, Stratospheric Apex That Owns Most of the World'-Exploring Geographies of Super-Rich." *Geographical Research*, 50(1), pp. 75–88.

Hürriyet (2007) "Gazyağı depose sanat eseri oldu." Available at: www.hurriyet.com.tr/gazyagi-deposu-sanatevi-oldu-6019196 [Accessed 28 November 2017].

İzmirbel (2015) "İzmir Belediyesi." Available at: www.İzmir.bel.tr/NewsDetail/13055/en [Accessed 11 August 2017].

İzmir Dergisi (2017) "İzmir'in ydın köyü Bademler." Available at: www.İzmirdergisi.com/tr/dergi-arsivi/32-8inci-sayi/1331-İzmir-in-aydin-koyu-bademler [Accessed 23 August 2017].

İzmir Kültür Turizm (2017a) "Urla." Available at: www.İzmirkulturturizm.gov.tr/TR,77470/urla.html [Accessed 29 August 2017].

İzmir Kültür Turizm (2017b) "Klazomenai." Available at: www.İzmirkulturturizm.gov.tr/TR,77421/klazomenai-urla.html [Accessed 28 November 2017].

Klazomeniaka (2017) "Zeytinyağı İşlikleri." Available at: www.klazomeniaka.com/15-KLAZOMENAI-ZEYTINYAGI-1.html [Accessed 28 November 2017].

Kondo, M.C., Rivera, R., and Rullman, S., Jr. (2012) "Protecting the Idyll But Not the Environment: Second Homes, Amenity Migration and Rural Exclusion in Washington State." *Landscape and Urban Planning*, 106, pp. 174–182.

Loeffler, R. and Stenicke, E. (2007) "Amenity Migration in the U.S. Sierra Nevada." *The Geographical Review*, 97(1), pp. 67–88.

McCarthy, J. (2008) "Rural Geography: Globalizing the Countryside." *Progress in Human Geography*, 32, pp. 129–137.

Milliyet (2015) "Urla Karantina Adası Hikayesi." Available at: http://blog.milliyet.com.tr/urla-karantina-adasi-hikayesi/Blog/?BlogNo=492027 [Accessed 29 November 2017].

Nufusmobi (2017) "Türkiye'de ilçe nüfusları 2016." Available at: http://nufus.mobi/turkiye/nufus/ilce/urla-İzmir [Accessed 10 August 2017].

Öner, A.C. and Vos, J. (2015) "Amenity Migration and Land Use Planning Conflicts." 2015 Association of Collegiate Schools of Planning, Houston, Texas, USA, Unpublished Conference Article, 22–25 October.

Smith, M.D. and Krannich, R.D. (2000) "'Culture Clash' Revisited: Newcomer and Longer-Term Residents' Attidtudes toward Land Use, Development, and Environmental Isues in Rural Communities in the Rocky Mountain West." *Rural Sociology*, 65(3), pp. 396–421.

Sönmez, İ.Ö. (2009) "Re-Emergence of Suburbia: The Case of İzmir Turkey." *European Planning Studies*, 17(5), pp. 741–763.

Taylor, P.J. (1999) *Modernities: A Geopolitical Interpretation*. Cambridge, MA: Polity Press.

Ticaret Gazetesi (2017) "İzmir'de konut fiyat artışıTürkiye ortalamasının iki katı." Available at: www.ticaretgazetesi.com.tr/İzmirde-konut-fiyat-artisi-turkiye-ortalamasinin-iki-kati [Accessed 26 November 2017].

Uras, G. (2016) "Urla'da 'Bağ Yolu' turizmi." Available at: www.milliyet.com.tr/yazarlar/gungor-uras/urla-da–bag-yolu–turizmi-2245467/ [Accessed 10 August 2017].

Urla Bağ Yolu (2017) "Urla Bağ Yolu." Available at: www.urlabagyolu.net/index.php?lang=2 [Accessed 10 August 2017].

Urla Egemen Haber (2016a) "Arasta ve Malgaca Pazarı, Urla'nın belleği." Available at: http://urlaegemenhaber.com/arasta-malgaca-pazari-urlanin-bellegi-2/ [Accessed 11 August 2017].

Urla Egemen Haber (2016b) "Kendilerini tarigi araştırmaya adamış bir grup insan." Available at: http://urlaegemenhaber.com/kendilerini-tarihi-arastirmaya-adamis-bir-grup-insan/ [Accessed 30 November 2017].

Urla Etkinlikleri (2017) "Urla etkinlik takvimi." Available at: https://urlaetkinlikleri.com/urla-etkinlik-takvimi/ [Accessed 23 August 2017].

Velibeyoğlu, K. (2004) "Institutional Use of Information Technologies in City Planning Agencies: Implications From Turkish Metropolitan Municipalities." Unpublished PhD Dissertation, City and Regional Planning. İzmir, Turkey: İzmir Institute of Technology.

Walker, P. and Fortmann, L. (2003) "Whose Landscape? A Political Ecology of the 'Exurban' Sierra." *Cultural Geographies*, 10, pp. 469–491.

Woods, M. (2007) "Engaging the Global Countryside: Globalization, Hybridity and the Reconstitution of Rural Place." *Progress in Human Geography*, 31(4), pp. 485–507.

Woods, M. (2011) "The Locational Politics of the Global Countryside: Boosterism, Aspirational Ruralism and the Contested Reconstitution of Queenstown, New Zealand." *GeoJournal*, 76, pp. 365–381.

Yarımadaİzmir (2015) "The Peninsula Project." Available at: www.yarimadaİzmir.com/en/home [Accessed 11 August 2017].

26

Cultural production in the suburban context

Alison L. Bain

Introduction

Around the world, cities have experienced intense suburbanization driven by rising incomes, innovations in transportation and communication, and the decentralization of employment. Decades of significant demographic growth and urban restructuring have rapidly transformed the form and use of suburbs, producing places of great structural and cultural complexity that can make distinctions between city and suburb increasingly unclear. Suburbs are no longer predominantly low-density and residential. While suburbs may have developed in ways that appear incoherent and formless, they contain a full range of urban land uses, some of which were abandoned by the central city and others that were never located there (Harris and Keil, 2017). Urban cultural policy and planning have yet to adequately engage with this complexity.

Through the 2000s, fast-policy models of neoliberal "cultural urbanism" have predominated globally (Peck, 2011, 2012). Such models have driven culture-led urban redevelopment projects, working in ways that repetitively transfer and reproduce within national and international policy networks a business-as-usual focus on economic development that privileges central cities as sites of creativity and cultural production, to the neglect of suburban contexts (Shaw, 2014). Culture-led urban redevelopment deploys culture as the primary driver within flagship- and amenity-driven urban revitalization strategies to secure competitive advantages (Grodach, 2010). In contrast, cultural redevelopment integrates cultural activity "into an area strategy alongside other activities in the environmental, social, and economic sphere" (Evans, 2005, p. 968). When culture is approached as a city's "software" and the built fabric as its "hardware," it can be difficult to meaningfully address the local specificity of social inequities as well as their variation across a wider city-region (Meyer, 2014). As Smith and von Krogh Strand (2011) emphasize, within urban neoliberalism "policy has been reconfigured to emphasize economic development (over wealth redistribution), targeted area interventions (over universal policies), and individual projects (over wider planning)." This chapter argues that when policymakers operate within a paradigm of prescriptive creative urban policy, socio-spatial inequities between city and suburb encourage the neglect and constraint of suburban cultural production.

Placing creativity in the cultural economy

International scholarship on culture and creativity as drivers of economic growth has grown rapidly in the decade following the publication of Richard Florida's (2002) book *The Rise of the Creative Class*. Much of the ensuing scholarship examining the role of knowledge-based and creative industries in a city's cultural economy has attended to the locational factors that attract young, mobile creative professionals to cities but not their suburban peripheries. A city's relative appeal to creatives correlates to reinvestment and regional economic advantage. This human capital approach to urban economic development à la Florida uses proxy indices of talent, technology, and tolerance to predict where creative people, jobs, and businesses will locate and where economic growth will occur. Florida's approach is metrocentric in its focus.

The amenities marketed to lure and service a labor pool of creative professionals – streetscapes animated by hipster coffee shops, boutiques, independent bars, live music venues, art galleries, festivals, and bike lanes – are in Florida's (2002) estimation concentrated in trendy, gentrifying downtown neighborhoods, not the suburbs. Ironically, however, Florida also asserts that high-income knowledge workers in the cultural economy are often more likely to live in suburbs rather than central cities, because they are perceived to provide a higher quality of life and the necessary amenities for a nuclear family. Scholarly critiques of creative city policy caution that although this "vision of urban potentialities" is alluring to policymakers, it is accompanied "by heavy social costs and disappointments" (Scott, 2014, pp. 566, 573). Moreover, creative city policy has a disproportionately negative impact upon racialized, low-income residents and local cultural producers (Munzner and Shaw, 2015). In North America, most people live in the suburbs, are not members of a policy-privileged creative class, and have little to gain from urban development based on creativity (Peck, 2005; Parker, 2008; Pratt, 2008, 2011; Oliveira and Breda-Vázquez, 2012; McLean, 2014).

The problem of creative cities

Florida (2002) popularized the concept of the creative city and introduced it to the global mainstream. He became a "trailblazer for what is now an industry of consultants receiving substantial fees to travel the world advising on how to best mobilize 'your city's creative (economic) potential" (Shaw, 2014, p. 142). Creative cities are designated as such because they use creative capital to generate new economies of innovation, research, and cultural production, and to generate unique place-identities (Sape, 2014). Creative cities are seldom, if ever, identified as suburban because the "nonurban," like people or classes who are "non-creative," have been largely relegated to "the scrap-heap of economic dynamism" (Shearmur, 2012, p. 9). Suburbs, until relatively recently, have been largely disregarded by academics, journalists, and central city residents as too young, homogeneous, and bland to be creative. Peter Hall (1998), for example, showcases European cities that have all accumulated knowledge, competence, and global recognition as creative places incrementally through history and across lengthy and uncomfortable development trajectories. For Hall, cities function as repositories of cultural assets, values, and practices. He does highlight Los Angeles and Memphis as cities that have capitalized on the fusion of cultural and industrial innovation; however, in his estimation, these urban articulations are not the culture of creative cities that have been around for millennia, but rather the culture of innovative cities that are the recent by-products of the industrial enlightenment. Regardless,

> [t]echnologically innovative cities and culturally creative cities are similar in that each would maintain a core of carefully developed competence in multiple areas of knowledge, be

relatively unregulated, attract talented outsiders, support the idea of the heroic individual and have infrastructures that promoted encounters.

(Brown, 2010, p. 121)

It is the metropolitan core of cities where culture is repeatedly asserted to be concentrated, with passing acknowledgment that "some [creative industry] enterprises are situated in suburban areas" (Hutton, 2009, p. 603).

Cities can provide infrastructure, stimuli, competition, market access, and dense commodity chains that enhance opportunities for innovation and collaboration (Lee and Rodríguez-Pose, 2014; Curran, 2010). Such cost- and risk-reducing affordances correlate with city size. Larger cities are attributed with spatially concentrating knowledge, creativity, and innovation (Stolarick and Florida, 2006). For cultural economists, creative cities are where analytical, synthetic, and symbolic knowledge is recombined in ways that can produce "new wealth" (Montgomery, 2007). But what kind of wealth and culture can suburbs produce if they are conventionally framed as disconnected from practices of urban creativity? It is necessary to reconsider creativity from the urban periphery – from the vantage point of suburbs and suburban cultural producers and practitioners. In so doing, greater consideration can be given to the micro-dynamics and networked complexities of local cultural development processes rather than to development formulas, benchmarking, and checklists for fast-tracking a city to creative status (Comunian, 2011). Thus, this chapter demonstrates how culture is embedded in the suburbs and intimately connected to suburban opportunities and challenges.

What about creative suburbs?

Creative city discourse has been critiqued by scholars for its metro-centricity and its neglect of the ways in which culture is produced in remote, rural, small, and suburban places (e.g., Markusen and King, 2003; Bell and Jayne, 2006, 2010; Gibson and Brennan-Horley, 2006; Brennan-Horley and Gibson, 2009; Mayes, 2010; Gibson, 2011; Bain and McLean, 2012, 2013a, 2013b; Van Heur and Lorentzen, 2012). Australian geography and creative industry scholars have led these poignant critiques, directing attention to the specificities of creativity outside of major cities, "in places that are physically and/or metaphorically remote, are small in population terms, or which because of socioeconomic status or inherited industrial legacies are assumed by others to be unsophisticated, or marginal in an imaginary geography of creativity" (Gibson, 2010, p. 3). An Australian Research Council Discovery grant supported a project entitled "Creative suburbia: a critical evaluation of the scope for creative cultural development in Australia's suburban and peri-urban communities" (2009–2011). Responding to an inner-urban bias in understanding sources of creativity, this research team investigated the experiences of creative industry practitioners working in the suburbs of Brisbane and Melbourne (Felton, 2013) and the role creative industries play in contemporary suburban development (Flew et al., 2012).

Taking inspiration from this leading research in the Global South and applying it to the Global North, I too have questioned the mistaken assumption embedded in creative city discourse that creativity is a central city phenomenon. In *Creative Margins: Cultural Production in Canadian Suburbs* (2013), I examine models of cultural production and enactments of creative practice through case studies of the inner and outer suburbs of Toronto and Vancouver. Using locality-based ethnographic analysis of the work of suburban cultural producers, I contest urban spatial hierarchies that position suburbs as the "sub-creative," and a denigrated spatial "Other" (Phelps, 2012). I demonstrate that suburbs are anything but featureless, homogeneous cultural wastelands by providing a detailed portrait of suburban creativity that shows the practical, hidden, temporary, and spontaneous dimensions of the everyday working lives of cultural workers in suburbia. The sprawling physical appearance of suburbs can be misleading, leading to the mistaken impression that because they do not physically resemble dense, clustered cultural districts that they are "'uncreative' zones – places of domestic consumption rather than sites of innovation, the arts and creativity" (Gibson and Brennan-Horley, 2006, p. 457). When some areas of a city-region are erroneously dismissed as less creative than others, that judgment and accompanying stigma can not only carry over into the realms of policy development and infrastructure investment, but also to the reputations of cultural institutions and the career trajectories of the cultural workers who live and work there.

North American suburbs were initially conceived with minimal artistic intention. Nevertheless, a new generation of suburban Bohemians is transforming cul-de-sacs, yards, and homes into counter-spaces of cultural production. Through creative re-invention and adaptive re-use of landscapes and built form, some suburban cultural workers have actively resisted social conventions and property norms while simultaneously conforming to middle-class consumer preferences in their locational decision-making. Homes that are affordable and spacious in family-oriented neighborhoods offering closeness to nature, domestic convenience, freedom, and privacy are all qualities of residential environments that appeal to suburban cultural workers. Car and home-ownership further opens up the possibilities of suburbs for cultural workers to interpret them as creatively emancipatory locales. However, an absence of formal, government-led investment in cultural infrastructure (especially that which supports creative processes rather than cultural products) and minimal public acknowledgment of suburban cultural work can detract from feelings of emancipation. Infrastructural disinvestment and relative invisibility are interpreted as both liberating and constraining. In suburbia, opportunities exist to create with limited financial and institutional support as well as reduced competitive and critical scrutiny. The lack of "third places" (Oldenburg, 1991) outside of the home to network with peers and receive critical feedback has motivated some cultural workers to selectively retrofit suburbia to better meet their professional needs.

Cultural workers have been active in the establishment of what I have termed "informal community cultural service hubs" (Bain, 2013) – grassroots workspaces that meet individual and collective needs for learning, exchange of ideas, and collaboration. In these hubs, people from different disciplinary backgrounds collaborate, sharing knowledge and equipment. This is a particularly suburban model of cultural production involving mutually supportive interactions that are "intermittent, irregular, informal, and not based on contractual agreements" (Shorthose, 2004, p. 153). In suburbia, creativity happens in an unplanned, unpredictable, and flexible way. There is often an explicitly non-economic dimension to suburban cultural work that can be variously motivated by community development, educational, social justice, or art-for-art's-sake agendas that prioritize social utility and also make it challenging to quantify in economic impact assessments (Shorthose, 2004). Nonetheless, it is important to value the range of intangible positive externalities that creative labor contributes to the quality of life in the suburbs and the potential it affords for reimagining communities from the bottom-up.

I have interpreted suburban cultural workers as essentially middle-class in their residential choices, but it needs to be acknowledged that this relative conventionality does not impede creative place-making. While suburban cultural workers may live within a middle-class status quo environment, they also have the creative capacities to critique it and re-imagine different uses and possibilities for suburban places. Such socio-spatial transformation is aided by "changes in communication technology" that "profoundly destabilize and create new opportunities in art and culture" (Peterson and Anand, 2004, p. 314). Innovations in computer-facilitated communication allow suburban cultural workers to create within and across extensive geographies, helping to alleviate feelings of isolation. For suburban cultural workers, the internet is a valuable site of social interaction, mobilization, and local cultural activity planning. The bridging of virtual and material worlds helps to animate suburban spaces through the transfer of cultural resources and intelligence within and through networks.

In their ground-up approach to innovation and sociability, suburban cultural workers are well positioned as suburban agents of change; however, their professional contributions to place-making are often significantly underappreciated by civic leaders. Lists of influential professionals and civic leaders usually include politicians, policymakers, urban planners, developers, architects, financiers, and real estate agents, but rarely do they also include cultural workers. Yet, the cultural workers, I argue, are key players in transforming suburbs and imagining suburban futures. Suburban cultural workers have valuable critical and analytical capacities that are underappreciated and under-resourced: they "are adventurous, and take off without warning for territories not yet familiar to us, returning with words, images, movements, and sounds that fascinate, concern, question, disturb, reveal, fill us with wonder, or prepare us for changes in how we perceive things" (Brault, 2010, pp. 33–34). Such creative provocations are invitations to participate in a different kind of suburbia, one that has the potential to be more socially inclusive and visibly expressive of different histories and lived experiences.

Learning from suburbia

Suburbs have an infrastructure that is more forgiving (financially and materially) than their urban equivalent, with more diverse standards, norms, and traditions (Rekers, 2012). This quality of infrastructural forgiveness creates spatial openings for creativity, experimentation, and the development of new ideas and products. The creative labor and interventions of suburban cultural workers emphasize the inherent value in small-scale and incremental changes to suburban built form through adaptive re-use at an ad hoc pace and unpredictable rhythm. Such gradual and synergistic adaptations are valuable dimensions of suburban place-making because they enable diverse citizen engagements in public life across an interdependent city-region.

Although suburban cultural workers have the potential to be significant agents of change in suburbia, that role needs much greater municipal acknowledgment and support. For starters, cultural sector managers need to direct greater attention to fostering cross-generational, cross-disciplinary, and cross-cultural ties. It is professional cultural workers rather than amateurs who experience underinvestment in resources and training opportunities. It is cultural production and not cultural consumption that needs greater suburban municipal infrastructure investment. In particular, it is decentralized nodes of affordable cultural workspace that require funding, along with operational support for the arts and culture organizations and networks that sustain them. Rather than continue to invest in flagship cultural mega-projects and festivals, which are often financially out of reach for local cultural workers, a greater impact can be made on the daily lives of suburban residents, I argue, by addressing the unspectacular. Drawing on their Australian research findings, Gibson et al. (2012, p. 299) agree; they too propose, "that unheralded and prosaic sites of suburban creativity" such as "community halls, writers' centers, youth music studios, and art spaces . . . deserve better and more sustained financial support." In Canadian suburbs, this financial support needs to go into modest safety upgrades and improvements to older buildings. In the inner and outer suburbs there are numerous aging, low-rise buildings that are not of heritage designation or restoration quality, but they could be retrofitted as informal community cultural service hubs. If these hubs had strong public transit connections, they could assist with decreasing spatial disparities in suburban cultural infrastructure provision. There are currently numerous examples of retrofitted storefronts, strip malls and schools being transformed into multi-purpose community cultural service hubs. This is a noteworthy dimension of the suburban geography of cultural production that merits greater recognition by urban cultural policymakers. It is local creative spaces and talent that warrant media attention and investment, rather than perpetuating a celebration of flagship cultural venues and outside creative and cultural consultant expertise.

My interviews with suburban arts administrators revealed similar insights for cultural hiring practices. In order to reinforce the connections between suburban cultural institutions, organizations, and local arts audiences, it would be helpful if a greater proportion of suburban arts managers live and work in the suburbs that they serve, instead of commuting to them. This increases opportunities to mentor local and diverse suburban residents for future roles as cultural leaders in their respective communities. To more effectively retain the local creative talent of suburban arts administrators and cultural workers, cultural policy strategies need to be developed that not only link individual creatives with each other but also with city-regional, national, and global professional networks (Bennett, 2010).

Given the extensive geographies of many suburban municipalities, suburban cultural activities are often invisible to the general public which is not "in the know." Consequently, increased media coverage and multiple outlets for communicating, advertising, and building public awareness of local cultural workers, cultural activities, and cultural spaces would be advisable. When awareness of suburban artistic and cultural scenes is augmented, community ties of trust and collaboration can be reinforced in ways that combat professional isolation and anchor local scenes within wider cultural networks.

Suburbs have a major role to play as creative places within urban systems. To enhance the suburbs' creative role, civic leaders should appreciate and act upon three key points. First, in North America and many other English-speaking countries, a significant number of a city-region's cultural workers live and work in the suburbs rather than the central city, largely due to the need for affordable and stable workspaces. Second, the geography of suburban cultural production is different from that of central cities, and an arts district model is not the most effective way to serve dispersed cultural workers. Third, unspectacular, community-based suburban cultural infrastructure requires sustained investment. If civic leaders continue to operationalize creative city policies that rely upon economic indicators to determine the value of culture, then cultural resources will continue to amass in the already culturally wealthy areas of city-regions and spatial inequalities will be exacerbated (Flew, 2011).

Cultural policymakers should strive to engage in "whole-of-city thinking" (Luckman et al., 2009, p. 82). When cultural infrastructure inequities are considered at the scale of the city-region, questions of equitable distribution and access can be considered from within a more fulsome context. There is a real need to provide more suburban cultural resources in the form of funding, services, and facilities that can inclusively support cultural expression, participation, and production across a spectrum of expertise from amateur to professional. Although it is inadvisable for cultural policymakers to generate policy from mechanistic economic impact assessments or modifications of city-centric strategies (O'Conner and Gu, 2010), many continue to do so. The obvious drawback of overlaying formulaic creative city agendas on places is that they run "roughshod over local needs, aspirations, and already existing or vernacular creative expressions" (Luckman et al., 2009, p. 72). Ideally, suburban cultural policy should be developed incrementally in ways that can more effectively respond to the specificities and unique textures of local needs. This more "organic" approach to cultural policy development is perhaps most effectively supported by scholarly ethnographic qualitative research that provides detailed documentation of the complexities and situatedness of suburban cultural work, and which gives voice to the diverse lived experiences of real people (Shorthose, 2004).

Conclusion

This chapter has argued that cultural and creative city policy from within a predominantly North American and Australian context has focused on the central city to the neglect of suburbs. Creative locales on the peripheries of city-regions in the Global North and Global South are marginalized in policy circles because of the hegemony of neoliberal urban logics of economic development. Policy priority is given to inter- and intra-urban competition, place marketing, and market-led redevelopment prescriptions that favor the attraction of the creative class. It is important to acknowledge, however, that creativity is not just a product of a place but also of the people who live there. While creativity needs to be geographically and socially contextualized in order for its nuances to be adequately understood, it is also mobile and can be produced anywhere. The local possibilities for creativity depend in significant part upon the institutional thickness of places – the policies, resources, infrastructures, networks, conventions, and norms – that shape how culture is locally produced. The suburbs are a vital and vibrant location of cultural production that lacks sufficient and sustained investment.

It should be clear from this chapter's discussion that contemporary suburbs are at a critical juncture in policy- and place-making. Within a context of continued global economic uncertainty, suburban populations continue to age and diversify along intersections of income, ethnicity, religion, migration, settlement, household structure, gender, and sexuality, producing pronounced and divergent tensions. Concomitantly, new transportation and communications technologies have also radically changed patterns of working, living, commuting, and consuming. Given these many and varied demographic, economic, and technological changes, suburbs could be where new social contracts are negotiated for sustainable, creative, and inclusive city-regions (Ingersol, 2006). I maintain that suburban cultural workers are socially well positioned (although less so economically, in light of the precarious employment standing of many non-profit-making cultural workers and their relegation to the service class, see Bain and McLean, 2013b) to assist with the development of such a social contract because of the effective embeddedness of their socio-material practices in suburban places. To this end, the mutual support that many suburban cultural workers informally provide between the arts and local communities requires more sustained municipal financial support and cultural policy recognition. When financial and infrastructural resources and policy attention are reliably directed to suburban cultural production, only then will the conditions exist for culture to reconstitute and socially cohere a city-region from the outside in.

Guide to further reading

Bain, A.L. (2013) *Creative Margins: Cultural Production in Canadian Suburbs.* Toronto: University of Toronto Press.

Felton, E. (2013) "Working in the Australian Suburbs: Creative Industries Workers' Adaptation of Traditional Work Spaces." *City, Culture and Society,* 4, pp. 12–20.

Flew, T., Gibson, M., Collis, C., and Felton, E. (2012) "Creative Suburbia: Cultural Research and Suburban Geographies." *International Journal of Cultural Studies,* 15, pp. 199–203.

Gibson, C. (ed.) (2011) *Creativity in Peripheral Places: Redefining the Creative Industries.* London and New York: Routledge.

Phelps, N. (2012) "The Sub-Creative Economy of the Suburbs in Question." *International Journal of Cultural Studies,* 15, pp. 259–271.

References

Bain, A.L. (2013) *Creative Margins: Cultural Production in Canadian Suburbs.* Toronto: University of Toronto Press.

Bain, A.L. and McLean, D. (2012) "Eclectic Creativity: Interdisciplinary Creative Alliances as Informal Cultural Strategy." In: Van Heur, B. and Lorentzen, A. (eds.) *Cultural Political Economy of Small Cities.* New York: Routledge, pp. 128–141.

Bain, A.L. and McLean, D. (2013a) "From Post to Poster to Post-Industrial: Cultural Networks and Eclectic Creative Practice." In: Breitbart, M. (ed.) *Cultural Economies in Post-Industrial Cities: Creating a (Different) Scene.* Aldershot, Hants.: Ashgate, pp. 97–121.

Bain, A.L. and McLean, H. (2013b) "The Artistic Precariat." *Cambridge Journal of Regions, Economy, and Society,* 6, pp. 93–111.

Bell, D. and Jayne, M. (eds.) (2006) *Small Cities: Urban Life Beyond the Metropolis.* London: Routledge.

Bell, D. and Jayne, M. (2010) "The Creative Countryside: Policy and Practice in the UK Rural Cultural Economy." *Journal of Rural Studies,* 26, pp. 209–218.

Bennett, D. (2010) "Creative Migration: A Western Australian Case Study of Creative Artists." *Australian Geographer,* 41, pp. 117–128.

Brault, S. (2010) *No Culture, No Future.* Translated by Kaplansky, J. Toronto: Cormorant Books.

Brennan-Horley, C. and Gibson, C. (2009) "Where Is Creativity in the City? Integrating Qualitative and GIS Methods." *Environment and Planning A,* 41, pp. 2595–2614.

Brown, G.M. (2010) "The Owl, the City and the Creative Class." *Planning Theory and Practice,* 11, pp. 117–127.

Comunian, R. (2011) "Rethinking the Creative City: The Role of Complexity, Networks, and Interactions in the Urban Creative Economy." *Urban Studies,* 48, pp. 1157–1179.

Curran, W. (2010) "In Defense of Old Industrial Spaces: Manufacturing, Creativity, and Innovation in Williamsburg, Brooklyn." *International Journal of Urban and Regional Research,* 34, pp. 871–885.

Evans, G. (2005) "Measure for Measure: Evaluating the Evidence of Culture's Contribution to Regeneration." *Urban Studies,* 42, pp. 959–983.

Felton, E. (2013) "Working in the Australian Suburbs: Creative Industries Workers' Adaptation of Traditional Work Spaces." *City, Culture and Society,* 4, pp. 12–20.

Flew, T. (2011) "Right to the City, Desire for the Suburb?" *M/C Journal,* 14, pp. 1–7. Available at: http://journal.media-culture.org.au/index.php/mcjournal/article/view/398.

Flew, T., Gibson, M., Collis, C., and Felton, E. (2012) "Creative Suburbia: Cultural Research and Suburban Geographies." *International Journal of Cultural Studies,* 15, pp. 199–203.

Florida, R. (2002) *The Rise of the Creative Class: And How It Is Transforming, Work, Leisure, Community, and Everyday.* New York: Basic Books.

Gibson, C. (2010) "Creative Geographies: Tales from the Margins." *Australian Geographer,* 41, pp. 1–10.

Gibson, C. (ed.) (2011) *Creativity in Peripheral Places: Redefining the Creative Industries.* London and New York: Routledge.

Gibson, C. and Brennan-Horley, C. (2006) "Goodbye Pram City: Beyond Inner/Outer Zone Binaries in Creative City Research." *Urban Policy and Research,* 24, pp. 455–471.

Gibson, C., Brennan-Horley, C., Laurenson, B., Riggs, N., Warren, A., Gallan, B., and Brown, H. (2012) "Cool Places, Creative Places? Community Perceptions of Cultural Vitality in the Suburbs." *International Journal of Cultural Studies,* 15, pp. 287–302.

Grodach, C. (2010) "Beyond Bilbao: Rethinking Flagship Cultural Development and Planning in Three California Cities." *Journal of Planning Education and Research,* 29, pp. 353–366.

Hall, P. (1998) *Cities in Civilization.* New York: Fromm International.

Harris, R. and Keil, R. (2017) "Globalizing Cities and Suburbs." In: Bain, A.L. and Peake, L. (eds.) *Urbanization in a Global Context.* Toronto: Oxford University Press, pp. 52–69.

Hutton, T. (2009) "The Inner City as Site of Cultural Production *Sui Generis*: A Review Essay." *Geography Compass,* 3, pp. 600–629.

Ingersol, R. (2006) *Sprawltown: Looking for the City on Its Edges.* New York: Princeton Architectural Press.

Lee, N. and Rodríguez-Pose, A. (2014) "Creativity, Cities, and Innovation." *Environment and Planning A,* 46, pp. 1139–1159.

Luckman, S., Gibson, C., and Lea, T. (2009) "Mosquitoes in the Mix: How Transferable Is Creative City Thinking?" *Singapore Journal of Tropical Geography,* 30, pp. 70–85.

Markusen, A. and King, D. (2003) "The Artistic Dividend: The Hidden Contributions of the Arts to the Regional Economy: Project on Regional and Industrial Economics." In: *Humphrey Institute*. Minneapolis, MN: University of Minnesota Press.

Mayes, R. (2010) "Postcards from Somewhere: 'Marginal' Cultural Production, Creativity and Community." *Australian Geographer*, 41, pp. 11–23.

McLean, H. (2014) "Digging into the Creative City: A Feminist Critique." *Antipode*, 46(3), pp. 669–690.

Meyer, T. (2014) "Exploring Cultural Urbanism." *Planning*, April, pp. 14–19.

Montgomery, J. (2007) *The New Wealth of Cities: City Dynamics and the Fifth Wave*. Aldershot: Ashgate.

Munzner, K. and Shaw, K. (2015) "Renew Who? Benefits and Beneficiaries of *Renew Newcastle*." *Urban Policy and Research*, 33, pp. 17–36.

O'Conner, J. and Gu, X. (2010) "Developing a Creative Cluster in a Post-Industrial City: CIDS and Manchester." *The Information Society*, 26, pp. 124–136.

Oldenburg, R. (1991) *The Great Good Place*. New York: Paragon House.

Oliveira, C. and Breda-Vázquez, I. (2012) "Creativity and Social Innovation: What Can Urban Policies Learn from Sectoral Experiences." *International Journal of Urban and Regional Research*, 36, pp. 522–538.

Parker, B. (2008) "Beyond the Class Act: Gender and Race in the 'Creative City' Discourse." In: DeSena, J. (ed.) *Gender in an Urban World: Research in Urban Sociology*. Vol. 9. Bingley, UK: Emerald Publishing, pp. 201–232.

Peck, J. (2005) "Struggling with the Creative Class." *International Journal of Urban and Regional Research*, 24, pp. 740–770.

Peck, J. (2011) "Neoliberal Suburbanism: Frontier Space." *Urban Geography*, 32, pp. 884–919.

Peck, J. (2012) "Recreative City: Amsterdam, Vehicular Ideas, and the Adaptive Spaces of Creativity Policy." *International Journal of Urban and Regional Research*, 36, pp. 462–485.

Peterson, R. and Anand, N. (2004) "The Production of Culture Perspective." *Annual Review of Sociology*, 30, pp. 311–334.

Phelps, N. (2012) "The Sub-Creative Economy of the Suburbs in Question." *International Journal of Cultural Studies*, 15, pp. 259–271.

Pratt, A. (2008) "Creative Cities: The Cultural Industries and the Creative Class." *Geografiska Annaler: Series B, Human Geography*, 90, pp. 107–117.

Pratt, A. (2011) "The Cultural Contradictions of the Creative City." *City, Culture, and Society*, 2, pp. 123–130.

Rekers, J. (2012) "We're Number Two! Beta Cities and the Cultural Economy." *Environment and Planning A*, 44, pp. 1912–1929.

Sape, M. (2014) "Urban Transformation, Socio-Economic Regeneration and Participation: Two Cases of Creative Urban Regeneration." *International Journal of Urban Sustainable Development*, 6, pp. 20–41.

Scott, A. J. (2014) "Beyond the Creative City: Cognitive-Cultural Capitalism and the New Urbanism." *Regional Studies*, 48, pp. 565–578.

Shaw, K. (2014) "Melbourne's Creative Spaces Program: Reclaiming the 'Creative City' (If Not Quite the Rest of It)." *City, Culture, and Society*, 5, pp. 139–147.

Shearmur, R. (2012) "Are Cities the Font of Innovation? A Critical Review of the Literature on Cities." *Cities*, 29, pp. S9–S18.

Shorthose, J. (2004) "Accounting for Independent Creativity in the New Cultural Economy." *Media International Australia*, 112, pp. 150–161.

Smith, A. and von Krogh Strand, I. (2011) "Oslo's New Opera House: Cultural Flagship, Regeneration Tool or Cultural Icon." *European Urban and Regional Studies*, 18, pp. 93–110.

Stolarick, K. and Florida, R. (2006) "Creativity, Connections, and Innovation: A Study of Linkages in the Montreal Region." *Environment and Planning A*, 38, pp. 1799–1817.

Van Heur, B. and Lorentzen, A. (eds.) (2012) *Cultural Political Economy of Small Cities*. New York: Routledge.

Part V
Conclusion and future prospects

27

The end of the suburbs

John Rennie Short

Introduction

Once upon a time there was a place called *Suburbia*. It was built on the edge of U.S. cities from 1945 to the early 1970s. Its emergence was driven by the demands of a growing white middle class, underwritten by the federal funding of roads and government support for the mortgage market. It was part cause and part effect of a post-war economic boom centered on domestic consumption, and the creation of a new and expanded middle class. It was predominantly white, and most homes were single-family dwellings with a marked separation between residential areas and commercial districts. It was the mythic counterpoint to the *Inner-City*, a place for immigrants and the poor, with mixed land use and lax morals; *Suburbia*, by contrast, was American and affluent, the embodiment of a new domestic economy and a nuclear family where a profane business world did not pollute the sacred site of family values. It was mile after mile of separate houses spread across the land. And the people were happy.

Well, not quite. *Suburbia* had its critics. There were the cultural critics who initially saw the suburbs as sites of conformity. Scholars such as Lewis Mumford were always a bit sniffy about *Suburbia* and could never hide their distaste; it was too new and decidedly not urban. The very prefix "sub" indicated its inferiority. He described it "as an asylum for the preservation of illusion" (Mumford, 1961, p. 464). Later, as *Suburbia* became more pervasive and less associated with a new, gauche middle class and more a stand-in for the now, it was criticized as a place of rampant individualism, the setting for the unraveling of civic society's fabric, a place where people retreated behind their remote-controlled garage doors, turned their backs on community, and held tight to the nexus of family connections (Putnam, 2000).

Then there was the equity argument. For those trapped in the Inner-City, *Suburbia* became something to aspire to, a bridgehead to the middle-class. To move to *Suburbia* was to be accepted as American. Opening up *Suburbia* became part of the struggle for civil rights (Downs, 1975).

There were the environmental criticisms. In the early years, the construction of houses across the landscape took place with limited environmental regulations and often less environmental sensitivity. Hills were razed, habitats were destroyed, and ecosystems disrupted as the geometry of the grid was imposed across the land. The increase in impermeable surfaces led to flooding. There was a loss of green space in the wake of low-density sprawl. The heavy reliance on motor vehicles increased air pollution. The large doses of pesticides and fertilizers necessary to maintain the suburban lawns increased pollution and hastened the deaths of streams and rivers. The suburbs were an environmental crisis. Rather than an Eden, the suburbs were an environmental calamity (Benton-Short and Short, 2013).

There were also the feminist critiques that suburbs became a way to discipline and contain women after their participation in the workplace during World War II. *Suburbia* hardened traditional gender roles after the plasticity of the war years. *Suburbia* was a place where women raised children and looked after the home, a domestic prison camp that maintained male hegemony (Friedan, 1963).

Suburbia was the vehicle for wider social criticisms. It was not so much a causal analysis but an equivalence that read "contemporary" or "now" from *Suburbia*. Both the supporters and critics were responding to the myth. And remember that a myth is an intellectual construction that embodies beliefs and values as much as information and facts. The real question is not whether a myth is true or not but whose truth it is (Short, 1991).

Surburbia goes global

Suburbia was specific to the post-war United States, but the myth was so compelling, and spread so widely, that its physical form was widely copied. Shorn of its historical and geographic specificity, the built form of *Suburbia* came to suggest global modernity and newfound affluence; it became a symbol of progress and an embodiment of contemporary success. It was copied in various forms around the world as developers employed the myth to meet the demand for elites and wannabe-elites. A new version of a more exclusive *Suburbia* took hold in foreign lands. Tracts of single-family homes sprouted in the cities of the Global South and in the former Soviet Empire. Often set in gated communities, these new suburbs embodied modernity and affluence and global connectivity. It was no longer a phenomenon limited to the United States; it became a physical-cultural form to house the established and rising elites.

From Suburbia to suburbia

The myth has also dominated commentary on the suburban experience beyond the time and place of *Suburbia*. It is as if the experience of the post-war United States, despite all its particularities and specificities, became the template for future understandings. *Suburbia*, an American myth, became a global yardstick to measure and explain suburban experiences and understand metropolitan dynamics around the world.

Suburbia has passed into legend. We now live in a global suburbia of greater heterogeneity and difference. Rather than a place of Being, it is still in the process of Becoming. So let us look in detail at some of the processes behind this global suburbia and note their differences from *Suburbia*.

Let us begin with movement. The journey to *Suburbia* became a classic foundational myth of the United States, a mythic quest, the search for an Eden that was equal in power and sweep to the mythic move west beyond the frontier. In global suburbia, in contrast, banishment and displacement, the removal from Eden, is just as important as the search for Eden.

Banishment

Whereas *Suburbia* was a desired destination, suburbia is as much a place of banishment. Post-war public housing projects in countries such as Scotland and France were built on the periphery of the major cities. In Edinburgh and Paris, for example, while the rich lived in the central core, it was the poor that lived on the city's edge. Today, there is marked revalorization of selected central cities around the world. In this urban resurgence, the suburban edge now contains the banished and the displaced (Short, 2017).

The breakneck urbanization of China, for example, has involved displacement on a massive scale. In Shanghai between 1992 and 1999, more than one million people were evicted under municipal plans to clear away older housing in the central city areas. At least 1.5 million people in Beijing were displaced for the 2008 Summer Olympics, and this figure does not include the impact on rural migrants to the city. While many improved their housing conditions, compensation levels were low and often involved the destruction of tight-knit communities and relocation to distant areas of the city. In Shanghai, the Xintiandi project consisted of 52 hectares, with 23 residential blocks housing 70,000 people. The poor were displaced to make way for upmarket expensive housing. Many were forced to move to the outer suburban areas, far from family and work opportunities (He and Wu, 2005).

Displacement is also a feature of cities with large or substantial populations whose semi- and illegal occupancy of prime sites provides an opportunity for the state and capital to displace them, often without the need for compensation. In India during the 2000s, private capital investment and state mega-projects led to massive evictions. Communities living alongside the strategically located River Yamuna were especially targeted. People who had migrated from the rural areas of Bihar, West Bengal, and Utter Pradesh established these communities over 30 years ago, but hosting the Commonwealth Games in 2010 provided the opportunity for an urban makeover that made 200,000 people homeless.

The displaced people of Delhi were pushed out to settlements on the city's edge. Bhalswa Colony was established in 1999 with more than 30,000 people from eight different sites in Delhi, but the mixing delayed a unified community response. The people in Bhalswa are housed in 21 acres beside one of the city's three main landfill sites, a rubbish dump 22 meters high with 2,200 tons added each day. It is on low-lying land subject to regular flooding. It is a bleak place, windswept, and frequently fire-swept. Contaminated water from the landfill regularly floods people's homes. Almost 80 percent of residents suffer from stomach problems. Chronic diarrhea and vomiting are common ailments. The area suffers from poor sanitation and unclean drinking water, situations exacerbated by high densities. Most live in poverty. Half of the children are malnourished (Bhan, 2009).

The resettlement is 30 kilometers from Delhi, a half-hour bus ride from the end of the metro system. The people are far from the old social networks that provided jobs and access to public services. Many people scavenge the landfill looking to recycle rags, plastics, glass, and copper wires. Women are often employed as domestic servants, and in addition to their difficult working conditions and long commutes, they are often subject to exploitation and abuse.

Since 2000, large-scale investment projects in Manila in the Philippines revalorized the informal housing areas. The informal settlers were forced out of the central areas. The old political link between government and settlers was replaced by the growing link between the government and private developers. Private capital, with help from the World Bank and the Asian Development Agency, promoted a modernization ensemble of railway and highway construction, mixed-use developments, and condo buildings. One project, the North-South Rail Modernization Project, runs through metro Manila, cost $120 million, and involved the demolition of slums and the relocation of 90,000 households. The clearance frontier moved out from the railway line to adjacent informal settlements and businesses as authorities secured the purity of the cleared site. In 2010, rents in surrounding areas increased from 1,000 pesos for one room to 2,500 pesos. Across the city there was a "a pernicious urban warfare against urban settlers" as the informal residents were "displaced, deconcentrated, and relocated to distant socialized housing projects" (Ortega, 2016).

When revalorization of central cities occurs, as in the case of Delhi or Manila, the poor are displaced to the suburbs.

Market displacement

If banishment is an overtly political process, then displacement is its economic equivalent.

Let us look at San Francisco, a high-density city covering 6 square miles. During the past 10 years, there has been something of a digital gold rush as high-tech, computer app companies grow and flourish. The city is like a hot house that provides the conditions for super growth of certain plants, in this case computer behemoths and start-ups. While Silicon Valley, down the interstate in Palo Alto, was the home of the early industry, its current heart is in San Francisco. The creative young labor, the sector's basic raw material, prefers the city to sunny but essentially boring Palo Alto. There is also a division between the "older" (in Silicon Valley historical terms) companies established in Silicon Valley such as Apple, Cisco, Google, and Facebook and the new start-ups that tend to cluster in the city. In recent years, even the large companies have shifted operations to San Francisco. Almost 50,000 new jobs were created in the city in the last decade, most of them very high-paying relative to the average U.S. worker or household. The growing wealth gap in the United States plays out in the housing market of San Francisco.

Tight urban space and rising demand has led to escalating housing prices and rents. Million-dollar homes, rare in the city before 2012, are now much more common. A 1,400 square foot apartment rents for $10,000 a month. The median monthly rent is $4,225. Lower income groups in the rental sector are pushed out to marginal areas of the city such as Hunter's Point. Some houses are purchased only to be knocked down to make room for newer, more expensive housing. There are also the evictions. Under California's Ellis Act, landlords can evict tenants "to go out of business," which in practice means that landowners can evict all of a property's tenants, in effect going out of business, and then turn the property into expensive units. In Los Angeles, a city ordinance requires rent control for five years after the change, but in San Francisco, it only requires that tenants receive just under $6,000 per person in compensation for the eviction. Since 2009, more than a hundred properties each year have used the Ellis Act to evict tenants; more than 150 people are evicted each month. This is not an organized, state-run project of displacement but rather a market-driven response to housing demand in a permissive regulatory environment.

New social formations

Suburbia was based on a specific social formation. The post-war United States was experiencing a demographic dividend and the growth of a new middle-class. It was the baby boomer era of growing families.

In global suburbia, in contrast, there are a variety of positions on the demographic transition, from the rapid population growth of sub-Saharan countries to the maturing demographic dividend of Brazil, India, and China to the later stages of limited population growth in North America and Europe. In countries at the earlier stages of the demographic transition, rapid population growth is reflected by the growth of informal settlements, both in the central city and the suburbs. In countries of the demographic dividend, sometimes a new middle class is reflected in classic suburbia but more often, smaller family size precludes the need for large family dwellings. High-rise housing remains popular.

Consider the case of the United States. The great suburban boom from 1950 to 1970 was based on an expanded demographic base of large families. In 1947, the average household contained

3.5 people. By 2016, this had decreased to approximately 2.5, part of the global demographic transition to lower birth and death rates. The larger suburban houses are now less attractive for smaller families, especially if they also require costly maintenance and commuting. Smaller family size encourages reevaluating urban housing markets, placing the emphasis on maximizing accessibility more than space.

There is an increase in single person and non-child households, so the move to suburbia is less driven by family dynamics. If anything, there is a shift to the higher density cities, as the more affluent single person and non-child person households seek to maximize employment and recreational opportunities.

Then there is the revalorization of time. We have to do more in less time in part because of our technology's punishing immediacy. In *Suburbia*, time was in greater abundance and so longer journeys to work were more acceptable. People relied on snail mail and interactions took place over days and weeks rather than seconds and minutes. Today, time is more valuable, and so longer journeys to work are imposed on the poor rather than embraced by the rich. The pressure on time revalorizes central city locations and devalorizes more distant places. If suburban living involves marked separation between work and home, it slides down the income scale.

New physical forms

The metropolitan regions of the world are complex mosaics. Even in the United States the old binary of Suburbs and Inner-City, with its superimposed racial and income differences, is being undercut and replaced by a new metropolitan reality containing suburban poverty as well as suburban exclusivity, and central cities of resurgence and gentrification as well as stubborn and enduring pockets of poverty (Hanlon et al., 2010).

This new suburbia of the global metropolitan has a number of characteristics that distinguish it from *Suburbia*. First, suburbia is no longer dominated by single-family homes at low densities. The variety of housing ranges from gated communities to informal settlements and villages now swallowed up by metropolitan expansion. High-rise apartment dwellings across the world look out on the peri-urban fringe.

Second, suburbia contains a variety of income, racial, and ethnic groups. *Suburbia* was white and middle class; suburbia houses a richer variety of different social demographics.

Third, the division between work and home, which was such an essential feature of *Suburbia*, no longer holds. Jobs and economic activities are more widely diffused. In some cases, the spread is due to market forces, industries, and offices moving to cheaper areas, and in others it is aided by the government. Economic activities in the central city are at times displaced to the city's edge. At the edge of Seoul, Paju Book City houses 250 publishers and almost 10,0000 workers. The Korean government and publishers supported the site, eager to create a creative cluster. It opened in 2001 as publishers relocated from central Seoul.

In global suburbia, there are more complex economies than a simple binary between home and suburbs on the one hand, and city and work on the other. New, more complex metropolitan economies are emerging. The domestic sites of suburbia are also places of economic activity as people work from home, establish commercial enterprises, and turn the purely domestic into the domestic economy.

In *Suburbia*, the main travel pattern was from home to work and back again. In global suburbia, there are convoluted circulations, as the journey to work, shop, and recreate takes on patterns more similar to Brownian motion, with movement in many different directions. Main roads and major public transport routes still guide developments but there is a wide dispersal of jobs, homes, stores, and recreational opportunities. Two standard models of urban studies, the Burgess and Hoyt models, assumed a mononuclear city with patterns and processes circulating around a strong, organizing center. Perhaps we need to look again at another classic model, the Harris-Ullman model, for an alternative with more contemporary relevance. In 1945, Chauncy Harris and Edward Ullman (1945) proposed a multiple nuclei model in which there were multiple centers in a metropolitan economy. Organizing growth points were distributed throughout the metropolitan regions. Their model is closer to today's reality than the more venerated Burgess and Hoyt models, which are staples of urban studies. Perhaps, however, the Harris-Ullman needs a fuller airing.

The end of suburbia

As metropolitan expansion increases, the division between city and suburb, while it may have legal and government reality, becomes more blurred. Can we legitimately speak then of suburbs as something different from the rest of the metro region? I think only in the very broadest terms. The metropolitan region still contains concentrated areas of centrality at one extreme and peripheral zones at the other. But the rest, the vast middle, is so amorphous, so variegated, and now so large, that it is now difficult to speak of them in the singularity of suburbia. The metropolitan areas are now one giant suburbia topped and tailed by extreme concentration at one end, and a half-urban, half-rural liminal zone at the other.

This suburbia of the vast middle is so complex and heterogeneous that the term has lost real meaning other than the broadest of categories. We need then to identify different types of metropolitan districts. These include the gated communities and informal settlements, the rich escaping the city, and the poor banished and displaced from the city.

Sometimes the two intermingle. The rise of the new middle class is part of an overall economic transformation that also involves marked and growing inequality. In China, for example, the inequality is also overlain by citizenship status, as poor rural migrants are restricted to the city's periphery and edged out of the formal state's services. Increased securitization is in part a response to the rising inequality, but it is also a marker of status, an entrenched requirement of more affluent consumers. Uniformed guards and checkpoints fit the growing list of required attributes that also includes marble top counters and metallic-sheened durables.

Suburbia the myth referred to a specific time and place. Global suburbia is now the huge liminal space between central cities and urban fringes. Its coverage is so large that it is perhaps time that we drop the term. Its usage always contains, in some measure, the myth of a *Suburbia* now rendered obsolete by the pace, scale, and sheer variety of metropolitan emergences. It should come as no surprise to readers of this book's previous chapters that perhaps it is time to abandon the term for the sake of clarity and sharper intellectual focus. To use the terms "suburb" and "suburbia" is to invoke, sometimes directly, often implicitly, the myth of *Suburbia* even as the myth fades from view. And even if we separate out suburbia from the mythic elements of *Suburbia*, the term now enfolds such a huge and diverse metropolitan landscape that its usage is limited. It is the end of the suburbs as a phenomenon and suburbia as a useful discursive device. Our task is now to build more sophisticated models and understandings of a complex metropolis without the crutch of these increasingly obsolete terms.

Guide to further reading

Forsyth, A. (2012) "Defining suburbs." *Journal of Planning Literature*, 27, pp. 270–281.
Gottlieb, R. and Ng, S. (2017) *Global Cities: Urban Environments in Los Angeles, Hong Kong, and China*. Cambridge, MA: MIT Press.

Hanlon, B., Short, J.R., and Vicino, T.J. (2010) *Cities and Suburbs: New Metropolitan Realities in the US.* New York: Routledge.
Rothstein, R. (2017) *The Color of Law: A Forgotten History of How Our Government Segregated America.* New York: Liveright.

References

Benton-Short, L. and Short, J.R. (2013) *Cities and Nature.* 2nd ed. London and New York: Routledge.
Bhan, G. (2009) "'This Is No Longer the City I Once Knew:' Evictions, the Urban Poor and the Right to the City in Millennial Delhi." *Environment and Urbanization,* 2, pp. 127–142.
Downs, R. (1975) *Opening Up the Suburbs: An Urban Strategy for America.* New Haven, CT: Yale University Press.
Friedan, B. (1963) *The Feminine Mystique.* New York: W.W. Norton.
Hanlon, B., Short, J.R., and Vicino. T. (2010) *Cities and Suburbs: New Metropolitan Realities in the US.* London and New York: Routledge.
Harris, C.D. and Ullman, E.L. (1945) "The Nature of Cities." *Annals of American Academy of Political and Social Science,* 242, pp. 7–17.
He, S. and Wu, F. (2005) "Property-Led Redevelopment in Post-Reform China: A Case Study of Xintiandi Redevelopment in Shanghai." *Journal of Urban Affairs,* 27, pp. 1–23.
Mumford, L. (1961) *The City in History.* London: Secker and Warburg.
Ortega, A.A.C. (2016) "Manila's Metropolitan Landscape of Gentrification: Global Urban Development, Accumulation by Dispossession and Neoliberal Warfare against Informality." *Geoforum,* 70, pp. 35–50.
Putnam, R. (2000) *Bowling Alone: The Collapse and Revival of American Community.* New York: Simon and Schuster.
Short, J.R. (1991) *Imagined Country: Environment, Culture and Society.* London: Routledge.
Short, J.R. (2017) *The Unequal City: Urban Resurgence, Displacement and the Making of Inequality.* London and New York: Routledge.

28

Conclusion and future research on global suburbs

Bernadette Hanlon and Thomas J. Vicino

The suburban context in an urbanized world

We live in a suburbanized world – yet, global socioeconomic changes have resulted in a world that is increasingly a blurry suburban landscape. It is evident that divergent realities characterize the suburbs in the twenty-first century. First, the suburban population and economy is growing in some favored metropolitan areas while it is declining in other disinvested regions. Suburban Baltimore, Buffalo, Cleveland, Pittsburgh, and Philadelphia – the Rust Belt – have experienced dramatic declines since the 1970s whereas suburban Atlanta, Dallas, Houston, and Phoenix – the Sun Belt – have experienced notable growth spurts at the same time. Second, the rapid growth in the Global South is an emergent characteristic whereas the growth of the Global North has stagnated. This is highlighted by the rapid urbanization of the metropolitan fringe of city-regions such as Lagos, Mexico City, Mumbai, and São Paulo. Third, migration flows have impact metropolitan areas around the world in two distinct ways. On the one hand, rural-to-urban migration flows have contributed to rapid urban change in countries like Brazil and India. On the other hand, in-migration flows from around the world have impacted the cities and suburbs of countries like the United States and Germany. Indeed, as the world "flattens," people and resources flow more freely (Friedman, 2005). The result of these processes is that the world's urbanized areas – a collection of cities and suburbs – are the sites of global social change.

As John Rennie Short asserts in this volume, "*Suburbia* is a myth that referred to a specific time and place . . . global suburbia is now the huge liminal space between central cities and urban fringes." The globalization of city-regions reinforces the differentiation and blurriness of the suburbs. The United Nations *World Cities Report* (2016) shows that half of the world's population now lives in urbanized areas – that is, cities and suburbs – and the future growth will increasingly occur at the fringe of existing development. In a reflection on the contemporary process of urbanization, Canadian Prime Minister Pierre Elliott Trudeau observes that,

> human settlements are linked so closely to existence itself, represent such a concrete and widespread reality, are so complex and demanding, so laden with questions of rights and desires, with needs and aspirations, so racked with injustices and deficiencies, that the subject cannot be approached with the leisurely detachment of the solitary theoretician.
>
> *(United Nations, 2016)*

Indeed, the urban system is a tangled web of cities and suburbs that presents many new challenges to policymakers and planners. Let us set the suburban context in an urbanized world by summarizing our major findings in this volume.

In Part I, "Suburban Descriptions and Definitions," we confront one of the enduring and central questions in the field: what is a suburb? There is no single definition of a suburb, and scholars continue to debate the utility of any given definition. Ann Forsyth provides us with an orientation of the criteria that suburbs tend to be defined: political unit; age of development; location near the central city; economic function; physical form and design; transportation infrastructure; and so forth. Such criteria and definitions vary dramatically around the world. We learn from Richard Harris about the positive and negative aspects of suburban stereotypes. In many cases, it is our very stereotypes about the suburbs that define how we think and conceive of a suburban definition. Moreover, it is also a common benchmark by which we analyze social, economic, and political trends in the suburbs. Then, Nicholas Phelps tells us about the post-suburban era and its role in shaping planetary urbanization. In considering the multiple of descriptions and definitions of the suburbs, it is clear that they matter for empirical analysis and discussion of future trends about urbanization.

In Part II, "Global Perspectives on the Suburbs," we examine the multiple ways that the process of suburbanization is expressed around the world. Drawing on their Major Collaborative Research Initiatives project on *Global Suburbanisms*, Pierre Hamel and Roger Keil introduce us to global perspectives by offering a theoretically grounded approach to the comparative study of suburbanization around the world. Then, we learn about these experiences throughout various regions in the world. In Latin America, for example, Lawrence Herzog demonstrates that mega-suburbs that are built on the periphery, often times segregating land uses through walled communities while disenfranchising residents living in informal settlements. In the case of Australian, Robert Freestone, Bill Randolph, and Simon Pinnegar show us that the suburban landscape is socially and economically diverse and houses the majority of metropolitan residents. Turning to the case of Europe, Ruth McManus traces the evolution of suburbanization of Dublin, illustrating that low-density garden suburbs for the working classes helped build the nation. In India, Annapurna Shaw frames the process of suburbanization as periurbanization, one that yields rapid growth on the metropolitan fringe by segregating land use and social classes. In Indonesia, Deden Rukmana, Fikri Zul Famhi, and Tommy Firman explain that the suburbanization of Jakarta, one of the world's largest cities, produced rapid urban growth and land use change, facilitated by new town and industrial estate developments. Last, in metropolitan Seoul, Chang Cyu Choi and Sugie Lee attribute the growth of the region to very strong and centralized growth management policy that favored dense neighborhood and abundant housing supplies. It is therefore evident that suburbanization is a global phenomenon, a process of growth and outward decentralization of people and economic activities. Although the form and function of suburbs varies by region of the world, the pattern of suburban life is a reality for a plurality of inhabitants of the planet.

In Part III, "Diversity, Exclusion, and Poverty in the Suburbs," we explore the impacts of suburbanization on people. The process of the decentralization of people and their activities (i.e., economic activities, public policies, etc.) produces a series of both positive and negative externalities on the city, the suburbs, and the metropolitan area. Andrew Gorman-Murray and Catherine J. Nash explore the growth of queer suburbia, demonstrating the suburbs have been home to a diverse group of people, including the LGBT population. Erina Iwasaki sheds new light on the inequality of suburbs through an analysis of economic inequality in suburban Cairo. Magnus Dahlstedt and David Ekholm make clear that multi-ethnic suburbs of Sweden institutionalize various forms of social exclusion. Anjuli N. Fahlberg similarly examines exclusion and division of poor residents of Rio de Janeiro, showing the informal development on the metropolitan fringe disenfranchises residents from formal economic and political opportunities in the city. Jon Teaford historicizes the content of local political incorporation in the United States, which led to tensions between self-governance and separation from the metropolis. Kyle Walker demonstrates the suburbs in the United States have become settlement hubs for immigrants from around the world. Katrin Anacker uncovers how impoverished suburbs confront the provision of a social welfare infrastructure in the United States. Willow Lung-Amam and Alex Schafran uncover the political, racial, and economic tensions in suburbs, demonstrating how social movements form in the suburbs. Last, Whitney Airgood-Obrycki and Cody Price reflect on the evolution of suburban stigma and how it shapes the capacity to respond to socioeconomic inequality in communities of the United States.

In Part IV, "Planning, Public Policy, and Reshaping the Suburbs," we consider the role of urban planning and public policy in confronting issues of growth, decline, and renewal in the suburbs. First, we look at the case of metropolitan Paris. Theresa Enright highlights the regional politics of urban and suburban issues how the Métropole du Grand Paris (MGP), a new metropolitan institution, governs the region effectively. Juliet Carpenter explains the pattern of growth of the suburbs of Paris, namely the French *banlieue*. Constructed from the mid-1950s in response to rapid economic growth and subsequent migration to the cities, the Parisian periphery has now become synonymous with marginalization and socioeconomic exclusion, and social isolation in suburban environments. Next, turning to the case of the United States, Justin Hollander, Colin Polsky, Dan Zinder, and Dan Runfola examine the consequences of suburban housing collapse during the Great Recession in 2008, which resulted in widespread housing vacancy and left suburban governments with an imperative to plan for suburban shrinkage. Then, Lisa Benton-Short argues that the growth of the suburbs is a continual process of evolution, influenced by new design ideas and resulting in different spatial forms. Drawing on a case study in Mission Viejo, California, Benton-Short documents the redesign of the suburbs and the rise of the master-planned communities. In suburban Chicago, Suzanne Lanyi Charles examines the phenomenon of teardown mansionization and shows how different physical, social, and economic characteristics change the inner-ring suburban landscape. Bahar Durmaz-Drinkwater, Jaap Vos, and Asli Ceylan Öner explore the concept of "amenity migration," which deals with movements of people based on a place's natural, cultural, or lifestyle characteristics, in a case study of the growth and redevelopment of a suburb in metropolitan Turkey. Last, Alison L. Bain reflects on the cultural production processes in the suburbs, arguing that neoliberal economic development policies further exacerbate the socio-spatial inequalities between city and suburb.

Future research

As we reflect on the future prospects for suburbia, let us consider the issues confronting a suburban world and put forth a research agenda.

Globalization and urbanization

It is clear that future urban growth will occur at the periphery of urbanized regions around the world. Cites – and now suburbs – are the sites of globalization (Glaeser, 2011). A connected and networked world of cities and suburbs will continue to fuel the urbanization of our population centers. An understanding about how and why this process occurs in the suburban context will be essential to our knowledge about twenty-first century suburbanization. Indeed, divergent realities of suburbs will likely persist; demographic transformations, deindustrialization, and housing issues will shape these patterns of divergence (Short et al., 2007). These realities will also vary regionally around the world, particularly in developing countries. It will therefore be imperative to extend our research agenda to consider how and why the process of urbanization (and growth along the urban periphery) differs around the world.

Migration

The globalization of cities and suburbs has been facilitated by the process of migration – the movement of people and capital around the world (Hanlon and Vicino, 2014). In fact, this free flow of people is one of the primary drivers of urbanization. Internal migration often occurs with the movement of people from the rural hinterland to the urban centers whereas international migration occurs when people move between countries. In 2016, there were some 250 million international migrants living abroad. For example, some 44 million immigrants, or approximately 14 percent of population, live in the United States. What is more, the number of migrants to the United States doubled since 1990, making the United States one of the largest destination settlements in the world. Similarly, in the Europe Union, some 33 million residents are migrants. Since 2015, refugee migration to Europe has soared, with 1.2 million estimated asylum applications reported. In contrast, in other regions like Japan still struggle to attract international migrants as its demographic structure transforms dramatically. These global conditions establish the context for the future challenges of urbanization. Questions abound about the role of public policy and planning in the regulation of migration. What will be the impact of governments that promote or discourage migration on societies? Where will migrants settle, and what impact will this have on the communities they leave behind and the new places where they settle? Finally, an understanding of the increased population heterogeneity will be essential for evaluating the impact of growth.

Social movements and the right to the city – and suburb

Cities – and now suburbs – are the sites that empower social, economic, and political change (Mitchell, 2003; Carpio et al., 2011). There is a long tradition of using the space of the city as a site to shape the political agenda of the city. The right to the city, David Harvey (2008, p. 23) asserts, is

> far more than the individual liberty to access urban resources: it is a right to change ourselves by changing the city. It is, moreover, a common rather than an individual right since this transformation inevitably depends upon the exercise of a collective power to reshape the processes of urbanization. The freedom to make and remake our cities and ourselves is, I want to argue, one of the most precious yet most neglected of our human rights.

During the 2000s, some of the world's largest cities and their surrounding suburbs witnessed unprecedented social movements. Consider a few of these movements:

- In the United States, residents of cities and suburbs mobilized to protest discrimination and police violence after various shootings of unarmed, black men by police, thereby sparking the "Black Lives Matter" social movement.

(Lebron, 2017)

- In Brazil, more than one million residents took to the streets to protest urban inequalities in health, education, transportation, as well as political corruption, igniting "Free Fare Movement" social movement.

  (Vicino and Fahlberg, 2017)

- In Egypt, millions of residents protested the authoritarian rule of law in cities across Egypt, beginning the "Arab Spring" social movement.

  (Ketchley, 2017)

- In Turkey, the Gezi Park protests began over the contestation of the urban redevelopment plans for public space. Drawing millions of Turkish residents to the streets, protests ignited a movement to make Turkey a more democratic nation.

  (Gürcan and Peker, 2015)

- In Hong Kong, millions of young people protested in the streets of Hong Kong for electoral reform, forming the "Umbrella Revolution."

  (Dapiran, 2017)

There are a few commanilties among these social movements. These protests were sparked by events that triggered many people to protest. The city and the suburb served as a public space to take collection action. These protests, like many others around the world, reflected the variegated consequences of urbanization – socioeconomic inequality, voice, representation, and justice. As the process of uncontrolled growth persists throughout the world, these issues are more likely to impact residents of the suburbs. Scholars should focus increased attention on these movements.

Resilience and security

Finally, the resilience of metropolitan communities to rebound after a disruption to the system is an emerging concept worth examining in the suburban context. What makes resilient regions? Scholars have begun to explore how issues such as economic insecurities, failing infrastructure, lack of regional governance, and demographic shifts impact the long-term sustainability of metropolitan areas – their residents, their economies, their housing, and their rights (Pagano, 2013). Like cities, the suburbs are increasingly at risk for natural disaster, terrorism, and an aging public infrastructure (Flynn, 2007). As a result, it will be imperative to understand our vulnerabilities, challenges, and solutions to overcoming threats like climate change, global insecurity, migration, and war.

Conclusion

Suburbanization is a global phenomenon. It is a pattern of development that dates to the beginning of human civilization, and it blossomed during the twentieth century. Today, suburbia encompasses a vast territory of land that spans from the central city to the fringes of metropolitan areas. This pattern is evident in metropolitan areas of the world – the Global North and the Global South – in developed and developing metropolitan areas. The processes of growth, decline, and renewal of the suburbs demonstrate that the life cycle of suburbia reflects the patterns of living of our civilization. The growth of urbanized areas of the twenty-first century will be in the suburbs – the far fringes of the city. Thus, it will be essential to understand the impact of this form of development on the social, economic, and political systems of the world.

Guide to further reading

Berger, A.M. and Kotkin, J. (eds.) (2017) *Infinite Suburbia*. New York: Princeton Architectural Press.

Phelps, N.A. (2015) *Sequel to Suburbia: Glimpses of America's Post-Suburban Future*. Cambridge, MA: MIT Press.

References

Carpio, G., Irazábal, C., and Pulido, L. (2011) "Right to the Suburb? Rethinking Lefebvre and Immigrant Activism." *Journal of Urban Affairs*, 33, pp. 185–208.

Dapiran, A. (2017) *City of Protest: A Recent History of Dissent in Hong Kong*. New York: Penguin Classics China.

Flynn, S. (2007) *The Edge of Disaster: Rebuilding a Resilient Nation*. New York: Random House.

Friedman, T. (2005) *The World Is Flat: A Brief History of the Twenty-First Century*. New York: Farrar, Straus and Giroux.

Glaeser, E. (2011) *Triumph of the City: How Our Greatest Invention Makes Us Richer, Smarter, Greener, Healthier, and Happier*. New York: Penguin Press.

Gürcan, E. and Peker, E. (2015) *Challenging Neoliberalism at Turkey's Gezi Park: From Private Discontent to Collective Class Action*. New York: Palgrave Macmillan.

Hanlon, B. and Vicino, T.J. (2014) *Global Migration: The Basics*. New York: Routledge.

Harvey, D. (2008) "The Right to the City." *New Left Review*, 53, pp. 23–40.

Ketchley, N. (2017) *Egypt in a Time of Revolution: Contentious Politics and the Arab Spring*. New York: Cambridge University Press.

Lebron, C.J. (2017) *The Making of Black Lives Matter: A Brief History of an Idea*. New York: Oxford University Press.

Mitchell, D. (2003) *The Right to the City: Social Justice and the Fight for Public Space*. New York: The Guilford Press.

Pagano, M.A. (ed.) (2013) *Metropolitan Resilience in a Time of Economic Turmoil*. Urbana-Champagne, IL: University of Illinois Press.

Short, J.R., Hanlon, B., and Vicino, T.J. (2007) "The Decline of Inner Suburbs: The New Suburban Gothic in the United States." *Geography Compass*, 1, pp. 641–656.

Vicino, T.J. and Fahlberg, A. (2017) "The Politics of Contested Urban Space: The 2013 Protest Movement in Brazil." *Journal of Urban Affairs*, 39, pp. 1001–1016.

United Nations (2016) *World Cities Report*. Nairobi: UN Habitat.

Index

Note: Italicized page numbers indicate a figure on the corresponding page. Page numbers in bold indicate a table on the corresponding page.

Printed and bound by CPI Group (UK) Ltd, Croydon, CR0 4YY

24/10/2024

01778291-0005